THE INTERNATIONAL ENCYCLOPEDIA OF PHYSICAL CHEMISTRY AND CHEMICAL PHYSICS

Topic 14. PROPERTIES OF INTERFACES

EDITOR: D. H. EVERETT

Volume 4

THE ELECTRICAL DOUBLE LAYER

BY

M. J. SPARNAAY

THE INTERNATIONAL ENCYCLOPEDIA
OF PHYSICAL CHEMISTRY AND CHEMICAL PHYSICS

THE INTERNATIONAL ENCYCLOPEDIA
OF PHYSICAL CHEMISTRY AND CHEMICAL PHYSICS

Editors-in-Chief
D. D. ELEY
NOTTINGHAM

F. C. TOMPKINS
LONDON

List of Topics and Editors

1. Mathematical Techniques	H. JONES, *London*
2. Classical and Quantum Mechanics	R. McWEENY, *Sheffield*
3. Electronic Structure of Atoms	C. A. HUTCHISON, JR., *Chicago*
4. Electronic Structure of Molecules	J. W. LINNETT, *Cambridge*
5. Molecular Structure and Spectra	Editor to be appointed
6. Kinetic Theory of Gases	E. A. GUGGENHEIM (*deceased*)
7. Classical Thermodynamics	D. H. EVERETT, *Bristol*
8. Statistical Mechanics	J. E. MAYER, *La Jolla*
9. Transport Phenomena	J. C. McCOUBREY, *Birmingham*
10. The Fluid State	J. S. ROWLINSON, *London*
11. The Ideal Crystalline State	M. BLACKMAN, *London*
12. Imperfections in Solids	J. M. THOMAS, *Aberystwyth*
13. Mixtures, Solutions, Chemical and Phase Equilibria	M. L. McGLASHAN, *Exeter*
14. Properties of Interfaces	D. H. EVERETT, *Bristol*
15. Equilibrium Properties of Electrolyte Solutions	R. A. ROBINSON, *Washington, D.C.*
16. Transport Properties of Electrolytes	R. H. STOKES, *Armidale*
17. Macromolecules	C. E. H. BAWN, *Liverpool*
18. Dielectric and Magnetic Properties	J. W. STOUT, *Chicago*
19. Gas Kinetics	A. F. TROTMAN-DICKENSON, *Cardiff*
20. Solution Kinetics	R. M. NOYES, *Eugene*
21. Solid and Surface Kinetics	F. C. TOMPKINS, *London*
22. Radiation Chemistry	R. S. LIVINGSTON, *Minneapolis*

THE ELECTRICAL DOUBLE LAYER

BY

M. J. SPARNAAY

Philips Research Laboratories, Eindhoven
and
Technological University Twente, Netherlands

PERGAMON PRESS
OXFORD · NEW YORK · TORONTO
SYDNEY · BRAUNSCHWEIG

Pergamon Press Ltd., Headington Hill Hall, Oxford

Pergamon Press Inc., Maxwell House, Fairview Park, Elmsford,
New York 10523

Pergamon of Canada Ltd., 207 Queen's Quay West, Toronto 1

Pergamon Press (Aust.) Pty. Ltd., 19a Boundary Street, Rushcutters Bay,
N.S.W. 2011, Australia

Vieweg & Sohn GmbH, Burgplatz 1, Braunschweig

First edition 1972

Library of Congress Catalog Card No. 72–81171

PRINTED IN GREAT BRITAIN BY BELL AND BAIN LTD., GLASGOW

ISBN 0 08 016852 3

CONTENTS

PREFACE

IT IS frequently said of a new book that it is obsolete the day it is published. An attempt is made in this book to meet this complaint. Views which in the opinion of the author are of general validity and are likely to resist the attacks of time are brought to the forefront, sometimes even at the expense of more fascinating ideas which may not be substantiated by future experiments. These motives can also be traced in the literature references. Reference is made to easily obtainable review papers rather than to papers dealing with details, however interesting they may be. Therefore, although the lists of references after each chapter are rather long for an introductory book, they are not intended to be complete bibliographies. For similar reasons details of experimental procedures are not given. In agreement with the general philosophy of this series, only an indication is given of the way in which the pertinent experimental data were obtained.

The author is indebted to Dr. Roger Parsons for reading and commenting on Chapters 1, 2 and 3, to Professor J. Th. G. Overbeek (Chapters 4 and 5), to Dr. H. J. van den Berg (Chapter 5) and Dr. F. Meyer and Dr. R. Memming (Chapter 6). The author is especially indebted to Professor D. H. Everett for his constant encouragement and his constructive criticism. Without his support this book would not have been written.

The writing of this book has been a time-consuming task, but fortunately my wife happily endured this long period of writing, only occasionally interrupted by our common observation that our three children grew faster than the book.

INTRODUCTION

The International Encyclopedia of Physical Chemistry and Chemical Physics is a comprehensive and modern account of all aspects of the domain of science between chemistry and physics, and is written primarily for the graduate and research worker. The Editors-in-Chief have grouped the subject matter in some twenty groups (General Topics), each having its own editor. The complete work consists of about one hundred volumes, each volume being restricted to around two hundred pages and having a large measure of independence. Particular importance has been given to the exposition of the fundamental bases of each topic and to the development of the theoretical aspects; experimental details of an essentially practical nature are not emphasized although the theoretical background of techniques and procedures is fully developed.

The Encyclopedia is written throughout in English and the recommendations of the International Union of Pure and Applied Chemistry on notation and cognate matters in physical chemistry are adopted.

CHAPTER 1

INTRODUCTION

Definition. The beginnings

An electrical double layer is a system, electrically neutral as a whole, in which a layer of positive charges opposes a layer of negative charges and in which, moreover, layers of oriented polar molecules and (or) polarized atoms are present. Such unequal charge distributions are generally found at interfaces and surfaces, but not in the homogeneous bulk of the substances involved. With this definition in mind the scope of the book is outlined below and reasons are given why the double layer systems considered in Chapters 3 to 6 are selected. The properties which tend to a unifying treatise are more stressed than those which tend to specialization. It will appear that in the early days of research on electrified interfaces no specialization was made at all. Later on a situation existed in which almost every important double-layer system had its own students. It was not until recently that the different groups of workers "discovered" each other.

A double layer consists, in its simplest form, of two charge layers whose thickness and distance apart are of atomic dimensions. This is the concept which Helmholtz[1] had in mind when he first used the name "double layer" in 1853 for the array of charges at the interface between two dissimilar metals (see Fig. 1.1). On the basis of electrostatic theory he has shown that

$$U_1 - U_2 = 4\pi m \tag{1.1}$$

where (Helmholtz' notation) U_1 and U_2 are the two potentials at the two sides of the interface, and m was called the moment of the double layer: $m = \sigma d$, where σ is the charge density of the positive charge layer and d is the distance between the two charge layers.

Thomson[2] (later Lord Kelvin) observed in 1870 that two dissimilar metals, kept in metallic connection with one another, exerted a mutual attraction according to a d^{-2} force law or a d^{-1} energy law. Here d is now a macroscopic distance. The energy law can be written as

$$w = -\tfrac{1}{2}\sigma(U_1 - U_2) = -\frac{(U_1 - U_2)^2}{8\pi d}. \tag{1.2}$$

1

We may note already now that, if the distance d becomes smaller than about 1 micron (10^{-4} cm), London–Van der Waals forces interfere with the electrostatic interaction. This will be explained in Chapter 4, where London–Van der Waals forces between colloidal particles are discussed. From estimates of the maximum value of the energy w Thomson deduced that the minimum distance d_m between two charge layers would be of the order of angstroms (1 Å = 10^{-8} cm).

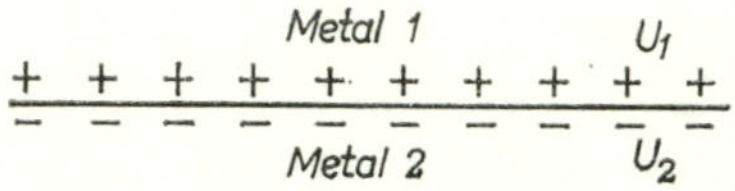

Metal 1 U_1
\+ + + + + + + + + +
– – – – – – – – – –
Metal 2 U_2

FIG. 1.1. Helmholtz' model of a double layer. Potential difference $U_1 - U_2$.

Helmholtz compared in 1879 the double layer of the metal/metal interface with that at the Pt/aq. soln. interface, for which capacitance data were available. He concluded correctly that the theoretical value of d_m (namely a few angstroms) of the metal/metal case was really attained in the Pt/aq. soln. case. Here the charge in the solution was provided by ions.

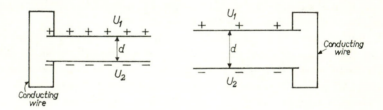

FIG. 1.2. Illustration of the Kelvin method for measuring a potential difference $U_1 - U_2$ which is independent of the distance d.

It is seen in Fig. 1.2 that at the same value of $U_1 - U_2$ the charge density σ is different if d is different, implying that a current has flowed through the connecting wire. The two situations in Fig. 1.2 have a bearing on the original experiments by Galvani[4] and by Volta[5]; in these experiments two discs of dissimilar metals were brought into contact, then isolated and finally separated. The discs then had acquired electric charges of opposite sign. A discharge was accomplished by connecting the discs through a frog's muscle. Animation of the muscle pointed to an electrical phenomenon: a discharge current. Volta observed in 1794 that it was essential to use dissimilar metals.

Molecular condensers. Galvani, Contact and Volta potentials

Helmholtz ascribed the formation of the double layer to a different affinity of the two adjoining media for the charge carriers. However, the situation is more complicated because the interface is a discontinuity and therefore the local electric fields are highly asymmetrical. The same applies to surfaces. For a single metal surface Bardeen[6] calculated, by wave mechanical methods, that the electrons have the tendency to protrude somewhat out of the surface and form a layer of negative charge, compensated by a positive layer of metal ions somewhat inside. In the case of Na for which the calculations were carried out, the potential drop is about $\frac{1}{2}$ volt. For semiconductors Tamm[7] was the first to give a quantum-mechanical treatment of the affinity of the surface atoms for electrons. This affinity is different from that of bulk atoms and may be larger or smaller. If it is larger, the surface atoms contain excess electrons. The surface is said to contain acceptor states, i.e. there are energy states at the surface which are favourable to receiving electrons. According to Tamm's calculations, the number of these states is of the same order of magnitude as the number of surface atoms. The excess electrons at the surface can be assumed to be released from bulk atoms, which have thus acquired a positive charge. The distribution of these positive charges will be discussed below. If the affinity for electrons at the surface is less than in the bulk, then the surface contains donor states, i.e. energy states which are unfavourable to electrons. The repelled electrons leave behind a positive layer of charges. The surface states, donors or acceptors, considered according to this concept are called Tamm states. At real surfaces both donors and acceptors will be present. This is so for germanium surfaces in high vacuum. The potential built up by the surface states is in this case about 200 mV, the surface being negative. In addition, both for metals and for semiconductors, the surface atoms may be polarized.

A potential drop will also be present at the surface of ionic crystals. For alkali halides Verwey[8] calculated that the anions probably tend to be pulled out of the surface. Moreover, they are polarized. The potential drop may be estimated to be of the order of 100 mV. For polar liquids, such as water, one may expect a preferential orientation of the polar molecules at the surface. However, neither experiments nor theoretical considerations have proved very conclusive, and it now seems probable that if a preferential orientation is present at all, it is weak.[9] At the Hg/water and the AgI/water interfaces there is reason to believe that the water molecules have turned their positive side to the aqueous phase.

The examples mentioned here provide evidence for the statement that as a general rule the electrostatic potential in the bulk of a substance is different from that at the surface or at the interface with an adjoining substance. The potential difference between the interior of two substances is the Galvani potential, denoted as $\Delta\varphi$. Part of this potential drop is ascribed to molecular condensers of one kind or another. This part is denoted as the χ-potential or also as the Lange potential (see below). The remaining part is the ψ-potential. For two substances separated by a finite distance d, one has (see Fig. 1.3)

$$\Delta\varphi = \Delta\psi + \Delta\chi. \tag{1.3}$$

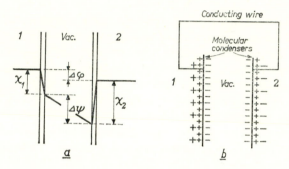

Fig. 1.3. Electrified state of two parallel surfaces. (a) Potential distribution. (b) Charge distribution.

The potential difference $\Delta\psi$ between the surfaces of the two substances is the contact potential. If the substances are electrically connected a current flows if d is varied. Use is made of this fact to determine the contact potential. Thus, two plates are brought close together and one plate is vibrated so that the distance d is periodically changed. An a.c. current flows which can be measured. The method just indicated is the Kelvin method. It is difficult if not impossible to measure the Galvani potential, not only by this method, but by any other method. The Volta potential is not essentially different from the contact potential. The quantity $\Delta\psi$ is the Volta potential when there is no space left between the two substances 1 and 2 of Fig. 1.3.

Reversible and ideally polarizable electrodes

To be able to discuss potential differences between two substances, an electrical connection of some kind must be established. This is simple enough for two metals. For the case of a metal in contact with a solution, however, two cases can be distinguished, which can be considered as extremes:

(a) Electrochemical equilibrium is set up at the interface. If this is so, the interface is a reversible interface. The metal can send (positive) ions into the solution and thereby *acquires* a negative charge, or alternatively if the ionic concentration is high, ions can be discharged, *making* the surface positive. In any case an electric field is set up which counteracts the dissolution or the discharge of the ions. An equilibrium is attained and in this situation during a given time interval as many ions arrive at the surface and are discharged as there are ions formed. The equilibrium reaction is

$$Me^{z+} + ze^- \leftrightarrows Me \qquad (1.4)$$

where z is the valency of the metal ion Me^{z+} and where e^- denotes the electrons. The minus sign here indicates that the electron has a negative charge ($z = -1$ for electrons). Through this reaction mechanism the electrical connection between and solution is established. Nernst[10] gave an expression of the Volta potential difference between metal and solution which we write here as follows:

$$\Delta\psi = \frac{RT}{F} \ln \frac{c}{c_0} \qquad (1.5)$$

where c is the concentration of the metal ions, R is the molar gas constant, F is the Faraday, the charge of a gramion, and T is the absolute temperature.† If $c = c_0$, the dissolution tendency of the ions just balances the discharge tendency. No excess electrons are present at $c = c_0$ at the metal surface and the double layer only consists of a molecular condenser. Nernst called ($p = RTc$) the dissolution pressure (Lösungstension) of the metal and wrote $\ln p/p_0$ instead of $\ln c/c_0$ where p is an osmotic pressure. The osmotic pressure concept was criticized by Van Laar.[11] The Nernst equation not only holds good when the ions proper to the metal are dissolved, but for many other systems as well. Thus, for the Pt/soln. interface the pertinent equation is

$$H^+ + e^- \leftrightarrows H. \qquad (1.6)$$

---

† $R = 8\cdot31$ J mol^{-1} °K^{-1} or $8\cdot31 \times 10^7$ erg mol^{-1} °K^{-1}. The product RT has the dimension energy per mol. We may write this product as $N_A kT$ where N_A is the Avogadro constant and k is the Boltzmann constant,

$$k = 1\cdot38 \times 10^{-23} \text{ J °K}^{-1} \text{ or } 1\cdot38 \times 10^{-16} \text{ erg °K}^{-1}.$$

We shall frequently use the product kT as an "energy unit". At room temperature $kT = 4\cdot1 \times 10^{-14}$ erg or $4\cdot1 \times 10^{-21}$ J. Often kT is called the thermal energy of a particle. A frequently used "unit" of the electrical potential is RT/F or kT/e, which, at room temperature, is about equal to 25 mV. Thus, 1 eV $= 1\cdot6 \times 10^{-12}$ erg $= 1\cdot6 \times 10^{-19}$ J, or: 1 eV $= 40 kT$ at room temperature.

An example of a widely studied reversible interface is the AgI/soln. interface. An Ag-electrode, covered with AgI, obeys the Nernst law with respect to variations of the Ag- and the I-ion concentration. These ions were called "potential-determining ions" by Lange.[12] There are good reasons to believe that during variations of their concentration the χ- (or Lange-) potential remains unaltered. Colloids of AgI particles have proved to be very useful for the study of the double layer surrounding each particle. We have chosen, in Chapter 4, the AgI/soln. interface as an example of a reversible interface.

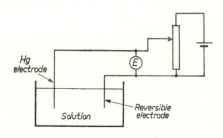

FIG. 1.4. Electric circuit in which participates a Hg electrode and a reversible electrode, both in contact with the same solution. E measures the applied potential difference.

(b) Electrochemical equilibrium cannot be achieved because no charges, electrons or ions, can cross the interface. The interface is then said to be ideally polarizable. The Hg/soln. interface can be made such that it can be considered as almost ideally polarizable. It is chosen as the topic of Chapter 3. Since the metal is a liquid, it is relatively easy in this case to prepare clean surfaces. To establish a real electrical contact between the metal and the solution, a circuit must be made as sketched in Fig. 1.4. An auxiliary electrode, which is reversible to at least one type of ion in the solution, is present. The potential difference at the Hg/ soln. interface can now be controlled by a potentiometer. The suitability of the Hg/soln. interface for studies of the electrified interfaces was first successfully pointed out by Lippmann, who derived the following equation:

$$\sigma = -\frac{\partial \gamma}{\partial E} \tag{1.7}$$

where σ is the charge density on the metal side of the interface; γ is the interfacial tension and E denotes the potentiometer reading. The differentiation of γ with respect to E must be carried out at constant temperature, pressure and chemical potentials of the components in the system (see Section 2.2). Changes of E are assumed to reflect changes of potential in polarizable interface only. Measurements of γ at different values of E provide values of σ. Since Hg is a liquid, measurements of γ can be carried out accurately. Lippmann designed an electrometer which bears his name. In it the value of γ was determined by measuring the height of a Hg-column in a glass capillary. Curves of γ vs. E, by whatever method they may be measured, are now generally called electrocapillary curves.

Space charges

Gouy,[14] in the beginning of this century, gathered a large amount of data concerning the Hg/soln. interface and he also made a major contribution to the theory of the double layer: if the Helmholtz picture of the double layer were correct, the double-layer capacity would be equal to $\varepsilon/4\pi\,\delta$ where ε is the relative dielectric constant of the medium between the two layers and δ the distance.† Defining the differential capacity (for integral capacities see Chapter 3) as

$$C = \frac{\partial\sigma}{\partial E} \qquad (1.8)$$

Gouy found that C was not a constant but depended both on E and on the ionic concentration. At low concentrations the E-dependence was especially marked and C was found to have a pronounced minimum. To explain this behaviour, Gouy assumed that the ions in the solution were not rigidly held in a two-dimensional array, but tended to diffuse away from the interface. This tendency was counteracted by the electrostatic forces arising from the imposed potential difference. The ions form a space charge, the Gouy layer, whose thickness lies between 10 Å and 100 Å. In concentrated solutions the diffusion tendency was less marked. There the Helmholtz picture is roughly correct. At the minimum value of the capacity (in dilute solutions) the electrostatic forces are weakest. If a z–z electrolyte is taken, then, in present-day notation,

$$C^{\mathrm{min}} = \frac{\varepsilon}{4\pi}\kappa \qquad (1.9)$$

where $\kappa = \sqrt{(8\pi n\,e^2z^2/\varepsilon kT)}$ is the reciprocal Debye–Hückel[15] length; n is the number of cations (charge $+ze$) and of anions (charge $-ze$) per cm³. The Gouy theory, which deals with the charge distribution in an ionic solution near a charged wall, and the Debye–Hückel theory,

---

† For reasons explained below we choose here δ as the symbol for distance. The symbol δ pertains to Helmholtz' model.

which deals with the charge distribution around a charged ion, have a common theoretical basis, as pointed out in Chapter 2. The importance of Gouy's work was stressed by Freundlich[16] and especially by Frumkin from 1923 onwards.[17] Space charges as discussed by Gouy now play an important role in colloid chemistry, notably in the stability theory

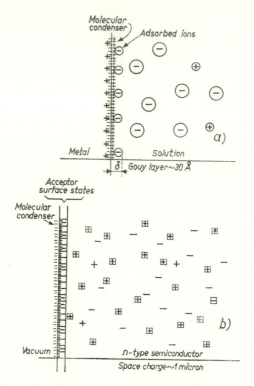

FIG. 1.5. Charge distributions at interfaces: (a) at a metal/solution interface. A space charge is present in the solution; (b) at a semiconductor surface. A space charge is present in the semiconductor. It is assumed that acceptor surface states are present, and that the semiconductor is of n-type conductivity. The squares indicate donor atoms (and one acceptor atom) at fixed positions.

of colloids as worked out by Derjaguin and by Verwey and Overbeek. They are present in ionic solutions bordering reversible and polarizable interfaces, at water surfaces covered with adsorbed ionized soap molecules, and also at semiconductor surfaces and interfaces where semiconductors are involved.

Semiconductors contain mobile electrons and "holes". As an example let us consider germanium Ge. A bulk Ge atom can be ionized, i.e. it may lose a valence electron, which then moves across the lattice. Such an electron has become a conduction electron. The ionized Ge atom has the tendency to saturate its demand for a valence electron at the cost of a neighbouring neutral atom which becomes positively charged. A "hole" may be defined as such a positively charged atom minus a neutral atom. According to the present concepts in solid state physics the above tendency can quite easily be followed, and it appears, that the displacements of holes are characterized by a mobility which is only slightly less than that of a conduction electron. The mobility of both a hole and an electron in Ge are of the order of 10^8 times higher than the mobility of an ion in water. A pure Ge crystal at room temperature contains about 5×10^{13} mobile charge carriers per cm^3 and has a specific conductivity of about 60 Ω-cm. This is much higher than that of an ionic solution in which nevertheless many more charge carriers per unit volume are present (a 10^{-3} M monovalent solution contains 6×10^{17} cations per cm^3 and the same number of anions).

To this quantitative difference a qualitative one should be added: in a semiconductor it is possible to introduce fixed positive or negative charges which are not subjected to thermal agitation. Thus, it is possible to replace a Ge atom in the lattice by a pentavalent impurity atom, which acts as a donor. Four electrons serve to saturate the chemical bonds, the fifth becomes a conduction electron. If the conductivity is mainly produced by electrons, it is said to be n-type (n stands for negative). On the other hand, a trivalent impurity atom replacing a Ge atom may act as an acceptor and lead to the formation of a hole. The conductivity is said to be p-type (p stands for positive) if holes are the majority charge carriers. Empty donors are positively charged, filled acceptors are negatively charged. The space charge theory for semiconductors has been developed by Schottky,[18] by Mott,[19] by Davydov[20] and, most completely, by Brattain and Garrett[21] while useful tables and curves were given by Kingston and Neustadter.[22] It is remarkable to note that the work of Gouy is hardly known in semiconductor research in spite of the fundamental similarities.

If the temperature of a semiconductor is increased, then more atoms are ionized in the bulk of the crystal, leading to an increased conductivity. According to Pearson and Brattain[23] this increase was first observed by Faraday[24] in 1833, on Ag_2S crystallites.

Electron emission. Work function. Surface structures

Electrons may be released[25] from the surface by heating (thermal emission). Another method of releasing electrons is by shining light on the crystal. If the wavelength is sufficiently small, atoms in the bulk (the light usually penetrates to a depth of several atomic diameters) are excited by the light waves and a number of electrons have sufficient energy to escape from the solid (photoelectric emission). A photocurrent can then be measured. The photoelectric threshold, which is the minimum energy required for the electrons to escape, lies between 4 and 6 electron volts (1 eV $= 1{\cdot}6 \times 10^{-12}$ erg). For metals the work function and the photoelectric threshold are the same. In general, however, the photoelectric threshold cannot be identified with the work function: the first quantity is an energy amount which is counted from the energy level of the valence electrons, while the second quantity is counted from the level of the electrochemical potential of the electrons in the semiconductor (see Chapter 6). For semiconductors the measurement of both quantities can provide information concerning the double-layer system at the surface. Usually a metal surface is involved in the measurement of the work function of a semiconductor surface by means of the Kelvin method. The Kelvin method provides the contact potential. This is the difference of the two work functions of the opposing plates. As one of these plates a metal is chosen whose work function is investigated by a photoelectric method.

Work functions have been the subject of intensive studies. These have been reviewed by Herring and Nichols in 1949.[26] Just as in the case of ionic crystals one of the factors which will be determining for the value of the potential drop across the surface is the atomic surface structure, i.e. the position of the outermost atoms with respect to their normal (bulk) lattice positions. A number of methods have become widely accessible since 1949 for the study of surface structures under various conditions (preparation of the surface, temperature, pressure of a given gas or vapour). We mention those of the field ion and field emission microscopes[27] and that of low-energy electron diffraction (LEED). Surface analysis with the aid of electron and ion probes is now in rapid progress. In particular the Auger electron spectroscopy (AES) is now becoming a widely available experimental technique. LEED and AES will be dealt with in Chapter 6. These methods are frequently used in conjunction with experiments leading to information on double layers. The determination of surface structures of base surfaces and of surfaces carrying adsorbed atoms offers an exceedingly complicated problem and so is the problem of a quantitative relationship between the χ-potential and a given surface structure.

Adsorption

Not all the ions participating in the double layer at the Hg/soln. interface, or at any other interface, are subject to thermal agitation. Stern[28] gave a theory in which he considered certain ions to be adsorbed at fixed sites at the interface, the adsorption being ascribed to a specific adsorption energy. This restored part of the original Helmholtz picture. Stern's theory is closely related to Langmuir's adsorption theory,[29] as is now generally recognized. Specifically adsorbed ions at the Hg/soln. have been the subject of numerous investigations, especially by Frumkin, by Grahame[31] and by Parsons.[32] Owing to the precision of the measurements at the "model" interface between Hg and an electrolyte, the study of the adsorption at the Hg/aq. soln. interface has revealed many general aspects and has served to elucidate the adsorptive properties of other systems.

The thermodynamic basis of this work was provided by the Gibbs adsorption equation[33]:

$$dγ = - \sum Γ_i \, dμ_i \qquad (1.10)$$

where $Γ_i$ is relative adsorption per cm^2 of the species i and where $μ_i$ is the chemical potential of this species (electrochemical potential if charged particles are involved). A derivation and a fuller definition of these quantities will be found in Chapter 2, but it may already be noted that, according to the equation, a positive adsorption leads to a decrease of $γ$ when $μ_i$ increases.

The electrical properties of interfaces and surfaces are often quite drastically altered by adsorption of ions and neutral molecules. Often only a minor amount of these compounds is needed for such an adsorption. Thus, an O_2 pressure of 10^{-7} torr is sufficient to increase the photoelectric threshold of Ge by 0·5 volt, and moreover there is a 100-fold decrease of the number of surface states (see Chapter 6). Entirely different examples of such a strong influence are the adsorption at a water surface or a Hg/water (aq. soln.) interface, of large organic molecules, long-chain aliphatic alcohol molecules, aromatic molecules and above all of soap molecules. Here also, small bulk concentrations are sufficient to cause major changes in the interfacial properties. Gouy[34] initiated the study of the adsorption of organic molecules at the Hg/soln. interface. He suggested that the adsorption energy of neutral adsorbed molecules was a function of the electrified state of the interface. This suggestion was developed by Frumkin,[35] who in 1926 gave a theory which is still a basis for the discussion of more recent experiments, and which is reproduced in Chapter 3. Gouy also suggested that the adsorbed molecules were oriented with their hydrophilic part toward the water side of the interface. For the adsorption at the water surface a similar suggestion was made by Harkins[36] and co-workers and, almost simultaneously, by Langmuir,[37,38] who had further developed the experimental techniques of Devaux[39] (see Chapter 5). Thus, as confirmed by many later experiments, the benzene ring, part of a surfactant molecule, will lie flat on the water surface and at the Hg/soln. interface. Aliphatic chains may, at high coverage, stand upright and form a close-packed two-dimensional array. This concept, further developed by Adam, Rideal and others, is not only important for understanding the properties of soap lamellae, but also for the study of biological membranes which therefore are also included in Chapter 5. Two-dimensional arrays, not unlike those in soap lamellae and in adsorbed films, are also assumed to constitute biological membranes, although the latter have a much more complicated structure. Their double-layer characteristics, such as the potential difference caused by them, are intimately connected with their rather anomalous diffusion properties for various ions, especially for Na^+- and K^+-ions and for small polar molecules. Although the physiological nature of the experiments of Galvani and Volta was never forgotten in the nineteenth century, it was only with the work of Overton[40] and of Bernstein[41] in 1902, who stressed the importance of Na^+- and K^+-ions, that modern research concerning the mutual influence of electrical and biological phenomena began.

Molecular structures in an aqueous medium

Studies of the interactions between water molecules and ions and molecules of widely different properties prompted a number of investigations into the molecular structure of water and the effects which various molecules and electrical charges may have upon this structure. Like other liquids close to the melting point, water[42] may be considered as a distorted solid rather than as a dense vapour. X-ray investigations have revealed a tetrahedral structure[43] with an orderliness which extends over a distance of about three molecular diameters, but certainly not more, because of the thermal movements of the strongly polar water molecules. If a noble gas atom or a non-polar foreign molecule is dissolved, the disturbing influence of the water molecules is locally absent and, according to Frank and Evans,[44] an "iceberg" is formed around the foreign atom or molecule. It is the entropy decrease, connected with the iceberg formation, rather than an energy increase, that makes the dissolution of non-polar species in water so difficult. If, however, ions and especially small ions are dissolved, energy is gained. The water molecules prefer to surround the ion and form a "hydration-shell", roughly containing between two water molecules (large size ions) and about ten (small size ions). Hydrated ions may have appreciable volumes, ranging to 100 $Å^3$ or more. The hydrated ions must not be viewed upon as static entities. Already Samoilov[45] pointed out that there may be a rapid exchange of water molecules immediately adjacent to a "bare" ion and water molecules some distance away. More quantitative indications to this effect were given by Hertz[46] who used nuclear magnetic relaxation methods, and by others.[47]

Between the hydration zone and the bulk of water there exists a distorted zone, which may be a few molecular diameters thick. This is the "molten zone". This concept arises again from entropy data. According to estimates based on thermodynamic data of dissolved salts, the K^+ ion, and larger cations, have the effect of increasing the entropy (structure breakers) whereas smaller cations lead to a decrease of the entropy of the system (structure formers). The F^- ion is a structure former, larger anions are structure breakers.[44]

The enforced iceberg formation around a non-polar, hydrophobic, molecule explains why two hydrophobic molecules in water may attract each other and form a "hydrophobic bond". The total number of water molecules involved in the formation of a joint iceberg is smaller than in the two icebergs taken separately. This leads to an entropy increase and therefore to a free energy decrease which may be estimated at one kT per hydrophobic bond. Hydrophobic bonding is therefore an important, though not decisive, factor affecting the stability of protein structures in biological systems. It may well have a bearing on biological membranes and their properties. Hydrophobic bonding may be considered as a first stage of demixing. Complete demixing occurs in micelles, clusters of about thirty surfactant molecules whose hydrophilic parts form, together with the ions in the solution, a double-layer system surrounding the cluster. The natural counterpart of "hydrophobic bonding" is "hydrophilic repulsion" which must exist between two ions surrounded by "molten zones". We shall return to "hydrophilic repulsion" in Chapter 7. There have been a number of theories on the structure of water, an early and in its simplicity attractive one, being the theory of Lennard-Jones and Pople[48] who assumed the hydrogen bonds to be bendable. Detailed models based on the concepts of an "iceberg" and a "molten zone" have been given[49,50] but the validity of these models has been criticized as being too speculative.[51] These speculations are now being superseded by discussions of results obtained by modern experimental techniques[46] whereas new theoretical work deserves attention.[52]

Non-equilibrium phenomena

Although we confine ourselves to equilibrium properties, a few words may be said about non-equilibrium properties and phenomena in order to define the limits of this treatise. We distinguish two classes of non-equilibrium properties. The first class is concerned with the tangential movements of a charged liquid along an interface. A "slipping plane" can then be defined (see Fig. 1.6). The potential at this slipping plane is called the zeta potential. There are numerous phenomena[53] where the two layers adjacent to the slipping plane can be made to move with respect to each other. Examples of these "electrokinetic phenomena" are electro-osmosis and electrophoresis. Electro-osmosis is obtained if a plug of porous material is placed in a liquid containing ions and two electrodes are present in the liquid. The potential difference between the electrodes causes the (charged) liquid to move. The volume of

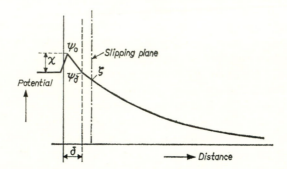

FIG. 1.6. Potential distribution at the metal/solution interface.

the liquid, displaced per unit time, is an indication of the value of the zeta potential. The relation between the *charge of the liquid* and the zeta potential is provided by the Gouy theory. Electrophoresis is observed if particles are present in a liquid and are subject to the influence of an electrostatic field. The double layer at and around the particles is distributed at either side of the slipping plane. The charge at the particle side of the slipping plane causes the particle to move. To a first approximation, the velocity is proportional to the value of the zeta potential. Electrokinetic phenomena have received a great deal of attention in the past, but their interpretation still presents many complications. One difficulty, however, has been definitively removed: the methods of irreversible thermodynamics have shown that the same physical quantity, the zeta potential, is fundamental in all these kinetic phenomena.[54] This was not always recognized earlier. The second class of non-equilibrium phenomena is provided by electrode processes. Here transport of electricity takes place across the interface and not along it. If positive charge carriers move from the electrode to the solution, there is a cathodic current, denoted by i_{cath}; if negative charge carriers move in that direction there is an anodic current to be denoted by i_{an}. The total current $i = i_{cath} + i_{an}$ If the cathodic and the anodic currents are separately zero, then the electrode is ideally polarizable. This was assumed under suitable conditions to be the case for the Hg-electrode. In assuming this, a group of phenomena such as that summarized under the name of polarography is disregarded. In polarography it is precisely the lack of polarizability of the (Hg-) electrode that is important. Given a well-defined potential difference, metal ions in the solution can be discharged, thus enabling a charge to be transferred across the interface. The fact that the ions do this at a well-defined potential is of the utmost importance for analytical purposes as first pointed out by Heyrovsky.[55] The theoretical considerations of electrode processes make use of the

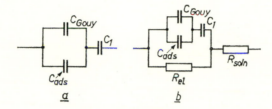

Fig. 1.7. Condenser representations of the double layer at the metal/ solution interface, where specifically adsorbed ions are present. (a) no leakage current assumed, (b) leakage assumed. C_1: inner region capacity; C_{ads}: capacity contribution due to specifically adsorbed ions; C_{Gouy}: contribution due to the Gouy layer; R_{el}: resistance across the double layer; R_{soln}: resistance of the solution.

concept of a potential barrier or a free energy of activation for the electrode reaction. For the ideally polarizable electrode this barrier has infinite height; for the reversible electrode the barrier is assumed to be so low that it can be neglected. As Frumkin[56] pointed out, electrode processes have a bearing on double-layer processes. Consider the most investigated case of the discharge or formation of hydrogen ions at an electrode. The reaction takes place at a small distance δ from the electrode, where the potential, ψ_δ, usually differs from that in the bulk of the solution (see Chapter 3). For the H^+-concentration $(H^+)_s$, which was directly involved in the reaction, Frumkin wrote

$$(H^+) \exp \frac{-e\psi_\delta}{kT}$$

where (H^+) is the bulk H^+-concentration and where the influence of the potential ψ_δ was given in the form of a Boltzmann factor. It was assumed that if a voltage V was applied in the cathodic direction, the potential barrier decreased by αV, where $0 < \alpha < 1$. Figure 1.8 gives a possible scheme for such an influence of V upon the barrier height. The cathodic current was written by Frumkin as

$$i_{cath} = k_1\,(\mathrm{H^+})\,\exp\frac{-e\psi_\delta}{kT}\exp\frac{\alpha V}{kT} \qquad (1.11)$$

where k_1 is a constant of the reaction $\mathrm{H^+} + e \to \mathrm{H}$. On the other hand, the anodic reaction was retarded, the barrier height for this reaction being increased by an amount $+(1-\alpha)V$, as is shown in Fig. 1.8. The anodic current is therefore

$$i_{an} = k_2\exp\frac{-(1-\alpha)\,V}{kT} \qquad (1.12)$$

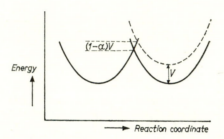

F<small>IG</small>. 1.8. Schematic energy diagram for electrode reaction. There are two energy minima. Application of a potential V leads to the dotted curve. The energy barrier or activation energy is hereby decreased by an amount αV, the fraction α depending on the shape of the curves.

where k_2 now is a rate constant of the reaction $\mathrm{H} \to \mathrm{H^+} + e$. The total current is

$$i = k_1(\mathrm{H^+})\exp\frac{-e\psi_\delta}{kT}\exp\frac{\alpha V}{kT} - k_2\exp\frac{-(1-\alpha)V}{kT}. \qquad (1.13)$$

At zero total current the potential is denoted as V_0. The quantity $V - V_0$ is the overpotential. As Frumkin noted, the potential was a function of the bulk concentration of $\mathrm{H^+}$-ions and of other ions adsorbed. The double-layer properties are reflected in the value of $\psi\delta$ and therefore V–i curves give information on double layers. It is seen that certain approximations of eq. (1.14) lead to

$$V = a + b\log i \qquad (1.14)$$

where a and b are constants ($b = kT/\alpha e$). Such an equation was proposed by Tafel[57] and straight V–log i curves are therefore called Tafel lines.

In the example of the H^+ discharge (H_2-formation) the product, made by one reaction, is the same as that consumed by the other, no new product being formed if $i = 0$. A similar situation is found at the (reversible) AgI/soln. interface. There are many cases, however, in which a new product *is* formed, namely if corrosion has taken place. One example is given in Section 6.4: the dissolution of Ge in O_2-containing water. The result of the corrosion is here the formation of germanic acids. A phenomenon, related to corrosion, is the passivity of iron and other metal surfaces.[58] A thin (~ 50 Å) layer of pore-free iron-oxide, probably γ-Fe_2O_3, is present on the Fe surface and the corrosion proceeds very slowly, the corrosion current density being about 10^{-7} amp/cm^2. Although passivity is only partly understood, and although it is difficult even to give a definition of it,[59] it seems certain that the small corrosion current is due to the low diffusion rate of ions across the oxide layer, which posesses good electronic conductivity.

As will be pointed out in Chapter 6, an important class of non-equilibrium phenomena is provided by semiconductor devices. The principles underlying these phenomena will be given, together with the aspects, which they have in common with some other non-equilibrium phenomena.

Finally, we mention the rather anomalous diffusion phenomena of ions across biological membranes. Numerous problems in that field await further study (see Chapter 5). It is probably here that most work can be expected in the future and that concepts and phenomena, known from the fundamentic work of double layers, can be put to good use.

REFERENCES

1. H. HELMHOLTZ, *Pogg. Ann.*, 1853, LXXXIX, 211.
2. W. THOMSON, *Nature,* 1870, **2**, 56.
3. H. HELMHOLTZ, *Ann. d. Physik u. Chemie (Wiedemann's Ann.)*, N.F., 1879, **7**, 337.
4. L. GALVANI, *De viribus electricitatis in motu musculari commentarius*, Prague, 1791. Galvani carried out his frog's muscle experiment for the first time in 1789. See *Handwörterbuch der Naturwissenschaften* IV, 462 Jena 1913.
5. *L'Opera di Alessandro Volta,* Milano, 1927 (ed. F. MASSARDI, Ass. Elettr. Ital.), p. 155. In a letter, written Feb. 1794, Volta ascribes the animation of the frog's muscle to the electrical difference of the two metals. He constructed his pile in 1799.
6. J. BARDEEN, *Phys. Rev.*, 1936, **71**, 717.
7. I. TAMM, *Phys. Zeits. Sowjetunion,* 1932, **1**, 733.
8. E. J. W. VERWEY, *Rec. trav. chim. Pays-Bas,* 1946, **65**, 521; G. C. BENSON, H. P. SCHREIBER and D. PATTERSON, *Can. J. Phys.,* 1956, **34**, 265.
9. A. N. FRUMKIN, *Russ. J. of Phys. Chem.* (Engl. transl.), 1961, **35**, 1064.

10. W. Nernst, *Theoretische Chemie,* 11th–15th ed., Stuttgart, 1926, p. 849; *Z. physik. Chemie,* 1889, **4,** 129.

11. J. J. Van Laar, *Zeits. Physik. Chemie,* 1893/4, **15,** 457; 1894/5, **18,** 245. *Die Thermodynamik einheitlicher Stoffe,* etc., Gron., 1935, p. 219. Also ref. 10, footnote p. 856.

12. E. Lange, *Handbuch d. Experimentalphysik,* 1933, XII, 2, 265, Akad. Ver. Leipzig.

13. G. Lippmann, *Ann. Chim. Phys.,* 1875, (5), **5,** 494; *J. Phys. Radium,* 1883, (2), **2,** 116.

14. G. Gouy, *Ann. Chim. Phys.,* 1908, (8), **8,** 294; a review in *Ann. Phys.,* 1917, (9), **9,** 129.

15. P. Debye and E. Hückel, *Phys. Zeits.,* 1923, **24,** 185, 305.

16. H. Freundlich, *Kapillarchemie,* 4th ed. Leipzig, 1930.

17. A. N. Frumkin, *Z. physik. Chemie,* 1923, **103,** 43 and, for example, *J. Electrochem. Soc.,* 1960, **107,** 461.

18. W. Schottky, *Zeits. für Physik,* 1939, **113,** 367.

19. N. F. Mott, *Proc. Roy. Soc.* A, 1939, **171,** 27.

20. B. Davydov, *J. Physics U.S.S.R.,* 1939, **1,** 167.

21. C. G. B. Garrett and W. H. Brattain, *Phys. Rev.,* 1955, **99,** 376.

22. R. H. Kingston and S. F. Neustadter, *J. Appl. Phys.,* 1955, **26,** 718.

23. G. L. Pearson and W. H. Brattain, *Proc. of the I.R.E.,* 1955, **43,** 1794.

24. M. Faraday, *Exp. Researches in Electricity,* I, 1839, in Everyman's Library, 1922, p. 44.

25. G. Hermann and S. Wagener, *Die Oxydkathode,* I, Leipzig, 1948.

26. C. Herring and M. H. Nichols, *Rev. Mod. Phys.,* 1949, **21,** 185.

27. G. Ehrlich, *Discussions Faraday Soc.,* 1966, **41,** 1; E. W. Müller and T. T. Tsong, *Field Ion Microscopy,* Elsevier, 1969.

28. O. Stern, *Z. Elektrochem.,* 1924, **30,** 508.

29. I. Langmuir, *J. Am. Chem. Soc.,* 1917, **38,** 1885; 1918, **40,** 1361.

30. A. N. Frumkin, general reviews on double-layer problems, which include specific adsorption, are given in: *J. Electrochem. Soc.,* 1960, **107,** 461; *Trans. Symp. Electrode Processes,* Philadelphia, Pa., 1959, p. 1 (publ. 1961. This symposium was dedicated to the memory of D. C. Grahame). D. M. Mohilner in *Electroanalytical Chemistry,* Vol. I, p. 241 (Marcel Dekker; N.Y., 1966). See also refs. Chapter **3.**

31. D. C. Grahame, general reviews on double-layer problems, which include specific adsorption, are given in: *Chem. Revs.,* 1947, **41,** 441; *Ann. Rev. Phys. Chem.,* 1955, **6,** 337; see also refs. Chapter **3.**

32. R. Parsons, general reviews on double-layer problems, which include specific adsorption, are given in: *Modern Aspects of Electrochemistry* (eds. J. O'M. Bockris and B. E. Conway), Vol. I, Butterworth, 1954; *Advances in Electrochemistry* (ed. P. Delahay), Vol. I, p. 1, Interscience, 1961.

33. J. W. Gibbs, *Collected Works,* Yale U.P., 1948, Vol. I, pp. 230, 234.

34. G. Gouy, *Ann. Chim. Phys.,* 1908, (8), **8,** 294; *Ann. Phys.,* 1917, (9), **9,** 129, a review.

35. A. N. Frumkin, *Z. Physik,* 1926, **35,** 792.

36. W. D. Harkins, *J. Am. Chem. Soc.,* 1917, **39,** 354, 541.

37. I. Langmuir, *J. Am. Chem. Soc.,* 1917, **39,** 1848; *Met. Chem. Eng.,* 1916, **15,** 469.

38. The first work on films was carried out by Miss Pockels, *Nature,* 1891, **43,** 437; 1892, **46,** 418; 1893, **48,** 152; 1894, **50,** 223; *Naturwissenschaften,* 1917, **5,** 137, 149. Lord Rayleigh, *Phil. Mag.,* 1899, **48,** 331.

39. H. Devaux. *J. Phys. Radium*, 1904, (4), **3**, 450; 1912, (5), **2**, 699, 891; *Repert. Smithsonian Inst.*, 1913, p. 261.

40. E. Overton, *Pflügers Archiv*, 1902, **92**, 346; 1904, **105**, 176.

41. J. Bernstein, *Pflügers Archiv*, 1902, **92**, 521; 1910, **131**, 589.

42. J. D. Bernal and R. H. Fowler, *J. Chem. Phys.*, 1933, **1**, 8, 515.

43. J. Morgan and B. E. Warren, *J. Chem. Phys.*, 1938, **6**, 666; G. W. Brady and W. J. Romanow, *J. Chem. Phys.*, 1960, **32**, 306.

44. H. S. Frank and M. W. Evans, *J. Chem. Phys.*, 1945, **13**, 507.

45. O. Ya. Samailov, *Structure of Aqueous Electrolyte Solutions*, Engl. transl. Consultants Bureau, 1965.

46. Symposium on Structures of Water and Aqueous Solutions, *J. Phys. Chem.*, 1970, **74**, no. 21.

47. See Proceedings of Conferences such as: *Hydrogen Bonding* (ed. D. Hâdzi), Pergamon Press, 1959; *The Structure of Electrolyte Solutions* (ed. W. J. Hamer), Wiley & Sons, New York, and Chapman & Hall, London, 1959, *Colloque C.N.R.S.*, 1953, no. 53 in: *J. Chim. Phys.*, 1953, **50**.

48. J. Lennard-Jones and J. A. Pople, *Proc. Roy. Soc. A*, 1951, **205**, 155; J. A. Pople, *ibid.*, p. 163.

49. H. S. Frank and Y. W. Wen, *Discussions Faraday Soc.*, 1957, **24**, 133.

50. G. Nemethy and H. A. Scheraga, *J. Chem. Phys.*, 1962, **36**, 3382, 3401; *J. Phys. Chem.*, 1962, **66**, 1773.

51. A. Holtzer and M. F. Emerson, *J. Phys. Chem.*, 1969, **73**, 26.

52. D. Eisenberg and W. Kauzmann, *The Structure and Properties of Water*, Oxford U.P., 1969.

53. J. Th. G. Overbeek in *Colloid Science*, ed. H. R. Kruyt, Vol. I, Elsevier, 1952, chap. V.

54. S. R. de Groot, P. Mazur and J. Th. G. Overbeek, *J. Chem. Phys.*, 1952, **66**, 1825.

55. J. Heyrovsky, *Chem. Listy Vedu Prumgsl*, 1925, **19**, 168; *Polarographie*, Wien, 1941. M. von Stackelberg, *Polarographische Methoden*, Berlin, 1950.

56. A. N. Frumkin, *Z. Phys. Chem.*, 1933, **164**, 121; reviews in *Advances in Electrochemistry* (ed. P. Delahay), Vols. I and III, Interscience, 1961 and 1963. See also R. Parsons in *Advances in Electrochemistry*, Vol. I.

57. J. Tafel, *Z. Phys. Chem.*, 1904, **50**, 641.

58. K. J. Vetter, *Elektrochemische Kinetik*, Springer, 1961.

59. K. J. Vetter, *J. Electrochem. Soc.*, 1963, **110**, 597.

Note added in proof

A number of interesting contributions in the field of double layer research are found in the Professor Overbeek Anniversary Volume of *J. Electroanal. Chem. and Interfac. Electrochem.*, 1972, vol. **37**.

CHAPTER 2

THE INTERFACE

As REMARKED in the previous chapter, double-layer properties are revealed by the interfacial tension (or surface tension, if a liquid or solid is exposed to vacuum). Some pertinent mechanical and thermodynamic aspects of the interfacial tension will therefore be recalled. If solids are bordering the interface, we make the assumption that they contain only isotropic stresses. With this assumption in mind it is justifiable for our purpose to ignore the difference between solids and liquids at interfaces.[1,2] For simplicity, however, we shall deal mainly with interfaces formed by liquids or by a liquid and a gas (vapour).

2.1. Mechanics

It has long been recognized that a liquid body tends to contract its surface and that, when unconstrained and in the absence of a gravitational field, it assumes the spherical shape. From a mechanical point of view, the tendency to contraction can be described in terms of the interfacial tension, which is located in the interface between the liquid and its surroundings, the interface here being considered as an infinitely thin layer which is homogeneous in two directions. It should be noted at once that this point of view is somewaht misleading because *physically* the interfacial tension is built up by intermolecular forces between molecules which are present in a transition region of finite thickness, but in this section we consider the interfacial tension from a macroscopic point of view. We stress here furthermore its vector character.

The interfacial tension is defined as follows[3]: Imagine a line of arbitrary shape, with length l, in the interface. At both sides of the line the interface is "pulling" at the line. At each line element δl the "pulling forces" are normal to the direction of the line and of magnitude $\gamma \delta l$. The quantity γ is called the interfacial tension or, if the liquid in question is surrounded by a vacuum: the surface tension. Now the liquid (or solid) is always assumed in our treatise, to be in equilibrium with its own saturated vapour. The definition of "surface tension" is therefore never fully obeyed, but we shall frequently use this term since it is common usage.

Techniques used to measure γ often depend on the properties of curved interfaces. Some mechanical properties of curved interfaces will therefore be considered briefly.

First: Assume an interface of spherical shape (radius r). The pressure at the concave side is p_{II} and that at the convex side is p_I. Mechanical equilibrium requires that the total force acting on each element of the interface is zero. Consider a ring on the surface bounded by two circles of radius a and $(a + \delta a)$ respectively, where δa can be assumed to be small. Two kinds of forces act on the ring: a force directed normal to the interface, due to the pressure difference $(p_I - p_{II})$ which is equal to $(p_I - p_{II}) \, 2\pi a (r/a) \, \delta a$, and a force due to the interfacial tension, acting in a direction tangential to the interface. At the upper boundary a force $\gamma 2\pi a$ is pulling in an upward direction and at the lower boundary a force $\gamma 2\pi (a + \delta a)$ is pulling in a downward direction. A requirement of mechanical equilibrium is that the components of the forces in the direction normal to the plane of the rings with radius a and $(a + \delta a)$ should balance each other. This means that

$$(p_I - p_{II}) \cdot 2\pi a \, \delta a = -\gamma \cdot 2\pi \left[\frac{a + \delta a}{r} (a + \delta a) - \frac{a}{r} a \right]$$

or

$$p_{II} - p_I = \frac{2\gamma}{r}. \qquad (2.1.1)$$

According to this equation the pressure inside a droplet is larger than the pressure outside. To estimate this effect, consider a water droplet with $r = 10^{-1}$ cm and a surface tension of 80×10^{-3} N m^{-1} ($= 80$ dyn cm^{-1}). The difference $(p_{II} - p_I)$ then becomes about 1·2 torr. For a mercury droplet of the same size, which has a surface tension about 6 times higher, this becomes about 7 torr.

An equation which is analogous to eq. (2.1.1) can be derived for small crystals obeying certain crystallographic relations: it should be possible (see Fig. 2.1b) to find a point 0 inside the crystal dividing the crystal in as many pyramids as there are crystal planes, all the pyramids having the point 0 as their apex. The different crystal planes have, as a general rule, different interfacial and surface tensions. If the interfacial tension of the ith pyramid is γ^i, then, denoting the height of this pyramid by h^i, the equation which is analogous to eq. (2.1.1), is written as

$$p_{II} - p_I = \frac{2\gamma^i}{h^i}. \qquad (2.1.2)$$

B

This equation is true for any pyramid

$$\frac{\gamma^1}{h^1} = \frac{\gamma^2}{h^2} = \dots = \frac{\gamma^n}{h^n} \quad (i = 1, 2, \dots n). \tag{2.1.3}$$

This result is due to Wulff[4] (Wulff's relations).

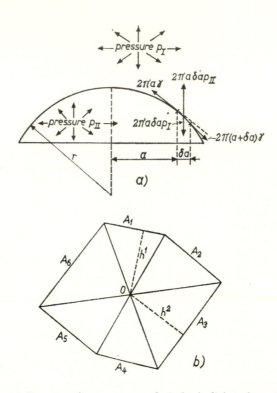

FIG. 2.1. (a) Forces acting on a curved (spherical) interface. (b) Cross section of a "Wulff" crystal.

Second: Consider a droplet on a solid surface. There is a characteristic angle, α, at the common boundary line of the droplet, the underlying surface and the surrounding gas (usually air). This is the contact angle. Its value is determined by the forces acting on the common boundary line. These forces are provided by the interfacial tension γ_{LG} (interface between droplet and gas), by γ_{SG} (interface between underlying surface and gas) and by γ_{SL} (interface between underlying surface and droplet). If these are the only forces, it is seen that (Fig. 2.2a)

$$\gamma_{SG} = \gamma_{SL} + \gamma_{LG} \cos \alpha \tag{2.1.4}$$

which is Young's equation.[5] There can be no equilibrium if

$$\gamma_{SG} < (\gamma_{SL} - \gamma_{LG}) \quad \text{and if} \quad \gamma_{SG} > (\gamma_{SL} - \gamma_{LG}).$$

In the first case the droplet is completely repelled by the underlying surface and will roll down the surface after tilting. In the second case there is complete wetting and the droplet spreads over the whole surface.

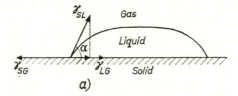

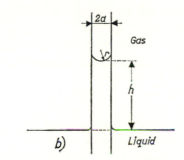

FIG. 2.2. (a) Droplet placed on a solid flat surface. Contact angle is α.
(b) Capillary rise method.

Third: Consider the case of an open capillary, placed vertically in a liquid. After mechanical equilibrium is established the liquid level *inside* the capillary will usually differ from that outside. When $\alpha < 90°$ the level inside will be higher than outside, when $\alpha > 90°$ it will be lower. Assume that the capillary is so narrow that the liquid surface inside, the meniscus, is spherical, with radius r. Then eq. (2.1.1) applies. The difference in height between the levels inside and outside is determined by the pressure difference $p_{II} - p_I$

$$p_{II} - p_I = h\rho g = 2\frac{\gamma}{r} \qquad (2.1.5)$$

where ρ is the density of the liquid and g is the gravitational acceleration. Here $p_{II} - p_I$ can also be interpreted as the pressure which should be applied to keep the levels inside and outside the capillary at the same height.

The cosine of the contact angle is here, in the case of a spherical surface, given by a/r, where a is the radius of the capillary. This, of course, need not always be so. Non-spherical surfaces can be treated by extension of the above ideas (Laplace[6]). Especially for small values of a ($a \leqq 1$ mm), eq. (2.1.5) is a good approximation and can be used for measurements of γ. A well-known example is the Hg/electrolyte interface. The replacement of the gas or vapour by a liquid (i.e. the electrolyte) does not greatly alter the considerations given here. Only the density ρ appearing in eq. (2.1.5) should be replaced by $(\rho_{Hg} - \rho_d)$ where ρ_{Hg} is the density of the mercury and ρ_d is the density of the electrolyte.

Variations of an applied potential difference bring about variations of h and consequently indicate variations of the interfacial tension at the Hg/electrolyte interface. Lippmann[9] was the first to carry out reproducible measurements. Two types of the Lippmann electrometer are shown in Fig. 2.3a and b.

The capillary rise method can be used to measure interfacial tensions down to the order of 10^{-1} N m^{-1} ($= 10^{-3}$ dyn cm^{-1}), as pointed out by Bungenberg de Jong.[10] Many other methods, such as the ring method, the Wilhelmy plate method, the sessile drop method, the drop-weight method, also make use of the mechanics of curved surfaces as outlined here. Descriptions are given in the literature.[10] A famous example of an instrument in which use is made of these methods is the Langmuir trough.[11] With this instrument one can also measure the surface pressure of an insoluble monolayer. Since the surface pressure is the difference between two surface (interfacial) tensions it is a differential method, the other methods mentioned being absolute ones.

2.2. Thermodynamics

The interfacial region between two bulk phases is the seat of the interfacial tension and will be considered as a separate phase. A bulk phase is homogeneous in three directions in space and has a characteristic composition whereas an interfacial phase is homogeneous in two directions and has a characteristic composition. For simplicity only flat interfaces will be considered. It is true that in measurements of the interfacial tension the (mechanical) properties of curved interfaces are often used, but the interfacial tension of a curved interface has the same value as that of the flat interface under the same physico-chemical conditions, unless the radius of curvature becomes of the order of molecular dimensions.

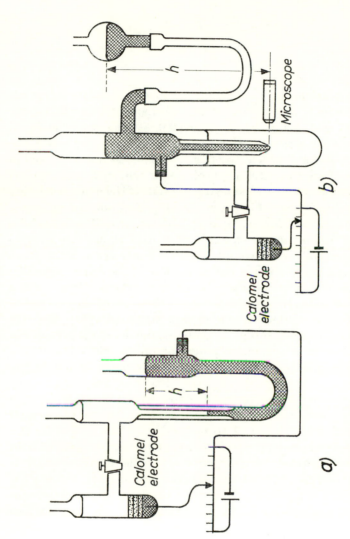

FIG. 2.3. Lippmann electrometers. Two versions. (a) leads to imprecise results and is now rarely used (the capillary should be extremely clean over a considerable length and during the time interval necessary for the experiments). (b) is more often applied. The position of the Hg/electrolyte interface in the conical capillary (lower end of the glass tubing) should remain the same at various values of the potential V. For observation of this position a microscope is used. A modification of version (b) serves to measure the capacity of the double layer at the Hg/electrolyte interface (see Chap. 3).

Thermodynamic equilibrium will be assumed throughout. This implies mechanical and chemical equilibrium and it means that the temperature has the same value throughout the system. The components in each phase must have a uniform chemical potential in that phase, and for a reaction

$$aA + bB + \ldots \leftrightharpoons pP + qQ + \ldots$$

one must have

$$a\mu_A + b\mu_B + \ldots = p\mu_P + q\mu_Q$$

where μ_A, μ_B, μ_P and μ_Q are the chemical or electrochemical potentials of the components A, B, P and Q. These components may be atoms, molecules, ions, electrons or holes. If reversible transport of material of a component i is possible from one bulk phase to another across the interface, then the chemical (electrochemical) potentials μ_i are equal in these phases and in the interfacial phase. Often such transport is impossible. Then the (electro-)chemical potentials of the non-permeating components need not be equal in the two bulk phases nor in some cases, in the interfacial phase. An example of the last case is provided by insoluble films. An important example of a case where the electrochemical potentials of certain components are equal in one bulk phase and the interfacial phase, but not equal to those in the other, adjoining bulk phase, is provided by ideally polarisable interfaces. Thus, for the Hg/electrolyte interface the electrochemical potentials of the ions in the electrolyte are equal to those in the interfacial phase where they are adsorbed, whereas no assumption needs be made about their values in the mercury. The electrochemical potential of the electrons at the mercury electrode will be different from that of the electrons in the metal connection to the reference electrode (Fig. 1.4). Another example of unequal electrochemical potentials in two adjoining phases occurs when a semipermeable membrane separates two solutions of unequal concentrations of the same salt. Two methods can be followed for the thermodynamic treatment of interfaces: that of Gibbs[20] and that of Verschaffelt as developed by Guggenheim.[21] The latter method is affiliated to that of Van der Waals and Bakker (Section 2.5).

Gibbs followed a subtraction procedure: the bulk phases were considered to be homogeneous up to a hypothetical dividing surface located somewhere in the interfacial region. To make the Gibbs model equivalent to physical reality it is necessary to associate a certain amount of matter with the dividing surface. Gibbs derived equations which were invariant with respect to the chosen location of the dividing plane. In the case of unequal chemical potentials in the adjoining phases it is natural to identify Gibbs' dividing plane with the dividing plane which is given by the physical situation. This is often possible and leads to simplification of the pertinent equations. Guggenheim defined an interfacial phase comprising all the inhomogeneities. It has a finite thickness which should extend from (the homogeneous part of) one bulk phase to (the homogeneous part of) another bulk phase. The equations obtained were invariant with respect to the location within the finite volume of the interfacial phase.

Guggenheim's treatment is less abstract than that of Gibbs, the applicability of which to curved interfaces has been challenged,[21] and it offers more physical insight. On the other hand, Gibbs' treatment leads to simpler mathematical expressions and we will use it for double-layer problems.

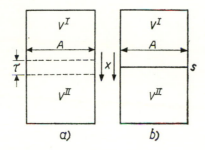

FIG. 2.4. Illustrating the definition of interfacial phase (a) according to Guggenheim, (b) according to Gibbs.

2.2.1. *Thermodynamic potentials*

The reversible change of the total energy or internal energy U^I of a bulk phase I is

$$dU^I = T\, dS^I - p\, dV^I + \sum_{i=1}^{i=c} \mu_i{}^I\, dN_i{}^I \tag{2.2.1}$$

in which S^I is the entropy, V^I is the volume, T the absolute temperature, p the pressure exerted on the phase, $N_i{}^I$ the number of molecules (atoms, ions, electrons, etc.) in phase I of component i $(i = 1 \ldots c)$ and $\mu_i{}^I$ the chemical potential of this component in phase I. Quantities with a superscript "I" are characteristic for phase I. Integration at constant T, p and the chemical potentials leads to

$$U^I = TS^I - pV + \sum_{i=1}^{i=c} \mu_i{}^I N_i{}^I. \tag{2.2.2}$$

The physical meaning of this integration is as follows: when temperature, pressure and chemical potentials are held constant, the entropy density s^{I} and the number densities of the components $n_i{}^{\mathrm{I}}$ also remain constant. The densities are defined by the equations $S^{\mathrm{I}} = s^{\mathrm{I}}V^{\mathrm{I}}$ and $N_i{}^{\mathrm{I}} = n_i{}^{\mathrm{I}}V^{\mathrm{I}}$. The differential dS^{I} can now be written as $s^{\mathrm{I}}\,dV^{\mathrm{I}}$ and the $dN_i{}^{\mathrm{I}}$ become $n_i{}^{\mathrm{I}}\,dV^{\mathrm{I}}$. The integration leading to eq. (2.2.2) amounts to the formation of a phase, with volume V^{I}, through bringing together identical parts with volumes dV^{I}. S^{I}, V^{I} and the $N_i{}^{\mathrm{I}}$ are extensive properties. Their value is proportional to the volume. In contrast, the intensive properties, such as T, p, the $\mu_i{}^{\mathrm{I}}$'s, the refractive index, can be defined in any part of the system, provided this part is large compared with molecular dimensions. A change in the total energy can always be expressed in terms of changes in the extensive properties. The enthalpy H^{I}, the Helmholtz free energy F^{I} and the Gibbs function G^{I} of the bulk phase I are:

$$H^{\mathrm{I}} = U^{\mathrm{I}}+pV^{\mathrm{I}} \qquad = TS^{\mathrm{I}} \; + \sum^{c} \mu_i{}^{\mathrm{I}}N_i{}^{\mathrm{I}}, \qquad (2.2.3)$$

$$F^{\mathrm{I}} = U^{\mathrm{I}}-TS^{\mathrm{I}} \qquad = -pV^{\mathrm{I}}+\sum^{c} \mu_i{}^{\mathrm{I}}N_i{}^{\mathrm{I}}, \qquad (2.2.4)$$

$$G^{\mathrm{I}} = U^{\mathrm{I}}+pV^{\mathrm{I}}-TS^{\mathrm{I}} = \qquad \sum^{c} \mu_i{}^{\mathrm{I}}N_i{}^{\mathrm{I}}. \qquad (2.2.5)$$

The quantities U^{I}, H^{I}, F^{I} and G^{I} are the thermodynamic potentials of phase I.

Consider now two bulk phases I and II and an interfacial phase s. According to the Gibbs definition

$$N_i{}^{s} = N_i-(N_i{}^{\mathrm{I}}+N_i{}^{\mathrm{II}}) \qquad (2.2.6)$$

and so on for the other extensive properties. Here N_i is the number of molecules or other elementary, charged or uncharged, particles of component i in the system, which consists of two bulk phases and an interfacial phase, $N_i{}^{\mathrm{I}}$ and $N_i{}^{\mathrm{II}}$ are these numbers for phases I and II which are assumed to extend homogeneously to a hypothetical dividing plane. As new characteristic determining the state of the system we have the interfacial tension γ and the interfacial area A. If the chemical potentials of all the components are uniform through the whole system, a reversible change dU^{s} of the total energy U^{s} of the interfacial phase can be written as†

$$dU^{s} = T\,dS^{s}+\gamma\,dA+\sum^{c} \mu_i\,dN_i{}^{s} \qquad (2.2.7)$$

---

† Whereas in Section 2.1 γ was expressed in force units per unit length, it is here more appropriate to express γ in energy units per unit area. In the c.g.s. system 1 dyn $\mathrm{cm^{-1}} = 1$ erg $\mathrm{cm^{-2}}$ but in the future many workers will use the Newton (N; 1 N $= 10^5$ dyn) as a force unit, the joule (J; 1 J $= 10^7$ ergs) as an energy unit and the metre instead of the centimetre. A number of workers may find it difficult to use Newtons and metres because the forces involved in their experiments are of the order of dynes rather than of Newtons and the surfaces used have a surface area of a $\mathrm{cm^2}$ rather than of a $\mathrm{m^2}$.

where the chemical potentials μ_i have the same value in the three phases. If this is not the case, i.e. if chemical potentials of the same component are different in different phases, an equation can be written down, which is formally the same as eq. (2.2.7) but it includes now also contributions with $\mu_i{}^I$, $\mu_i{}^s$ and $\mu_i{}^{II}$. We return to this point if the phase rule is considered (Subsection 2.2.2). Integration of eq. (2.2.7) at constant T, γ and the μ_i can be carried out in a way similar to the integration of eq. (2.2.1) and gives

$$U^s = TS^s + \gamma A + \sum^c \mu_i N_i{}^s. \qquad (2.2.8)$$

Extensive properties here are proportional to A. The interfacial analogues of enthalpy, Helmholtz free energy and Gibbs function are:

$$H^s = U^s - \gamma A \qquad = TS^s + \sum^c \mu_i N_i{}^s, \qquad (2.2.9)$$

$$F^s = U^s - TS^s \qquad = \gamma A + \sum^c \mu_i N_i{}^s, \qquad (2.2.10)$$

$$G^s = U^s - TS^s - \gamma A = \qquad \sum^c \mu_i N_i{}^s. \qquad (2.2.11)$$

We write the Gibbs function as a sum of products $\mu_i N_i{}^f$, where f denotes the phase. The Gibbs function of the whole system under consideration here is $G = \sum^c \mu_i N_i$ where G is the sum of G^I, G^{II} and G^s. A potential which is sometimes used is $G^I + G^{II} + F^s = U - TS + pV = G'$ and is called the Gibbs function of the system.[11] However, G' cannot be expressed as a sum of products $\mu_i N_i$.

Introduction of an interfacial phase according to Verschaffelt and Guggenheim gives

$$V = \mathbf{V^I} + \mathbf{V^{II}} + \tau A \qquad (2.2.12)$$

where τ is the thickness of the interfacial phase and where bold-face letters denote quantities relating to Guggenheim's treatment. Now

$$\mathbf{N}_i{}^s = N_i - (\mathbf{N}_i{}^I + \mathbf{N}_i{}^{II}) \qquad (2.2.13)$$

where $\mathbf{N}_i{}^s$ is the full content of component i present in the interfacial phase. The thermodynamic potentials can easily be written down. For example, the Helmholtz free energy $\mathbf{F}^s$ is

$$\mathbf{F}^s = \gamma A + p\tau A + \sum^c \mu_i \mathbf{N}_i{}^s. \qquad (2.2.14)$$

Usually $p\tau \ll \gamma$, and the term $p\tau A$ can be dropped. Thus for the water surface at room temperature and atmospheric pressure one has $\gamma = 80 \times 10^{-3}$ N m^{-1} ($= 80$ dyn cm^{-1}), $p = 10^6$ N m^{-2} ($= 10^7$ dyn cm^{-2}) and τ will be about 10^{-1} cm. Then $p\tau = 10^{-3}\gamma$. At elevated temperatures and pressures, i.e. closer to the critical point, τ is increased and γ is decreased. Under these circumstances it is no longer permitted to neglect $p\tau$ with respect to γ. The theory of Van der Waals on the origin of the interfacial tension (see Section 2.5) deals with liquids close to the critical point.

We now discuss the chemical potentials of bulk phases to give some necessary definitions. The chemical potentials are often written as

$$\mu_i = kT \ln \lambda_i = \mu_{i0}{}^* + kT \ln a_i \qquad (2.2.15)$$

where λ_i is the absolute activity, $\mu_{i0}{}^*$ is a standard chemical potential of the pure species i at given T and p, and a_i is the relative activity. The quantity $\mu_{i0}{}^*$, which is equal to the molecular free energy of the pure species i, is independent of the nature of the other components of the bulk phase. If the bulk phase is an ideal mixture then $a_i = N_i/\sum N_k = x_i$ is equal to the mole fraction. If for *all* values of x_i, large or small, the following equation holds:

$$\mu_i(= kT \ln \lambda_i) = \mu_{i0}{}^* + kT \ln x_i, \qquad (2.2.16)$$

then the bulk phase under consideration is called a perfect solution. If only for small values of x_i a linear relationship between μ_i and $\ln x_i$ holds:

$$\mu_i = \mu_{i0} + kT \ln x_i, \qquad (2.2.17)$$

then we have an ideal dilute solution. As pointed out by Everett[18] the standard chemical potentials μ_{i0} and $\mu_{i0}{}^*$ are often unequal, μ_{i0} often being dependent upon the nature of the other components (i.e. the solvent).

For dilute solutions $N_i/\sum N_k$ approaches N_i/N_1 where the subscript 1 denotes the solvent molecules. The mole ratio N_i/N_1 is closely related to the molality which is defined as the number of moles of solute per kg of solvent. For aqueous solutions 1 kg of solvent contains $10^3/18$ moles. The volume concentration, c_i, is the number of moles per unit volume of the solution. The volume concentration has the disadvantage of being temperature dependent. This complication can be met by defining a reference concentration, $c_i{}^b$ say. Then

$$\mu_i = (\mu_{i0} + kT \ln N_i{}^b/N_1) + kT \ln c_i/c_i{}^b \qquad (2.2.18)$$
$$= \mu_{i0}{}^b + kT \ln c_i/c_i{}^b$$

where $N_i{}^b = c_i{}^b V N_0$, V being the volume and N_0 being Avogadro's number.

If the bulk phase is a one-component ideal gas with pressure p one has, denoting the chemical potential by μ^g,

$$\mu^g = \mu_0{}^g + kT \ln p/p_0 \qquad (2.2.19)$$

where $\mu_0{}^g$ is a standard chemical potential at given temperature T and pressure p_0. Usually $p_0 = 1$ at is chosen. If the gas is in equilibrium with the same component in liquid or solid form, the pressure p is the equilibrium pressure of the liquid or the solid at the given temperature T. We have in this case $p < p_0$. If T is increased then also p rises. At $p = p_0$ we have the boiling point of the liquid or the temperature of vaporization of the solid. At $p > p_0$ the liquid or solid phase is no longer stable. If we have an ideal gas mixture then the partial pressure p_i of component i can be written as px_i. Instead of eq. (2.2.19) we have[18]

$$\mu_i{}^g = [\mu_{i0}{}^g + kT \ln p/p_0] + kT \ln x_i. \qquad (2.2.20)$$

The bracketed part is the chemical potential of the pure component i at temperature T and pressure p.

Fugacities should replace the pressures if non-ideal gases are considered. However, in our applications we shall mainly use ideal one-component gases or vapours.

2.2.2. *The Gibbs–Duhem equation. The phase rule*

It can easily be derived from eqs. (2.2.1) and (2.2.5) that a reversible change dG of a bulk phase can be written in two ways:

$$dG = -S\,dT + V\,dp + \sum^c \mu_i\,d_iN_i, \qquad (2.2.21a)$$

$$dG = \sum^c \mu_i\,dN_i \qquad + \sum^c N_i\,d\mu_i. \qquad (2.2.21b)$$

(We have dropped the superscript "I"). Combination of eqs. (2.2.21a) and (2.2.21b) gives

$$N_0 \sum^c c_i\,d\mu_i = -s\,dT + dp \qquad (2.2.22)$$

where $s = S/V$ is the entropy per unit volume of the phase. Equation (2.2.22) is the Gibbs–Duhem equation. It is a relation between $c + 2$ variables. A relation of this kind can be written down for each phase. We now consider r phases and assume that all the μ_i's have the same value through the whole system. If, furthermore, it is assumed that all the c components are independent, the system contains

$$k = c + 2 - r \qquad (2.2.23)$$

degrees of freedom. Equation (2.2.23) is the Gibbs phase rule. Thus, a one-phase system containing c components has $c + 1$ degrees of freedom: the temperature, the pressure and the composition which is characterized by $c - 1$ parameters x_i.

However, it would be dangerous to apply eq. (2.2.23) without inspection of the particular system under consideration. This is already evident from the fact that we have made assumptions in the derivation.

Five points are worth mentioning here:

1. A new variable, γ, must be considered if the system contains an interface. If the interface is homogeneous over the whole interfacial area then one new phase, the interfacial phase, must be taken care of. We return to this point in Subsection 2.2.8 (Clausius–Clapeyron equation). We notice that the omission in, for example, a liquid–vapour system, of both the interfacial phase and the interfacial tension does not affect the calculated number of degrees of freedom of the system.

2. Usually the condition of electrical neutrality will be imposed upon the system. If charged components are present then this condition reduces the number of degrees of freedom by one.

3. If chemical reactions are possible then at least some of the components are no longer independent. There exist relations between the chemical potentials of some of the components (see the introduction of Section 2.2) and this means again a reduction of the number of degrees of freedom.

4. If between two phases no reversible exchange is possible of a number (say c') of components, then the number of degrees of freedom is increased (by c'). The number of independent components can then be given as $c + c'$.

5. If a space charge is present, then the density of one or more components depends on the location. In this case the definition of "phase" (a homogeneous part of the system) is not fully obeyed. For the case of the terrestrial atmosphere where, just as in a space charge, the density of some components is different from place to place, Van der Waals and Kohnstamm[22] introduced the concept of an infinite number of phases. However, there is no conflict with the phase rule, because for each newly introduced phase a new variable can be introduced and k remains constant. Therefore we loosen somewhat the definition of a "phase" and allow for density variations such as caused by space charges.

2.2.3. *The Gibbs adsorption equation*

This equation is the two-dimensional analogue of the Gibbs–Duhem equation. It can be derived by considering the differential dG^s:

$$dG^s = -S^s \, dT - A \, d\gamma + \sum^c \mu_i \, dN_i{}^s = \sum^c \mu_i \, dN_i{}^s + \sum^c N_i{}^s \, d\mu_i. \quad (2.2.24)$$

The Gibbs adsorption equation then is

$$dy = s^s \, dT - \sum^c \Gamma_i \, d\mu_i \qquad (2.2.25)$$

in which $\Gamma_i = N_i{}^s/A$ is the interfacial excess concentration in particles per unit area, and $s^s = S^s/A$ is the interfacial entropy per unit area. At positive adsorption of component i ($\Gamma_i > 0$) an increase of μ_i leads to a decrease of γ.

The excess quantities s^s and the Γ_i are defined with respect to a certain dividing surface and accordingly vary with variations of the position of the chosen dividing surface. However, at equilibrium not all

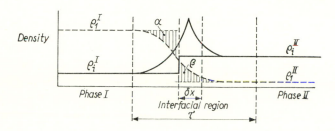

FIG. 2.5. Densities in interfacial region of component I (main component of phase I) and dissolved, adsorbed component i. Two Gibbs dividing surfaces are indicated, a distance δ apart.

μ_i's can be varied independently because of the Gibbs–Duhem equation in the bulk phases. We can eliminate one of the chemical potentials, μ_1 say, where the subscript "1" denotes the main constituent of one of the bulk phases. At the same time the dividing surface can be given a well-defined, fixed, position. To show this, consider the Gibbs–Duhem equations for the two adjoining bulk phases I and II. We have

$$dp = s^I \, dT + N_0 \sum^c c_i{}^I \, d\mu_i = s^{II} \, dT + N_0 \sum^c c_i{}^{II} \, d\mu. \qquad (2.2.26)$$

To eliminate μ_1 write

$$d\mu_1 = -\frac{\Delta s}{\Delta c_1} dT - \sum^c \frac{\Delta c_i}{\Delta c_1} d\mu_i \qquad (2.2.27)$$

where

$$N_0 \, \Delta s = s^I - s^{II} \quad \text{and} \quad \Delta c_i = c_i{}^I - c_i{}^{II} \quad (i = 2 \ldots c).$$

Inserting eq. (2.2.27) into eq. (2.2.25) gives

$$dy = -\left(s^s - \Gamma_i \frac{\Delta s}{\Delta c_1}\right) dT - \sum^c \left(\Gamma_i - \Gamma_1 \frac{\Delta c_i}{\Delta c_1}\right) d\mu_i. \qquad (2.2.28)$$

The quantities $s^s - \Gamma_1(\Delta s / \Delta c_1)$ and $\Gamma_i - \Gamma_1(\Delta c_i / \Delta c_1)$ are invariant with respect to the location of the dividing surface. This may be illustrated by Fig. 2.5, where two dividing surfaces are drawn, a distance δx apart. Inspection shows that $\Gamma_1^\delta = \Gamma_1^0 + \delta x \Delta c_1$ and $\Gamma_i^\delta = \Gamma_i^0 + \delta x \Delta c_i$ the superscripts "δ" and "0" indicating excess concentrations with respect to the two dividing planes. A similar reasoning can be given for the entropy. It is easy to show that

$$\Gamma_i{}^0 - \Gamma_1{}^0 \frac{\Delta c_i}{\Delta c_1} = \Gamma_i{}^\delta - \Gamma_1{}^\delta \frac{\Delta c_i}{\Delta c_1}, \qquad (2.2.29)$$

$$(s^s)^0 - \Gamma_1{}^0 \frac{\Delta c_i}{\Delta c_1} = (s^s)^\delta - \Gamma_1{}^\delta \frac{\Delta s}{\Delta c_1}.$$

The quantities in this equation are the relative adsorption and the relative entropy. They are the same quantities as those appearing in eq. (2.2.28). Their invariance with respect to the location of the dividing plane allowed Gibbs to choose the location such that $\Gamma_1 = 0$. Excess concentrations referring to this dividing surface are denoted by $\Gamma_i(0)$. Gibbs noted that $\Gamma_1/\Delta c_1$ is the distance between the surface where the surface excess is Γ_1 and the surface where $\Gamma_1 = 0$. In many instances the transition between two bulk phases is quite abrupt and the position of the dividing surface can easily be defined. This position can often be chosen such that the interfacial excesses of the main constituents of the bulk phases are both practically zero. Then we are left with the "truly" adsorbed compounds. This is the case with the interfaces Hg/aq. soln. (although even here some difficulties have arisen as to the interpretation of experimental results[23]), the AgI/aq. soln., the semiconductor/aq. soln. interfaces and with most solid/gas interfaces. Less abrupt transitions may occur at the interface between a liquid and its own vapour near the critical point (see Section 2.5).

By using the treatment of Verschaffelt and Guggenheim an adsorption equation can be derived along the same lines as followed here and its invariance with respect to the location of any excess material within the defined interfacial region can be proved.

2.2.4.   *Interfacial tension, interfacial free energy, interfacial energy*

The Gibbs treatment of one-component systems leads to very simple equations because $\Gamma_1(0)$ can be taken zero. Thus

$$F^s = \gamma A. \qquad (2.2.30)$$

Furthermore,

$$S^s \, dT = -A \, d\gamma, \tag{2.2.31}$$

$$U^s = A\left(\gamma + T\left(\frac{\delta\gamma}{\delta T}\right)_A\right). \tag{2.2.32}$$

However, the Guggenheim approach leads to equations in which the physical origin of the interfacial tension becomes more apparent. Thus consider a c-component system and compare $\sum N_i{}^s$ molecules in a bulk phase I with the same number in the interfacial phase s with which phase I is in equilibrium. One has

$$\sum_{}^{c} N_i{}^s \, d\mu_i = -S^I \, dT + V^I \, dp, \tag{2.2.33}$$

$$\sum_{}^{c} N_i{}^s \, d\mu_i = -S^s \, dT + V^s \, dp - A \, d\gamma. \tag{2.2.34}$$

Assuming that $(V^I - V^s)p \ll \gamma A$, the combination of these two equations gives

$$S^s - S^I = A\left(\frac{\delta\gamma}{\delta T}\right)_A. \tag{2.2.35}$$

The same subtraction procedure applied to free energies leads to

$$F^s - F^I = \gamma A. \tag{2.2.36}$$

Owing to the relation $U = F + TS$ one has

$$U^s - U^I = A\left(\gamma + T\left(\frac{\delta\gamma}{\delta T}\right)_A\right). \tag{2.2.37}$$

The fact that the interfacial tension and its derivative with respect to T are obtained by comparing bulk and interfacial properties is more clearly brought out by Guggenheim's approach than by the Gibbs approach.

2.2.5. *The electrochemical potential*[15a]

In a number of cases, and indeed in those which are of importance to double layers, the thermodynamic potentials so far given can not account for charge distributions in the system and it is necessary to consider those parts in the system which have a non-zero average charge. Thus assume a subsystem with volume V' (V' being small, i.e. of small dimensions, compared with the volume of the system) containing a charge Q'. A contribution $\psi' \, dQ'$ must be added to the differential of the total energy dU' of the subsystem, ψ' being the electrostatic potential in the volume V'. Here ψ' is defined with respect to some reference level, preferably that in the bulk of a volume phase where no space charge is expected. Owing to the presence of the charge Q' in the volume V' there must be a non-zero grad ψ'. The electric field in V' has a field strength $E' = -\text{grad } \psi'$ and it polarizes the medium in the small considered volume V'. This polarization leads to the addition to dU' of a contribution $(1/4\pi)E' \, dD' V'$ where the electric displacement $D' = \varepsilon E'$, ε being the dielectric constant of the medium. We consider $D'V'$ as an extensive property of the subsystem. The charge Q' is

$$Q' = e \sum_{}^{c} z_i N_i' \qquad (2.2.38)$$

where $z_i e$ is the charge on an ion of component i and valency z_i. For an electron $z_i = -1$, for monovalent cations $z_i = +1$ and so on. For the volume V', dU' becomes

$$dU' = T \, dS' - p \, dV' + \sum_{}^{c} (\mu_i' + z_i \, e\psi') \, dN_i' + \frac{1}{4\pi} E' \, dD' V'. \qquad (2.2.39)$$

It is now no longer the chemical potential μ_i' which should have the same value everywhere, but the electrochemical potential $\mu_i' + z_i \, e\psi'$. This quantity is denoted as $\tilde{\mu}_i$:

$$\tilde{\mu}_i = \mu_i' + z_i \, e\psi'. \qquad (2.2.40)$$

The electrochemical potential takes account of influences whose origin is to be found outside the considered volume (V' in our case) or even outside the phase. It is evident that we assume our volume V' located somewhere in a double-layer system, either in the space charge region or at the boundary between two phases which is charged owing to the peculiarities of the whole system. Whenever it is possible to distinguish energies derived from space charge properties, from energies derived from interatomic or intermolecular interactions it will be sensible[24] to consider separate terms $\psi' \, dQ'$ although care is needed as has been pointed out by Gibbs and by Guggenheim. It may be noted that the electrochemical potential is an example of what was called by Van der Waals and Kohnstamm[27] the total potential. They considered a number of examples, one of them being that of a small volume in the terrestrial atmosphere where the effect of gravitation was included in the expression of the total potential, another being that of a small volume in the liquid/vapour transition region near the critical point. Here long-range Van der Waals interactions were included.

There is a (usually small) effect of the polarization of the medium upon the electrochemical potential. To derive this polarization contribution, consider the differential of the Helmholtz free energy dF' which is equal to $d(U'-TS')$. This is a total differential and the application of the cross-differentiation procedure gives

$$\left(\frac{\partial^2 F'}{\partial N_i' \, \partial D'}\right)_{T,V',N'_{k \neq i}} = \left(\frac{\partial \tilde{\mu}_i}{\partial D'}\right)_{T,V',N'_i} = V'\left(\frac{\partial E'}{4\pi \, \partial N_i'}\right)_{T,V',D',N'_{k \neq i}}.$$

$$(2.2.41)$$

Defining a concentration c_i' by the equation $N_i' = c_i' V' N_0$ the right-hand side of eq. (2.2.41) can be written as

$$\left(\frac{\partial \varepsilon}{\partial c_i'}\right)\frac{E'}{4\pi \varepsilon N_0}.$$

We stress that the differentiation of E' with respect to N_i' must be carried out at constant T, V', D' and numbers N_k' of the other components. The contribution to $\tilde{\mu}_i$ which arises from the polarization is found by integration to be

$$-\frac{1}{8\pi N_0}\left(\frac{\partial \varepsilon}{\partial c_i'}\right)E'^2.$$

As will be pointed out in Chapter 7 the effect of the addition of component i upon the dielectric constant is often small and accordingly the polarization contribution can often be ignored.

Omitting this, and other effects of similar magnitude, it is often permissible to write

$$\mu_i' = \mu_{i0} + kT \ln c_i' \quad \text{(see Chapter 7)}. \qquad (2.2.42)$$

This equation is approximately valid for ions in solution and for electrons and holes in a semiconductor. Thermodynamic equilibrium requires that the total average force acting on each particle is zero. The total average force, $\bar{f}$, is

$$\bar{f} = kT \operatorname{grad} \ln c_i' + z_i e \operatorname{grad} \psi' = 0. \qquad (2.2.43)$$

Here $kT \operatorname{grad} \ln c_i'$ is the force due to the concentration gradient and $z_i e \operatorname{grad} \psi'$ is the force exerted on a charge $z_i e$ in a field $-\operatorname{grad} \psi'$. The potential ψ' is the potential of the average force. Application of eqs. (2.2.40) and (2.2.42) to two volumes V' and V_0' in the same phase in which the potentials are respectively ψ' and 0, leads to

$$c_i' = c_{i0} \, e^{-(z_i e \psi / kT)} \qquad (2.2.44)$$

where c_{i0} is the bulk concentration of component i. This (Boltzmann-type) equation is used in the theory of the diffuse double layer. We drop the dashes in the future. Possible modifications of this equation are considered in Chapter 7. The most important effects which we have so far omitted are:

the effect of the "self-atmosphere" of the ions (electrons, holes) in a space charge;

the effect of the pressure gradient and of the finite size of the ions;

the polarization effect.

In the following subsections three expressions of the electrochemical potential of a charged component i are of special importance, depending on the location of the particles i. These locations are:

1. the interface (potential $\psi_0 + \chi$), where particles i participate in the interfacial charge $+\sigma$;
2. the Gouy or space charge layer;
3. the bulk of the solution or semiconductor, where grad $\psi = 0$. If there is a reversible exchange possible of particles i between these three locations one has:

$$\tilde{\mu}_i = \underset{\text{(interface)}}{\mu_i{}^s + z_i\,e(\psi_0 + \chi)} = \underset{\text{(space charge)}}{\mu_{i0} + kT \ln c_i + z_i\,e\psi} = \underset{\text{(bulk)}}{\mu_i}. \quad (2.2.45)$$

In other cases the three locations are:

1. a plane close to the interface where the charged particles are specifically adsorbed. Here they contribute to the countercharge $-\sigma$;
2. the Gouy or space charge layer;
3. the bulk of the solution or semiconductor.

This situation exists at the Hg/solution interface, which we shall consider as ideally polarizable. Then, the expression $\mu_i{}^s + z_i\,e(\psi_0 + \chi)$ must be replaced by $\mu_i{}^s + z_i\,e\psi_{\text{ads}}$ where ψ_{ads} is the potential at the plane of the specifically adsorbed ions. Expressions of $\mu_i{}^s$ as a function of coverage are discussed in Section 2.3.

2.2.6. *Gibbs functions of systems containing double layers*

(a) REVERSIBLE INTERFACE: THE AgI/aq.soln. INTERFACE

Assume that this system consists of solid AgI in contact with an aqueous solution containing $AgNO_3$, KNO_3, KI, Ag^+, I^-, K^+, $NO_3{}^-$, and furthermore H^+ and OH^- ions. There are eleven components of which six are ionic. The system should obey the electrical neutrality condition and five conditions of chemical equilibrium:

$$
\left.
\begin{aligned}
e(z_{Ag}N_{Ag} + z_I N_I + z_K N_K + z_{NO_3}N_{NO_3} + z_H N_H + z_{OH}N_{OH}) &= 0 \quad \text{(a)} \\
\mu_{AgI} \;\; &= \tilde{\mu}_{Ag} + \tilde{\mu}_I, \quad\quad\quad\quad\quad\quad\; \text{(b)} \\
\mu_{H_2O} \;\; &= \tilde{\mu}_H + \tilde{\mu}_{OH} \quad\quad\quad\quad\quad\;\;\; \text{(c)} \\
\mu_{KI} \;\; &= \tilde{\mu}_K + \tilde{\mu}_I, \quad\quad\quad\quad\quad\quad\;\; \text{(d)} \\
\mu_{KNO_3} \;\; &= \tilde{\mu}_K + \tilde{\mu}_{NO_3}, \quad\quad\quad\quad\quad\;\; \text{(e)} \\
\mu_{AgNO_3} \;\; &= \tilde{\mu}_{Ag} + \tilde{\mu}_{NO_3}. \quad\quad\quad\quad\quad\;\; \text{(f)}
\end{aligned}
\right\} \quad (2.2.46)
$$

Therefore there are five independent neutral components. These are AgI, H_2O, $AgNO_3$, KNO_3 and KI. The Gibbs function is

$$
\begin{aligned}
G = \; & \mu_{AgI}N_{AgI} + \mu_{H_2O}N_{H_2O} + \mu_{AgNO_3}N_{AgNO_3} \\
& + \mu_{KNO_3}N_{KNO_3} + \mu_{KI}N_{KI}.
\end{aligned}
\quad (2.2.47)
$$

Next, consider the interface. Without much error the Gibbs dividing surface can be placed such that both N_{AgI}^s and $N_{H_2O}^s$ are zero. This limits the number of independent components in the interfacial phase to three.

The interfacial charge $+\sigma$ is provided by Ag^+ and I^- ions

$$
\sigma = e(\Gamma_{Ag} + \Gamma_I) = eA^{-1}(N_{Ag}^s + N_I^s) \quad (2.2.48)
$$

where Γ_{Ag} is the excess, or if negative, of the deficiency per unit area of the Ag^+ ions at the interface. A similar definition can be given for Γ_I. The countercharge $-\sigma$ is provided by Ag^+, I^-, K^+ and NO_3^- ions. We disregard the H^+ and the OH^- ions. As a general rule it is possible to arrange the ionic components such that the charge ($-\sigma$ in our case) is ascribed to only one kind of ion, the ions a say, whereas the remaining ions, and among them also ions a, are arranged in neutral salt molecules b. This way of dealing with double-layer problems will also be followed in the treatment of the ideally polarizable electrode.

The charge $-\sigma$, compensating the interfacial charge $+\sigma$ can now be written as

$$
-\sigma = z_a e \Gamma_a. \quad (2.2.49)
$$

The Gibbs function of the interfacial phase becomes

$$
\begin{aligned}
G^s = \; & \mu_{Ag}^s N_{Ag}^s + \mu_I^s N_I^s + \sigma A[(\psi_0 + \chi) - (z_a e)^{-1}\mu_a] \\
& + \sum \mu_b N_b^s.
\end{aligned}
\quad (2.2.50)
$$

(b) WEAKLY IONIZED MONOLAYERS

Consider the interface between air and an aqueous solution containing a weakly ionized acid HA and an almost completely ionized acid HB. Assume that only the acid HA is able to adsorb at the solution/air interface. There are four ionic components: H^+, OH^-, A^- and B^-, and there are three neutral components: H_2O, HA and HB. At equilibrium the system obeys three chemical conditions arising from the reactions $H_2O \leftrightarrows H^+ + OH^-$, $HA \leftrightarrows A^- + H^+$ and $HB \leftrightarrows H^+ + B^-$. These are

$$\mu_{H_2O} = \tilde{\mu}_H + \tilde{\mu}_{OH}, \tag{a}$$
$$\mu_{HA} = \tilde{\mu}_H + \tilde{\mu}_A, \tag{b}$$
$$\mu_{HB} = \tilde{\mu}_H + \tilde{\mu}_B. \tag{c}$$

$$(2.2.51)$$

Furthermore, there is the condition of electroneutrality:

$$e(z_H N_H + z_{OH} N_{OH} + z_A N_A + z_B N_B) = 0, \tag{d}$$

Therefore there are three independent, neutral, components: HA, HB and H_2O. The Gibbs function is

$$G = \mu_{HA} N_{HA} + \mu_{HB} N_{HB} + \mu_{H_2O} N_{HOO}. \tag{2.2.52}$$

The interfacial charge $+\sigma$ is assumed to consist solely of adsorbed A^- ions and the countercharge $-\sigma$ is provided by A^- ions, B^- ions and (above all) by H^+ ions. Denoting by α^s the degree of ionization at the interface and by N^s the sum of the non-ionized molecules HA and ions A^- adsorbed, the interfacial charge is written as

$$\sigma = z_A \, e N^s (\alpha^s) A^{-1} \tag{2.2.53}$$

whereas the compensating charge $-\sigma$ can be written as

$$-\sigma = z_a \, e N_a{}^s A^{-1} \tag{2.2.54}$$

where it is understood that the ions denoted as "a" are here the H^+ ions. The Gibbs function of the interfacial phase is

$$G^s = \mu_{HA} N^s (1 - \alpha^s) + \mu_A{}^s N^s \alpha^s + \sigma_A [(\psi_0 + \chi) - (z_a \, e)^{-1} \mu_a)] + \sum \mu_b N_b{}^s. \tag{2.2.55}$$

In this case the neutral components "b" are HA and HB.

The value of α^s depends on the value of the electrostatic potential and of course on the values of the ionization constants of the equilibrium reaction $HA \leftrightarrows H^+ + A^-$. Thus, assuming ideal laws, one has:

$$\mu_{HA} = (\mu_{HA})_0 + kT \ln c'(1 - \alpha') \tag{2.2.56}$$

where we have again introduced for the moment dashes to indicate that we are considering a certain volume (V') in the Gouy layer region. In this volume V' the degree of ionization is α' and the concentration is $c'(1 - \alpha')$. In the same volume one has for $\tilde{\mu}_A$

$$\tilde{\mu}_A = (\mu_A)_0 + kT \ln c'\alpha' + z_A \, e\psi'. \tag{2.2.57}$$

These equations, inserted in eq. (2.2.51a), give

$$kT\left[\ln\frac{1-\alpha'}{\alpha}-\frac{z_A\,e\psi'}{kT}\right]=\tilde{\mu}_H-(\mu_{HA})_0+(\mu_A)_0=kT\ln K \quad (2.2.58)$$

in which K is the ionization constant.

Thus, α' varies with variations of ψ'. By virtue of eq. (2.2.56) if α' varies c' also varies. It is seen that variations of ψ' bring about fairly complicated redistributions of molecules HA and ions A^-.[30]

In a similar way an equation for the ionization constant K^s at the interface can be written down. It contains α^s instead of α':

$$kT\ln K^s=kT\left[\ln\frac{1-\alpha^s}{\alpha^s}-\frac{z_A\,e(\psi_0+\chi)}{kT}\right]. \quad (2.2.59)$$

It is not a priori to be expected that K and K^s are equal because the environment of the HA molecules at the interface is different from that in the Gouy layer or in the (space charge free) bulk of the solution.

(c) SEMIPERMEABLE MEMBRANES. DONNAN EQUILIBRIUM

We wish to consider two types of semipermeable membrane. The first type is permeable only to some kinds of dissolved ions and molecules; but not to the solvent molecules; the other type is permeable to some kinds of dissolved ions and molecules and also to the solvent molecules.

Consider a membrane of the first type, which separates two aqueous solutions I and II of a completely ionized salt CA ($z_C=-z_A=1$). The salt concentrations are $C_{CA}{}^I$ and $C_{CA}{}^{II}$. The solutions I and II are two different phases. Assume that C^+ ions can diffuse through the membrane and that A^- ions and water are unable to do so. This situation was considered by Van der Waals and Kohnstamm.[30] In discussions of Donnan equilibria a somewhat more complicated situation is often considered. The ions A^- are large and may contain many elementary charges. A salt CD is assumed to be present of which both ions C and D have low valency and can permeate through the membrane. Our simple system gives

$$\mu_{H_2O}{}^I\neq\mu_{H_2O}{}^{II}\quad\text{and}\quad\tilde{\mu}_A{}^I\neq\tilde{\mu}_A{}^{II} \quad (2.2.60)$$

but

$$\tilde{\mu}_C{}^I=\tilde{\mu}_C{}^{II}=\tilde{\mu}_C. \quad (2.2.61)$$

Assuming that initially $C_{CA}{}^I>C_{CA}{}^{II}$ cations have diffused from I to II. Since they are charged, a potential difference, the Donnan potential, is set up, which counteracts the diffusion of the cations across the membrane. The excess cations in II form a (positive) space charge extending some distance away from the membrane. A corresponding space charge forms at the other side of the membrane, where it has of course an opposite sign.

Denoting the value of the Donnan potential by $\psi^{\text{I}} - \psi^{\text{II}}$, and assuming ideal behaviour one has

$$kT \ln c_C{}^{\text{I}}/c_C{}^{\text{II}} = -z_c\, e(\psi^{\text{I}} - \psi^{\text{II}}). \qquad (2.2.62)$$

The number of diffused cations will be denoted by $N_c{}^d$. The excess positive charge in II is written as $\sigma = z_c\, eN_c{}^d$ and that in I is $-z_c\, eN_c{}^d$. The Gibbs function in I is

$$G^{\text{I}} = \mu_{\text{H}_2\text{O}}{}^{\text{I}}N_{\text{H}_2\text{O}}{}^{\text{I}} + \mu_{\text{CA}}{}^{\text{I}}N_{\text{CA}}{}^{\text{I}} - \tilde{\mu}_C N_c{}^d \qquad (2.2.63)$$

and that in II is

$$G^{\text{II}} = \mu_{\text{H}_2\text{O}}{}^{\text{II}}N_{\text{H}_2\text{O}}{}^{\text{II}} + \mu_{\text{CA}}{}^{\text{II}}N_{\text{CA}}{}^{\text{II}} + \tilde{\mu}_C N_c{}^d. \qquad (2.2.64)$$

Therefore

$$\begin{aligned} G = G^{\text{I}} + G^{\text{II}} = {} & \mu_{\text{H}_2\text{O}}{}^{\text{I}}N_{\text{H}_2\text{O}}{}^{\text{I}} + \mu_{\text{H}_2\text{O}}{}^{\text{II}}N_{\text{H}_2\text{O}}{}^{\text{II}} \\ & + \mu_{\text{CA}}{}^{\text{I}}N_{\text{CA}}{}^{\text{I}} + \mu_{\text{CA}}{}^{\text{II}}N_{\text{CA}}{}^{\text{II}}, \end{aligned} \qquad (2.2.65)$$

i.e. it is again a collection of terms of neutral components.

The second type of membrane differs from that of the first type in that now also water can permeate. Then

$$\mu_{\text{H}_2\text{O}}{}^{\text{I}} = \mu_{\text{H}_2\text{O}}{}^{\text{II}} = \mu_{\text{H}_2\text{O}} \quad \text{and} \quad \tilde{\mu}_c{}^{\text{I}} = \tilde{\mu}_c{}^{\text{II}} = \tilde{\mu}_c \qquad (2.2.66)$$

but

$$\tilde{\mu}_{\text{A}}{}^{\text{I}} \neq \tilde{\mu}_{\text{A}}{}^{\text{II}}. \qquad (2.2.67)$$

In addition to the double layer, now also an osmotic pressure difference is set up. Owing to this pressure difference the value of $(\mu_c)_0$ of Subsection 2.2.2 (the component i of that subsection is here the cation, the quantity μ_{i0} is here $(\mu_c)_0$) may be different in I and in II. We neglect this difference and G is now

$$G = \mu_{\text{H}_2\text{O}}N_{\text{H}_2\text{O}} + \mu_{\text{CA}}{}^{\text{I}}N_{\text{CA}}{}^{\text{I}} + \mu_{\text{CA}}{}^{\text{II}}N_{\text{CA}}{}^{\text{II}}. \qquad (2.2.68)$$

(d) THE INTERFACE Hg/AQUEOUS SOLUTION

Consider an ideal polarizable Hg electrode in contact with an aqueous electrolyte. For thermodynamic purposes the presence will also be assumed of an electrode which is reversible to one of the ionic components. The electric circuit is completed by a potentiometer which is able to vary the potential difference E.

For simplicity an electrode of the first kind will be selected as the reversible electrode. It is denoted by A and is reversible to anions A⁻. Then:

$$e^- + A \rightleftharpoons A^-. \qquad (2.2.69)$$

We could also have assumed the presence of an electrode of the second kind, made of the slightly soluble salt DA. We should have had to consider then also a relation of the kind $\mu_{DA} = \tilde{\mu}_D + \tilde{\mu}_A$. Assume that there are k charged components. One of these consists of the electrons the others are ions, including A⁻. The electroneutrality condition can be completely satisfied by arranging the ions into $(k-2)$ kinds of salt molecules such that in the solution one single ionic species remains for which charge compensation is not complete. This ionic species then serves to balance the charge of the electrons at the Hg electrode. The procedure is analogous to the one described for the reversible case under (a).

One form of the condition of electroneutrality is

$$z_e \, eN_e{}^s + z_A \, eN_A{}^s = 0 \qquad (2.2.70)$$

where $N_e{}^s$ is the number of excess electrons at the Hg-side of the electrode and $N_A{}^s$ is that of the A⁻ ions at the aqueous side. If $N_e{}^s$ is positive, then since z_e and z_A have the same sign, $N_A{}^s$ is negative. Instead of the anions A⁻ we can also choose cations C⁺ for charge compensation. Then

$$z_e \, eN_C{}^s + z_C \, eN_C{}^s = 0. \qquad (2.2.71)$$

The difference between eqs. (2.2.70) and (2.2.71) is that in the latter equation the additional adsorption is assumed of salt molecules CA. We take $z_C = 1$ and $z_A = -1$ and consider a salt CA. It is possible[32] to extend the treatment to more complicated salts C_xA_y (with $x/y = -z_A/z_C$) but this will not be undertaken since no essentially new features are revealed. The importance of considering in this way the adsorption of a salt CA was first stressed by Grahame.[31] We want to retain here a term relating the chemical potential of CA and the surface excess of one kind of ions in the final form of the Gibbs function. To this purpose we shall need the relation

$$\mu_{CA} = \tilde{\mu}_C + \tilde{\mu}_A. \qquad (2.2.72)$$

Furthermore, the neutrality condition eq. (2.2.71) will be observed. For simplicity it will be assumed that only one kind of salt molecules namely CA is present. The Gibbs function of the system Hg/aqueous solution/A is

$$G = \tilde{\mu}_e N_e{}^s + \tilde{\mu}_C N_C{}^s + \mu_{CA} N_{CA} + \mu_{H_2O} N_{H_2O}$$
$$+ \mu_{Hg} N_{Hg} + \mu_A{}^{rev} N_A{}^{rev} \qquad (2.2.73)$$

where $\tilde{\mu}_e$ is the electrochemical potential of the electrons in the mercury and where the superscript "rev" refers to the reversible electrode. The other symbols have their usual meaning. The potential difference E can be written as:[33]

$$E = \frac{1}{z_e\, e}\,(\tilde{\mu}_e - \tilde{\mu}_e{}^{\text{rev}}).\qquad(2.2.74)$$

In semiconductor physics the electrochemical potential of the electrons is usually called Fermi energy. We can adopt this usage here too. Then eq. (2.2.74) can be said to indicate the difference between two **Fermi** energies. By virtue of eq. (2.2.69) one has

$$\tilde{\mu}_e{}^{\text{rev}} + \mu_A{}^{\text{rev}} = \tilde{\mu}_A.\qquad(2.2.75)$$

We note that the subscript "A" in $\tilde{\mu}_A$ refers to the *ions*, whereas the subscript "A" in $\mu_A{}^{\text{rev}}$ refers to *atoms* A in the reversible electrode. Then G can be written as:

$$G = (z_e\, eE - \mu_A{}^{\text{rev}} + \tilde{\mu}_A)N_e{}^s + \tilde{\mu}_C N_C{}^s + \sum_{j=1}^{j=4} \mu_j N_j.\qquad(2.2.76)$$

The subscript "j" denotes neutral components, namely CA, H_2O, Hg and A. We rewrite eq. (2.2.76) by observing that we have assumed $z_e = -z_C = -1$. Therefore $N_e{}^s = N_C{}^s$. Then

$$G = \sigma A\left(E - \frac{\mu_A{}^{\text{rev}}}{e}\right) + \mu_{CA} N_C{}^s + \sum_{j=1}^{j=4} \mu_j N_j \qquad(2.2.77)$$

where $\sigma A = z_e\, e N_e{}^s$.

This Gibbs function for the whole system differs from those in the previous cases (a), (b) and (c), in that it contains explicitly a term referring to the charge σ in the double layer. The Gibbs function of the interfacial phase becomes (we assume the dividing surface to be situated such that $N_{H_2O}{}^s = 0$ and $N_{Hg}{}^s = 0$):

$$G^s = \sigma A\left(E - \frac{\mu_A{}^{\text{rev}}}{e}\right) + \mu_{CA} N_C{}^s.\qquad(2.2.78)$$

(e) GAS ADSORBED AND IONIZED ON A SEMICONDUCTOR SURFACE

Let us assume the following case: oxygen gas is adsorbed in a reversible way on a semiconductor surface. Part of the adsorbed molecules is ionized according to

$$O_2 + e^- \leftrightarrows O_2{}^-.\qquad(2.2.79)$$

The required electrons are withdrawn from the bulk of the semiconductor either because constituent molecules of the semiconductor are ionized or because excess conductivity electrons were already available. There are two neutral components, O_2 and the semiconductor denoted by SC. There are three electrically charged components: holes which are formed if a constituent molecule of the semiconductor is ionized, conductivity electrons and, thirdly, oxygen ions O_2^-. The holes are denoted as h^+ and the conductivity electrons as e^-. There are three conditions which should be obeyed: the condition of electrical neutrality and two chemical conditions, arising from eq. (2.2.79) and from the equilibrium

$$h^+ + e^- \leftrightarrows 0. \tag{2.2.80}$$

This condition will be considered more closely in Chapter 6. For the moment it be sufficient to state that a hole is an electron deficiency around a constituent molecule or, rather, atom of the semiconductor. The arrow pointing left in eq. (2.2.80) indicates the creation of a hole–electron pair and the arrow pointing right indicates the annihilation of such a pair. The electrochemical potential of the conductivity electrons or Fermi energy (see case (d)) is usually denoted by E_F:

$$E_F \equiv \tilde{\mu}_e. \tag{2.2.81}$$

The conditions can be written as:

$$
\begin{aligned}
e(z_e N_e + z_{O_2} N_{O_2} + z_h N_h) &= 0, &\text{(a)}\\
\mu_{O_2} + E_F &= \tilde{\mu}_{O_2}, &\text{(b)}\\
E_F + \tilde{\mu}_h &= 0. &\text{(c)}
\end{aligned}
\tag{2.2.82}
$$

We could have introduced as further components the (partly ionized) impurity atoms in the semiconductor. However, each such impurity brings with it a condition of chemical equilibrium $Imp + e^- \leftrightarrows (Imp)^-$ and the number of independent components would have remained the same. Thus there are two neutral independent components: O_2 and SC. The Gibbs function is

$$G = \mu_{O_2} N_{O_2} + \mu_{SC} N_{SC}. \tag{2.2.83}$$

The interfacial charge $+\sigma$ is provided by adsorbed O_2 ions:

$$\sigma = z_{O_2} e N^s \alpha^s, \tag{2.2.84}$$

whereas the compensating charge $-\sigma$ is provided by electrons and holes. As before, the countercharge will be formally attributed to one component. For a semiconductor this is, in view of eq. (2.2.80), particularly simple:

$$-\sigma = z_e e N_e{}^s.\tag{2.2.85}$$

The Gibbs function of the interfacial phase is ($N_{SC}{}^s$ is chosen zero):

$$G^s = \mu_{O_2} N_{O_2}{}^s (1-\alpha^s) + \tilde{\mu}_{O_2} N_{O_2}{}^s \alpha^s - E_F N_e{}^s.\tag{2.2.86}$$

With $z_e = z_{O_2} = -1$ this can be written as

$$G^s = \underset{\text{(non-ionized)}}{\mu_{O_2} N_{O_2}{}^s (1-\alpha^s)} + \underset{\text{(double layer)}}{\sigma A[(\psi_0 + \chi) - E_F]} + \underset{\text{(chem)}}{\mu_{O_2}{}^s N_{O_2}{}^s \alpha^s}.\tag{2.2.87}$$

There are three contributions: the first term of eq. (2.2.87) gives the contribution of the non-ionized adsorbed oxygen molecules, the second term is the contribution directly referring to the double layer and the third term is the "chemical" contribution of the adsorbed ionized oxygen molecules. It is seen that the Gibbs function here has much in common with that discussed in case (b) (weakly ionized adsorbed monolayers).

2.2.7.   *The Helmholtz free energy and the integral* $-\int_0^{\psi_0} \sigma \, d\psi_0$

We want to stress the relationship between the Helmholtz free energy and the interfacial tension. The latter quantity will play an important role in Section 2.3. It can be expressed in a number of ways:

$$\gamma = A^{-1}(F^s - G^s),\tag{2.2.88}$$

$$\gamma = A^{-1}(F - G + pV).\tag{2.2.89}$$

These two equations follow immediately from definitions given above (eqs. (2.2.10) and (2.2.11)). The third relation which we want to write down is an integrated form of the Gibbs adsorption equation, eq. (2.2.24):

$$\gamma = -A^{-1} \int S^s \, dT - \sum \int \Gamma_i \, d\mu_i + \text{constant}.\tag{2.2.90}$$

We are not interested in the total value of γ, but only in that part which is directly connected with the double layer. This part will be denoted by $\Delta\gamma$, and is defined as the difference of the interfacial tension before and after the double layer has formed. This definition implies that the system should be available in two states and that the formation of the double layer can take place in a reversible manner. Similar definitions as for $\Delta\gamma$ will be given for ΔF, ΔF^s, ΔG and ΔG^s.

The same examples will be discussed as in the previous subsection. We assume that the temperature remains constant. Thus in our examples the integrals $\int S^s \, dT$ are zero.

(a) AgI/AQUEOUS SOLUTION

We consider the system in the following two states: first both σ and ψ_0 are zero and secondly σ and ψ_0 have a finite non-zero value. The first state is identified as the state before the double layer has formed and the second state is the state after double-layer formation. In the underlying case, AgI/aqueous solution, it is not difficult to obtain both states in a reversible manner. The addition of $AgNO_3$ makes the surface charge more positive, the addition of KI makes the surface charge more negative. At a well-defined concentration, $10^{-5\cdot6}$ M of Ag^+ ions, we just have $\sigma = 0$. We note that in the "state before the double layer has formed", the χ-potential is not zero. A reversible increase or decrease of the surface charge in the way described leaves the χ-potential unaltered. This part of the double layer will be left out of the discussion.

In order to obtain an expression for $\Delta\gamma$ the Gibbs adsorption equation (2.2.90) will be used. In this equation the μ_i's must be specified. A convenient way to do this is to reconsider the special form which G^s assumes in the underlying case. This form is given by eq. (2.2.50). We obtain for $\Delta\gamma$

$$\Delta\gamma = -\int \Gamma_{Ag}\, d\mu_{Ag}{}^s - \int \Gamma_I\, d\mu_I{}^s$$
$$-\int \sigma\, d[\psi_0 - (z_a\, e)^{-1}\mu_a] + \int \Gamma_b\, d\mu_b \qquad (2.2.91)$$

in which we have assumed a constant temperature and a constant χ-potential. The limits of the integration are given by the formation process of the double layer. The first two integrals of eq. (2.2.91), i.e. the "chemical" part of $\Delta\gamma$, can often be taken zero for the following reason: The surface charge is in practical cases of the order of 10^{10}–$10^{12}e$ per cm^2. This means that the excess Ag^+ or I^- ions is of this order of magnitude. It is of little importance to the values of $\mu_{Ag}{}^s$ or $\mu_I{}^s$ that this excess amount is incorporated in a lattice which contains more than 10^{14} of these ions per cm^2. Only for very high surface charges the "chemical" part can play a role. This extreme case is considered in Section 4.2. Also the role of the component "a" can often be neglected in eq. (2.2.91). This may be seen as follows: the component a is an ionic component for which we can often choose NO_3^- or K^+. Now assume that the solution contains KNO_3 to the amount of 10^{-3} M. A value of ψ_0 of 60 mV requires a Ag^+ concentration of about $10^{-4\cdot6}$ M. This concentration can be obtained by the addition of a slight amount of $AgNO_3$ to a system, which is in the state prior to the formation of the double layer. It is seen that this addition has only a minor effect upon the concentration of the NO_3^- ions and therefore upon μ_{Ag}. Finally the components "b" are neutral components. They may be adsorbed as a consequence of the formation of the double layer, but this will hardly change their chemical potential. Therefore also this contribution to $\Delta\gamma$ will be very small in most cases. Thus, we are usually left (at constant T and χ) with the integral $-\int_0^{\psi_0} \sigma \, d\psi_0$. For the Helmholtz free energies ΔF and ΔF^s we have

$$\Delta F = -A \int_0^{\psi_0} \sigma \, d\psi_0 + \Delta G + \Delta(pV), \qquad (2.2.92)$$

$$\Delta F^s = -A \int_0^{\psi_0} \sigma \, d\psi_0 + \Delta G^s. \qquad (2.2.93)$$

We can usually take $\Delta(pV) = 0$. We stress the difference between ΔG and ΔG^s. The Gibbs function of the whole system, G, is the sum of quantities related to neutral components (see eq. (2.2.47)). The expression for ΔG consists of the difference of two such sums. It does not contain a term in which appears explicitly σ or ψ_0. In contrast ΔG^s contains a contribution explicitly pertaining to the double layer. This contribution will be denoted by $\Delta G_e{}^s$. We see from eq. (2.2.50) that $\Delta G_e{}^s$ can be written as:

$$\Delta G_e{}^s = A\sigma\psi_0 + A\sigma\chi - A(z_a e)^{-1}\mu_a. \qquad (2.2.94)$$

Especially the first term, $A\sigma\psi_0$, deserves the attention. Its presence means that actually ΔF^s in eq. (2.2.93) contains the integral $+A \int_0^\sigma \psi_0 \, d\sigma$. Thus, comparing eqs. (2.2.92) with eq. (2.2.93) we see that the Helmholtz free energy of the whole system contains a contribution $-A \int_0^{\psi_0} \sigma \, d\psi_0$ whereas the Helmholtz free energy of the interfacial phase contains the contribution $+A \int_0^\sigma \psi_0 \, d\sigma$.

(b) WEAKLY IONIZED MONOLAYERS

Almost the same procedure can be carried out as in case (a). The experiment, which should lead the system from the first to the second state, consists here of the reversible addition of the acid HA to an aqueous solution of the completely ionized acid HB. The components i are now found specified in eq. (2.2.55) which gives G^s for the present case. Thus $\Delta\gamma$ becomes

$$\Delta\gamma = -\int \Gamma^s(1-\alpha^s) \, d\mu_{HA} - \int \Gamma\alpha^s \, d\mu_A{}^s - \int \sigma \, d\psi_0 - \sum \int \Gamma_b \, d\mu_b$$
$$- (z_a e)^{-1} \int \sigma \, d\mu_a. \qquad (2.2.95)$$

The ionic component "a" can be specified here as the H^+ ion. If the addition of HA takes place at constant pH (this requirement can of course not completely be met), then the last integral of eq. (2.2.95) is zero. For reasons, similar to those mentioned under (a), contributions arising from components b will be neglected. The "chemical" contributions, given in the first two integrals, can now become important. Assuming the validity of ideal laws, one has

$$\Delta\gamma = -kT\Gamma - \int_0^{\psi_0} \sigma \, d\psi_0. \tag{2.2.96}$$

The degree of ionization is implicit in the second term. As we shall see in Section 2.3 and in Chapter 5, the first term of eq. (2.2.96) need not be small compared to the second.

For ΔF^s one has

$$\Delta F^s = +A \int_0^{\sigma} \psi_0 \, d\sigma - A\Gamma kT + A\Gamma[(1-\alpha^s)\mu_{HA} + \alpha^s \mu_A^s]$$

$$+ \sigma A[\chi - (z_H e)^{-1}\mu_H]. \tag{2.2.97}$$

For ΔF one has

$$\Delta F = -A \int_0^{\psi_0} \sigma \, d\psi_0 + \Delta G. \tag{2.2.98}$$

For ΔG one usually simply has $\mu_{HA} N_{HA}$. This simple expression involves the approximation, that the addition of HA to the system leaves μ_{HB} and μ_{H_2O} unchanged.

(c) SEMIPERMEABLE MEMBRANES

The same systems are considered as in the previous subsection, i.e. it is assumed that the first membrane considered is permeable only to C^+ ions and impermeable to A^- ions and to water. Then no osmotic pressure difference is set up if the ionic concentration at both sides of the membrane become different. In contrast, in the case of the second membrane considered, which is permeable to water and to the C^+ ions, such an osmotic pressure difference is set up and it will be shown that the equation for it contains the integral $-A \int_0^{\psi_0} \sigma \, d\psi_0$. For both membranes considered it is assumed that there is no specific ionic adsorption at the walls of the membrane.

We compare again two states of the system. First, both solutions separated by the membrane are identical, the salt concentration being $C_{CA}{}^{II}$. The salt is added reversibly to I until a concentration $c_{CA}{}^I$ is reached. This addition is assumed to occur without change of volume.

For the first membrane considered one simply has $\Delta F = \Delta G$ where ΔG is given by the difference of two expressions (2.2.65).

The osmotic pressure difference for the second membrane may be calculated as follows:

If the concentration of CA in I is reversibly increased by steps dc_{CA}, then an osmotic pressure difference in molar units

$$\Pi = \int_{II}^{I} c_{CA}\, d\mu_{CA} \qquad (2.2.99)$$

is set up under equilibrium conditions. Here c_{CA} is a concentration in between $c_{CA}{}^{II}$ and $c_{CA}{}^{I}$. We not only have $\mu_{H_2O}{}^{I} = \mu_{H_2O}{}^{II} = \mu_{H_2O}$ but also $\tilde{\mu}_C{}^{I} = \tilde{\mu}_C{}^{II} = \tilde{\mu}_C$. In order to show the approximations involved[34] in arriving at the well-known classical Van't Hoff result (the proportionality between osmotic pressure and concentration difference) we first assume that the ions are chargeless. Then eq. (2.2.99) becomes

$$\Pi = \int_{II}^{I} c_A\, d\mu_A \qquad (2.2.100)$$

and $d\mu_A$ may be written as (see also Subsection 7.1.1):

$$d\mu_A = \left(\frac{\partial \mu_A}{\partial c_A}\right)_{T,p,c_k} dc_A + \sum \left(\frac{\partial \mu_A}{\partial c_k}\right)_{T,p,c_1} dc_k + \left(\frac{\partial \mu_A}{\partial p}\right)_{T,c} dp \qquad (2.2.101)$$

where c_k indicates all components except A; c_1 indicates all components except k and c indicates all components. For $(\partial \mu_A / \partial p)_{T,N}$ one finds by considering a total differential:

$$\left(\frac{\partial^2 G}{\partial p\, \partial c_A}\right)_{T,c_k} = \left(\frac{\partial \mu_A}{\partial p}\right)_{T,c} = \left(\frac{\partial V}{\partial c_A}\right)_{T,c_k,p} = \bar{V}_A. \qquad (2.2.102)$$

Identifying the pressure p here with the osmotic pressure Π, eq. (2.2.101) becomes:

$$(1 - c_A \bar{V}_A)\, d\Pi = c_A \left(\frac{\partial \mu_A}{\partial c_A}\right)_{T,p,c_k} dc_A + c_A \sum \left(\frac{\partial \mu_A}{\partial c_k}\right)_{T,p,c_1} dc_k. \qquad (2.2.103)$$

If the solution is dilute, then $c_A \bar{V}_A \ll 1$ and can be neglected. If the solution is ideal, then $\partial \mu_A / \partial c_k = 0$. With these two approximations the osmotic pressure difference finally becomes:

$$\Pi = RT(c_A{}^{I} - c_A{}^{II}). \qquad (2.2.104)$$

This procedure can be repeated after the ions are given their charge. Although $d\tilde{\mu}_C = 0$ during the whole process, we wish to replace eq. (2.2.99) by:

$$\Pi = \int_{II}^{I} c_A\, d\tilde{\mu}_A + c \int_{II}^{I} c_C\, d\tilde{\mu}_C. \qquad (2.2.105)$$

This equation becomes, after writing in molar units: $d\tilde{\mu}_A = RT\, d\ln c_A$ $+ z_A\, e\, N_0 \psi'$ and $d\tilde{\mu}_C = RT\, d\ln c_C + z_C\, e\, N_0 \psi$

$$d\Pi = RT(dc_A' + dc_C') + eN_0(z_A c_A' + z_C c_C')\, d\psi'. \qquad (2.2.106)$$

Some of the C^+ ions have diffused from I to II. This makes $c_C{}^I$ and $c_C{}^{II}$ unequal. Assuming for simplicity that the volumes of the two reservoirs are the same and equal to V, say, then $eN(z_C c_C + z_A c_A) = \sigma' A/V$. Integration of eq. (2.2.106) gives

$$\Pi = 2RT(c_{CA}{}^I - c_{CA}{}^{II}) - A/V \int_0^\psi \sigma'\, d\psi' \qquad (2.2.107)$$

where the upper limit of the integration is given by eq. (2.2.62). Since the Gibbs function of this system and therefore also ΔG contains only terms referring to neutral components, it is seen that ΔF contains the integral $-A/V \int_0^\psi \sigma\, d\psi$.

For charged particles it is dangerous to use the concept of ideal solutions especially if one of the particles contains many elementary charges. For such (colloidal) systems, Stigter and Terrell Hill,[35] omitting the integral in eq. (2.2.107), have given equations and indicated how electrostatic properties of colloidal particles can be obtained by osmotic pressure data. In this case, virial terms were added to the ideal Van't Hoff term.

Also for colloidal particles Langmuir[36] argued that osmotic forces may be responsible for their repulsion. These forces, driving the particles apart, were due to the presence of ions between the particles. Use was essentially made of the integral in eq. (2.2.107). The osmotic method will be considered more closely in Section 4.3.

(d) THE Hg/AQUEOUS SOLUTION INTERFACE

The specification of the components i, which is necessary to find an expression for $d\gamma$ in this case is provided by eq. (2.2.78). For $d\gamma$ we find

$$d\gamma = -\Gamma_C\, d\mu_{CA} - \sigma\, d(E - e^{-1}\mu_A{}^{rev}). \qquad (2.2.108)$$

At constant μ_{CA} (and T) this equation reduces to the Lippmann equation. Then usually $d\mu_A$ is held zero, i.e. changes are often brought about without affecting the reversible electrode A.

At constant E (and T) eq. (2.2.108) reduces to

$$d\gamma = -\Gamma_C\, d\mu_{CA}. \qquad (2.2.109)$$

Thus, as pointed out by Grahame, it is possible, by measuring the interfacial tension at different values of μ_{CA}, to find the adsorption of one kind of ions. It is essential that the second electrode is reversible to one of the kinds of ions present in the solution.

For $\Delta\gamma$ one obtains:

$$\Delta\gamma = -\int \Gamma_C \, d\mu_{CA} - \int \sigma \, d(E - e^{-1}\mu_A). \qquad (2.2.110)$$

Strictly speaking, the first term of this equation is a "chemical" term and it is possible to have it zero during the process of the formation of the double layer. Expression for ΔF and ΔF^s can be obtained in the usual way. However, there is a difference with the "reversible" cases (a), (b) and (e). In the cases (a), (b) and (e) the Gibbs function of the whole system contained only terms referring to neutral components. This is no longer so in the present case of an ideally polarizable electrode. Thus both G and G^s (see eqs. (2.2.77) and (2.2.78)) contain the contribution $\sigma A\{E - (\mu_A/e)\}$ and therefore both ΔF and ΔF^s contain the integral $+A \int \psi_0 \, d\sigma$.

(e) GAS ADSORBED AND IONIZED ON A SEMICONDUCTOR SURFACE

Equation (2.2.87) provides the necessary specification of the components. The expression for $d\gamma$ becomes

$$d\gamma = -\Gamma_{O_2}(1-\alpha^s) \, d\mu_{O_2} - \sigma \, d\psi_0 + \sigma \, dE_F - \Gamma_{O_2}\alpha^s \, d\mu_{O_2}{}^s \qquad (2.2.111)$$

where, as usual, we have taken the χ-potential (and T) as constants. Furthermore, the Fermi energy must be considered as a constant of the given semiconductor material. Gas adsorption leaves E_F unchanged. Therefore the term $\sigma \, dE_F$ can be dropped. The case is not unlike that of an acid HA adsorbed on the surface of an aqueous solution of a given pH (case (b)). It also bears resemblance with case (a) where the process of the formation of the double layer was carried out through the addition of $AgNO_3$ to a system containing the AgI/aq. soln. interface and a solution with an NO_3^- concentration which is unaffected by the addition of $AgNO_3$. The expression for $\Delta\gamma$ becomes

$$\Delta\gamma = -\int \Gamma_{O_2}(1-\alpha^s) \, d\mu_{O_2} - \int_0^{\psi_0} \sigma \, d\psi_0 - \int \Gamma_{O_2}\alpha^s \, d\mu_{O_2}{}^s. \qquad (2.2.112)$$

If the adsorbed molecules, ionized or non-ionized, show ideal behaviour, i.e. if $d\mu_{O_2} = d \ln \Gamma_{O_2}(1-\alpha^s)$ and $d\mu_{O_2}{}^s = d \ln \Gamma_{O_2}\alpha^s$, then

$$\Delta\gamma = -kT\Gamma_{O_2} - \int_0^{\psi_0} \sigma \, d\psi_0. \qquad (2.2.113)$$

This equation is quite similar to eq. (2.2.96) derived for the case of ionized adsorbed monolayers.

2.2.8. *The Clausius–Clapeyron equation*

If material can be reversibly exchanged between two phases, the Clausius–Clapeyron equation is a useful relation between various thermodynamic quantities. It has a bearing on adsorption phenomena, because the interfacial phase which contains the adsorbed species can be taken as one of the considered phases. The Clausius–Clapeyron equation was originally derived for a one-component system, for instance for the water–ice equilibrium. Although it may formally be simple to derive analogous equations for multi-component systems, their applicability to experimental data is usually very limited.

We first consider a one-component liquid in equilibrium with its own vapour. Then, the peculiarities with respect to the equilibrium between a bulk phase and an interfacial phase will be pointed out.

For a one-component liquid phase L one has

$$dG^L = -S^L \, dT + V^L \, dp + \mu^L \, dN^L. \qquad (2.2.114)$$

Since $G^L = \mu^L N^L$ one has:

$$N^L d\mu^L = -S^L \, dT + V^L \, dp \qquad (2.2.115)$$

or, introducing $s^L = S^L/N^L$ and $v^L = V^L/N^L$,

$$d\mu^L = -s^L \, dT + v^L \, dp. \qquad (2.2.116)$$

For the vapour phase (V) one has in a similar way

$$d\mu^V = -s^V \, dT + v^V \, dp. \qquad (2.2.117)$$

The equilibrium condition is, apart from uniformity of temperature and pressure through the whole system,

$$\mu^L = \mu^V \quad \text{and} \quad \mu^L + d\mu^L = \mu^V + d\mu^V. \qquad (2.2.118)$$

Therefore, from eqs. (2.2.116) and (2.2.117),

$$(s^L - s^V) \, dT = (v^L - v^V) \, dp. \qquad (2.2.119)$$

The vapour can often be considered as an ideal gas and the molecular volume v^V can be written as kT/p. Usually $v^L \ll v^V$ and eq. (2.2.119) can be written as

$$s^L - s^V = -kT \frac{d \ln p}{dT} \qquad (2.2.120)$$

This is the Clausius–Clapeyron equation for a one-component liquid/vapour system. Denbigh[37] pointed out that according to the phase rule this system has only one degree of freedom. At each temperature the pressure is given unambiguously.

c

Consider now an adsorption system consisting of a one-component gas, an adsorbent, and the adsorbed gas molecules (atoms). As is often allowed this system will be reduced to a gas phase and an interfacial phase consisting of the adsorbed molecules or atoms. The interfacial phase introduces the interfacial tension as a new variable and therefore a comparison of a one-component gas/adsorbed phase system with a one-component vapour/liquid system leads to the result that in the former case the number of degrees of freedom is one more than in the latter case. The pertinent Clausius–Clapeyron equation is similar to eq. (2.2.120) but now the right-hand side of this equation becomes a partial differential: the p–T relationship is only determined if one parameter of the system is held fixed, for instance the surface coverage Γ. Everett has considered this condition.[38] In this case we get:

$$s^{\text{ads}} - s^V = -kT\left(\frac{\partial \ln p}{\partial T}\right)_\Gamma. \qquad (2.2.121)$$

Instead of Γ also γ, the interfacial tension, may be held constant. This may be shown by a more formal treatment than the one given so far. This treatment is based on Everett's analysis, in particular on his discussion of the use of the various thermodynamic potentials. The treatment is suitable for the extension to more than one component. Thus, if the system contains c independent components, then c parameters must be held constant. Of special interest are systems containing charged components. These systems are studied by Parsons,[33] his discussion of Clausius–Clapeyron equations being an extension of Everett's analysis.

For a one-component gas/adsorbed-phase system the differential dG^s is

$$dG^s = -S^s\, dT - A\, d\gamma + \mu^s\, dN^s \qquad (2.2.122)$$

and the differential dF^s is

$$dF^s = -S^s\, dT + \gamma\, dA + \mu^s\, dN^s. \qquad (2.2.123)$$

Cross-differentiation of eq. (2.2.122) gives

$$-\left(\frac{\partial S^s}{\partial N^s}\right)_{T,\gamma} = \left(\frac{\partial \mu^s}{\partial T}\right)_{N^s,\gamma} \qquad (2.2.124)$$

and cross-differentiation of eq. (2.2.123) gives

$$-\left(\frac{\partial S^s}{\partial N^s}\right)_{T,A} = \left(\frac{\partial \mu^s}{\partial T}\right)_{Ns,A}. \qquad (2.2.125)$$

We interpret the right-hand side of these equations as partial differentiations of μ^s (an intensive property) with respect to T respectively at constant γ and Γ.

The equilibrium condition can be chosen in two ways: Either μ^s is equal to μ^V at all temperatures but given γ

$$\mu_\gamma{}^s = \mu_\gamma{}^V; \quad \left(\frac{\partial \mu^s}{\partial T}\right)_\gamma = \left(\frac{\partial \mu^V}{\partial T}\right)_\gamma \tag{2.2.126}$$

or this equality must exist at given Γ

$$\mu_\Gamma{}^s = \mu_\Gamma{}^V; \quad \left(\frac{\partial \mu^s}{\partial T}\right)_\Gamma = \left(\frac{\partial \mu^V}{\partial T}\right)_\Gamma. \tag{2.2.127}$$

For the vapour phase we have

$$dG^V = -S^V\,dT + V\,dp + \mu^V\,dN^V. \tag{2.2.128}$$

Cross-differentiation gives

$$-\left(\frac{\partial S^V}{\partial N^V}\right)_{T,p} = \left(\frac{\partial \mu^V}{\partial T}\right)_p. \tag{2.2.129}$$

We observe the condition that any change in the system must take place at constant Γ. Then, since $\mu^V = \mu^V(p, T)$ we have

$$\left(\frac{\partial \mu^V}{\partial T}\right)_\Gamma = \left(\frac{\partial \mu^V}{\partial T}\right)_{p,\Gamma} + \left(\frac{\partial \mu^V}{\partial p}\right)_{T,\Gamma}\left(\frac{\partial p}{\partial T}\right)_\Gamma. \tag{2.2.130}$$

To obtain the Clausius–Clapeyron equation, eq. (2.2.129) must be subtracted from eq. (2.2.125). We also use the equilibrium condition eq. (2.2.127) and we apply eq. (2.2.130). Then

$$\left(\frac{\partial S^s}{\partial N^s}\right)_{T,A} - \left(\frac{\partial S^V}{\partial N^V}\right)_{T,p} = -\left[\left(\frac{\partial \mu^s}{\partial T}\right) - \left(\frac{\partial \mu^V}{\partial T}\right)_p\right]_\Gamma = -\left(\frac{\partial \mu^V}{\partial p}\right)_{T,\Gamma}\left(\frac{\partial p}{\partial T}\right)_\Gamma. \tag{2.2.131}$$

With $\mu^V = \mu_0{}^V + kT \ln p$ it is easily seen that the right-hand side of eq. (2.2.131) becomes equal to $-kT(\partial \ln p/\partial T)_\Gamma$.

Finally the left-hand side of eq. (2.2.131) can be written in terms of thermodynamic potentials. Thus, dU^s is given by

$$dU^s = T\,dS^s + \gamma\,dA + \mu^s\,dN^s. \tag{2.2.132}$$

Differentiation at constant T and A gives

$$\left(\frac{\partial U^s}{\partial N^s}\right)_{T,A} = T\left(\frac{\partial S^s}{\partial N^s}\right)_{T,A} + \mu^s. \tag{2.2.133}$$

The differential dH^V for the vapour phase is

$$dH^V = T\,dS^V + V\,dp + \mu^V\,dN^V. \tag{2.2.134}$$

Differentiation at constant T and p gives

$$\left(\frac{\partial H^V}{\partial N^V}\right)_{T,p} = T\left(\frac{\partial S^V}{\partial N^V}\right)_{T,p} + \mu^V. \tag{2.2.135}$$

Subtraction of eq. (2.2.135) from eq. (2.2.133) shows that under equilibrium conditions (namely $\mu^s = \mu^V$) we have

$$\left(\frac{\partial U^s}{\partial N^s}\right)_{T,A} - \left(\frac{\partial H^V}{\partial N^V}\right)_{T,p} = T\left[\left(\frac{\partial S^s}{\partial N^s}\right)_{T,A} - \left(\frac{\partial S^V}{\partial N^V}\right)_{T,p}\right]. \tag{2.2.136}$$

The left-hand side is the (isosteric) heat of adsorption. (Quantities referring to a constant coverage or a constant interface composition are called isosteric.)

Consider now as a simple extension of the one-component system a gas AB which is in equilibrium with its (gaseous) components A and B. In that case we have as a general rule non-zero values of Γ_{AB}, Γ_A and Γ_B. Since $\mu_{AB} = \mu_A + \mu_B$, the system is completely determined when, apart from pressure and temperature, two more parameters are given. Restricting ourselves to the interfacial phase these parameters may be chosen between Γ_{AB}, Γ_B, Γ_A and γ. Next, consider the case of adsorbate molecules (atoms) which upon adsorption are ionized: $A_{\text{ads}} \rightleftarrows A^+_{\text{ads}} + e^-$. We assume an electron–hole equilibrium in the adsorbent: $\tilde{\mu}_e + \tilde{\mu}_h = 0$. The parameters which must now be held constant may be chosen between Γ_A, Γ_{A^+}, Γ_{e^-} and γ. A linear combination, for example $(\Gamma_{A^+} - \Gamma_{e^-})$, together with, for instance, Γ_A and of course temperature and pressure also provides the parameters which according to the phase rule must be necessarily given to obtain a Clausius–Clapeyron equation. In this way the requirement of a constant interfacial charge $\sigma = e(\Gamma_{A^+} - \Gamma_{e^-})$ enters. In a similar way the other cases dealt with in the previous subsections can be shown to lead to a similar requirement. Parsons pointed out that Clausius–Clapeyron equations can also be obtained when ψ_0 and not σ is held constant. Thus at least four Clausius–Clapeyron equations can be obtained in each case depending on the differential we wish to start with: dG^s, dF^s, $d(G^s - \sigma\psi_0)$ or $d(F^s - \sigma\psi_0)$.

Selecting the interfacial charge σ together with the interface composition (such as is in agreement with the phase rule, see above) denoted by Γ_j, as the parameters to be held constant, the equilibrium condition which replaces eq. (2.2.127) is for component i:

$$(\mu_i^s)_{\Gamma_j, \sigma} = (\mu_i^b)_{\Gamma_j, \sigma}; \quad \left(\frac{\partial \mu_i^s}{\partial T}\right)_{\Gamma_j, \sigma} = \left(\frac{\partial \mu_i^b}{\partial T}\right)_{\Gamma_j, \sigma}. \tag{2.2.137}$$

As a general rule one can write $\mu_i^b = \mu_i^b(p, T, x_k)$, where x_k indicates the composition of the bulk phase b and where the subscript k denotes the independent components. Then

$$\left(\frac{\partial \mu_i^b}{\partial T}\right)_{\Gamma_j,\sigma} = \left[\left(\frac{\partial \mu_i^b}{\partial T}\right)_{p,xk} + \left(\frac{\partial \mu_i^b}{\partial p}\right)_{T,xk}\left(\frac{\partial p}{\partial T}\right)_{xk}\right.$$

$$\left. + \sum \left(\frac{\partial \mu_i^b}{\partial x_k}\right)_{p,T,x_1 \neq k}\left(\frac{\partial x_k}{\partial T}\right)_{p,x_1 \neq k}\right]_{\Gamma_j,\sigma}. \qquad (2.2.138)$$

If the bulk phase is an ideal gas, eq. (2.2.20) applies for μ_i and, if i is an independent component, the heat of adsorption is given by $-kT^2(\partial \ln px_i/\partial T)_{\Gamma_j,\sigma}$. If the bulk phase is an ideal solution then eq. (2.2.18) applies and the heat of adsorption is given by $-kT^2(\partial \ln c_i/\partial T)_{\Gamma_j,\sigma}$. If the component i is not an independent component, the situation is more complicated. Thus, consider the adsorption of ions A$^-$ provided by an ideal solution of the completely ionized salts CA and CD.[39] Here eq. (2.2.138) leads to the result, that the heat of adsorption is given by

$$-\alpha kT^2\left(\frac{\partial \ln c_A}{\partial T}\right)_{\Gamma_j,\sigma}$$

where $\alpha = 2$ for small concentrations of CD, such that $c_A \cong c_C$, and where $\alpha = 1$ for large concentrations of CD. As Parsons concluded, a rather detailed knowledge of the interfacial composition is needed, before heats of adsorption can be determined by means of Clausius–Clapeyron equations. However, Clausius–Clapeyron equations have been used for a wide variety of systems, ranging from gas adsorption and adsorption from liquids to heats of formation of micelles.

2.3.   The theory of Gouy (space charge theory)

As explained in Chapter 1, the theory of Gouy is of primary importance for a number of double-layer systems, notably in those considered in the following chapters (3–6). Therefore the main aspects of the theory are given here. Possible corrections are discussed in Chapter 7.

Gouy realized that a relatively small value of the capacity at the Hg/soln. interface (he developed his theory with this case in mind) could be explained by a relatively large value of the mean distance of the ions from the Hg surface. He assumed that the ions were subject to thermal agitation which led to diffusion and which counteracted the electrostatic interaction with the charged interface. Gouy's theory is closely related to the theory of Debye and Hückel for dilute electrolytes. Both theories use the so-called Poisson–Boltzmann equation as their basis. The Poisson equation relates the potential ψ at a certain position, with the charge density ρ at the same position:

$$\nabla^2 \psi = -\frac{4\pi}{\varepsilon} \rho \qquad (2.3.1)$$

in which ∇^2 is the Laplace operator. The dielectric constant ε is assumed to have the same value everywhere. For the flat plate approximation one has

$$\nabla^2 = \frac{d^2}{dx^2} \qquad (2.3.2)$$

in which x is the distance from the flat plate. Debye and Hückel considered the potential around a selected ion. Then

$$\nabla^2 = \frac{d^2}{dr^2} + \frac{2}{r}\left(\frac{d}{dr}\right)^2 = \frac{1}{r^2}\frac{d}{dr}\left(r^2 \frac{d}{dr}\right) \qquad (2.3.3)$$

in which r is the distance from the centre of the ion. If the system considered has cylindrical symmetry, then

$$\nabla^2 = \frac{d^2}{ds^2} + \frac{1}{s}\left(\frac{d}{ds}\right)^2 = \frac{1}{s}\frac{d}{ds}\left(s \frac{d}{ds}\right) \qquad (2.3.4)$$

in which s is the distance from the core of the cylinder.† The flat plate approximation is nearly always appropriate for electrode/solution interfaces, because the radius of curvature will nearly always be large compared with the thickness of the double layer. Equation (2.3.3) has also been used for the description of diffuse layers around spherical colloidal particles.[42] Equation (2.3.4) has been used for cylindrical colloidal particles.[43]

The charge density ρ can be written as $e \sum z_i n_i$, where n_i is the concentration (we take the concentration here in numbers per cm³) of charge carrier i with charge $z_i e$ at the considered position. The charge carriers consist of only ions if solutions are considered and of electrons, holes and ionized atoms at fixed positions if semiconductors are considered. We first deal with the solution case. Owing to the thermal movements of the ions, the charge density fluctuates somewhat. The expression $e \sum z_i n_i$ to be used is an averaged quantity. Consequently

---

† In the expressions (2.3.2), (2.3.3) and (2.3.4) of the Laplace operator, three coordinates should appear instead of one. However, we consider here ψ only as a function of the distance from plate, sphere or cylinder and disregard its dependence upon variations in other directions.

also the potential in eq. (2.3.1) is an averaged potential. For z–z electrolytes, to which we restrict our attention, one has

$$\rho = z\,e(n_C - n_A) \tag{2.3.5}$$

in which the subscripts C and A indicate cations and anions. The ions are considered as point charges. They are assumed to be distributed according to the Boltzmann equations:

$$n_C = n\,e^{-ze\psi/kT} \quad \text{and} \quad n_A = n\,e^{+ze\psi/kT} \tag{2.3.6}$$

in which, strictly speaking, ψ is the potential of the average force $z\,e\,\mathrm{grad}\,\psi$ (see Subsection 2.2.4) and not the averaged potential used in eq. (2.3.1). We neglect the difference, however. Combination of eqs. (2.3.1), (2.3.2), (2.3.5) and (2.3.6) leads to the Poisson–Boltzmann equation for flat diffuse layers, i.e. for the flat Gouy layer:

$$\frac{z\,e}{kT}\frac{d^2\psi}{dx^2} = \kappa^2 \sinh\frac{z\,e\psi}{kT}. \tag{2.3.7}$$

Here

$$\kappa^{-1} = \sqrt{\left(\frac{\varepsilon kT}{8\pi n(z\,e)^2}\right)} \tag{2.3.8}$$

is the Debye–Hückel length. For a 10^{-3} normal 1–1 electrolyte in water at room temperature $\kappa^{-1} = 100$ Å.

So far three assumptions are made: the volume of an ion is zero; the value of ε is the same at every value of x (i.e. ε is assumed independent of the ionic concentration and of the field strength in the diffuse layer); the potentials appearing in the Boltzmann equation and in the Poisson equation are identical. We take here their validity for granted and proceed to the solution of eq. (2.3.7). One notes that

$$2\int \frac{d^2y}{dx^2}\,dy = \left(\frac{dy}{dx}\right)^2 - C \tag{2.3.9}$$

in which C is a constant, to be determined by the boundary conditions of the problem. Identifying y with $(z\,e/kT)\psi$, one obtains, if eq. (2.3.7) is inserted into eq. (2.3.9),

$$\left(\frac{dy}{dx}\right)^2 = 4\kappa^2 \sinh\frac{y}{2} + C \quad [y = (z\,e/kT)\psi]. \tag{2.3.10}$$

For $x = \infty$ (far in the solution) we require $y = 0$ and $dy/dx = 0$. This makes $C = 0$. There are two roots of eq. (2.3.10). The negative root is needed here, because at an increase of x, the absolute value of y should decrease.

$$\frac{dy}{dx} = -2\kappa \sinh \frac{y}{2}.$$ (2.3.11)

This equation serves to find the charge in the Gouy layer in the following way. The ions are assumed to be in thermal movement up to a plane $x = \delta$ close to the interface. It was already realized by Gouy that the position of this plane and that of the interface would not coincide, owing to the finite hydration volumes of the ions. However, this idea was not fully worked out by him. We take here the plane $x = \delta$ at a distance from the interface which is of the order of the thickness of the Helmholtz layer. If all the ions are in thermal movement and none is located at the interface then the charge per unit area of the Gouy layer is $-\sigma$ where

$$-\sigma = \int_{\delta}^{\infty} \rho \, dx = -\frac{\varepsilon}{4\pi} \int_{\delta}^{\infty} \frac{d^2\psi}{dx^2} \, dx = \frac{\varepsilon}{4\pi} \left(\frac{d\psi}{dx} \right)_{x=\delta}.$$ (2.3.12)

The value of ψ at $x = \delta$ is denoted as ψ_δ and y at this plane is denoted as z.† The combination of (2.3.11) and (2.3.12) leads to

$$\sigma = \frac{\varepsilon\kappa}{2\pi} \sinh \frac{z}{2}.$$ (2.3.13)

The integration of eq. (2.3.11) leads to an explicit relationship between x and y $(=(ze/kT)\psi)$. After some algebraic manipulations one finds from eq. (2.3.11)

$$-\kappa \int_{x}^{0} dx = \kappa x = \ln \tanh \frac{z}{4} - \ln \tanh \frac{y}{4}$$ (2.3.14)

or

$$y = 2 \ln \frac{1 + e^{-\kappa x} \tanh z/4}{1 - e^{-\kappa x} \tanh z/4}.$$ (2.3.15)

For high values of z, $\tanh z/4$ approaches unity. Therefore one has always $e^{-\kappa x} \tanh z/4 < 1$ and a good approximation is

$$y = 4 e^{-\kappa x} \tanh z/4.$$ (2.3.16)

This approximation has proved to be very useful in colloid chemistry.

---

† There should be no confusion between the use here of z as a dimensionless expression for the potential at $x = \delta$ and its use elsewhere for charge number of an ion.

Another approximation is found for small z ($z \ll 1$). Then eq. (2.3.11) becomes

$$dy/dx = -\kappa y \qquad (2.3.17)$$

and its solution is for our case

$$y = z\, e^{-\kappa x}. \qquad (2.3.18)$$

This approximation, also very useful in colloid chemistry, is essentially the same as that used by Debye and Hückel. They found a $(\kappa r)^{-1}\, e^{-\kappa r}$ law for the potential. The potential around a charged cylinder in an electrolyte follows approximately the $(\kappa s)^{-\frac{1}{2}}\, e^{-\kappa s}$ law, the approximation being invalid at low values of κs, where a logarithmic law is found.

From eq. (2.3.11) the (Gouy) differential capacity C_G is found:

$$C_G = \frac{\partial \sigma}{\partial \psi_\delta} = \frac{\varepsilon \kappa}{4\pi} \cosh \frac{z}{2}. \qquad (2.3.19)$$

Its minimum value ($|z| \to 0$) is

$$C_G = \frac{\varepsilon \kappa}{4\pi}. \qquad (2.3.20)$$

This is the expression of the capacity of a "flat-plate" condenser with "thickness" κ^{-1}.

For $|z| \gg 1$ one has

$$C_G = \frac{\varepsilon \kappa}{8\pi}\, e^{\frac{1}{2}|z|}. \qquad (2.3.21)$$

In Subsection 2.2.7 the importance was stressed of the integral $-\int_0^{\psi_\delta} \sigma\, d\psi$. The integration had to be carried out over all values of the interfacial potential between zero and a final value ψ_0 which was given by the physical situation, ψ_0 being the potential at the interface where the charge σ was located.

We consider here the integral $-\int_0^{\psi_\delta} \sigma\, d\psi$ as the (Helmholtz) free energy of the diffuse part of the double layer and denote it by ΔF_G. By using eq. (2.3.13) the integration gives

$$\Delta F_G = -8n\kappa^{-1}kT\left(\cosh \frac{z}{2} - 1\right). \qquad (2.3.22)$$

This is a negative quantity, both for positive and for negative values of z. Two extreme cases of this integral will now be considered. For convenience we assume 1–1 electrolytes. For small values of z, positive or negative, one has

$$\sigma = \frac{\varepsilon \kappa}{4\pi} z \qquad z = (e/kT)\psi_\delta \qquad (2.3.23)$$

and the integral becomes

$$\Delta F_G = -\frac{2\pi}{\varepsilon\kappa}\,\sigma^2\,kT/e = -\frac{2\pi\,e}{\varepsilon\kappa}\,(\sigma/e)^2 kT. \qquad (2.3.24)$$

For large absolute values of z, such that

$$\sigma = +\frac{\varepsilon\kappa}{4\pi}\,e^{\frac{1}{2}|z|} \quad \text{for positive values of } z,$$

$$\sigma = -\frac{\varepsilon\kappa}{4\pi}\,e^{\frac{1}{2}|z|} \quad \text{for negative values of } z, \qquad (2.3.25)$$

we have

$$\Delta F_G = -2(\sigma/e)kT \quad \text{for positive values of } z,$$

$$\Delta F_G = +2(\sigma/e)kT \quad \text{for negative values of } z. \qquad (2.3.26)$$

If the adsorbed charge $+\sigma$, which compensates the charge $-\sigma$ of the Gouy layer, consists of Γ particles per unit area with each a charge $+e$ (for positive σ) or $-e$ (for negative σ) then

$$\sigma = +e\Gamma \quad \text{or} \quad \sigma = -e\Gamma \qquad (2.3.27)$$

and it appears that

$$\Delta F_G = -\frac{2\pi\,e}{\varepsilon\kappa}\,\Gamma^2 kT \quad \text{(small } z\text{-values)}, \qquad (2.3.26a)$$

$$\Delta F_G = -2\Gamma kT \qquad \text{(large absolute } z\text{-values).} \qquad (2.3.26b)$$

2.4. Equations of state. Adsorption isotherms

Equations of states of adsorbed particles relate the interfacial pressure π to the number Γ_i per unit area of adsorbed particles. They are based on a, preferably simple, physical model. Two classes of such models can be distinguished:

(i) the particles move freely or almost freely over a structureless interface;

(ii) the particles are located at "sites" on the interface which are connected with the atomic structure of the interface. Surface migration is possible and may be described by a desorption–adsorption sequence.

The best-known models are enumerated below and some electrostatic aspects will be mentioned. The adsorption of only one component, component i, will be considered. We define the interfacial pressure π as

$$\pi = -(\gamma - \gamma^0) \qquad (2.4.1)$$

where γ is the interfacial tension of the system in the presence of component i and γ^0 is this quantity in the absence of component i. For an ideal polarizable interface the differential $d\pi$ can be written as

$$d\pi = A^{-1}(S^s - S^{s0})\, dT + (\sigma - \sigma^0)\, dE + \sum_{j \neq i}(\Gamma_j - \Gamma_j{}^0)\, d\mu_j + \Gamma_i\, d\mu_i \quad (2.4.2)$$

where we have taken as a starting point the system in the absence of component i at the same T, E and μ_j as afterwards. Here the electrons are an independent component. On the other hand, for an interface in true equilibrium this is not the case. Thus for the AgI/soln. interface, the term $+(\sigma - \sigma^0)\, dE$ should be replaced by $e^{-1}(\sigma - \sigma^0)\, d\mu_{Ag}$ or by $(\sigma - \sigma^0)\, d\psi_0$. For the other special cases, such as those enumerated in Subsections 2.2.2–2.2.5, the relevant expressions for dG^s can be used to derive expressions for $d\pi$.

If the temperature and all the chemical potentials except μ_i are kept constant, then

$$d\pi = \Gamma_i\, d\mu_i. \quad (2.4.3)$$

Since we consider only one component, we omit the subscript "i". In equilibrium one has

$$\mu^s = \mu^b = \mu. \quad (2.4.4)$$

The superscripts "s" and "b" indicate interface and bulk quantities respectively. For the bulk phase we write:

$$\mu^b = \mu_0{}^b + kT \ln \frac{a}{a_0}. \quad (2.4.5)$$

It will prove convenient to introduce a standard activity a_0 here. If the bulk phase is an ideal gas, then the activity a can be replaced by the pressure p and if it is an ideal solution then it can be replaced by the concentration c.

Adsorption equations give the relation between the activity a (or p or c respectively) and the amount adsorbed $N^s = A\Gamma$. Once this relation is known for a particular system, eqs. (2.4.3), (2.4.4) and (2.4.5) lead to an equation of state, i.e. to a π–Γ relationship. Such an equation of state can only occasionally be identified with a theoretical one. In fact as pointed out by Trapnell[44] and by Everett[45] often more than one theoretical equation of state fits the same adsorption isotherm.

For class 1 models we distinguish four cases:

(Ia) The adsorbed particles behave as a two-dimensional ideal gas. The equation of state is

$$\pi = \Gamma kT. \quad (2.4.6)$$

The chemical potential is

$$\mu = \mu_0{}^s + kT \ln \frac{\Gamma}{\Gamma_0} \quad (2.4.7)$$

where Γ_0 is a chosen reference value. The adsorption equation is

$$\Gamma = a\frac{\Gamma_0}{a_0}e^{-\Delta\mu_0/kT} \tag{2.4.8}$$

where $\Delta\mu_0 = \mu_0{}^s - \mu_0{}^b$ is the standard free energy of the adsorption. It is often denoted as ΔG_0.

Equation (2.4.8) is Henry's law. Other isotherms often approach this law for low Γ-values but this is not a general rule.

The chemical potential, and also $\Delta\mu_0$, can be written as the difference of two terms, an enthalpy term and an entropy term. Even for the one-component system considered in this section various definitions of these terms are possible and it is important to know which definition is observed. Thus, μ^s can be written in at least three ways:

(1) The relation

$$G^s = \mu^s N^s = H^s - TS^s \tag{2.4.9}$$

leads to

$$\mu^s = \frac{H^s}{N^s} - T\frac{S^s}{N^s} = \hbar^s - T\bar{s}^s. \tag{2.4.10}$$

(2) The differential

$$dU^s = T\,dS^s + \gamma\,dA + \mu^s\,dN^s \tag{2.4.11}$$

leads to

$$\mu^s = \left(\frac{dU^s}{\partial N^s}\right)_{T,A} + T\left(\frac{\partial S^s}{\partial N^s}\right)_{T,A} = u^s - Ts_{T,A}{}^s. \tag{2.4.12}$$

(3) The differential

$$dH^s = T\,dS^s - A\,d\gamma + \mu^s\,dN^s \tag{2.4.13}$$

leads to

$$\mu^s = \left(\frac{\partial H^s}{\partial N^s}\right)_{T,\gamma} - T\left(\frac{\partial S^s}{\partial N^s}\right)_{T,\gamma} = h^s - Ts_{T,\gamma} \tag{2.4.14}$$

whereas for the two-dimensional ideal gas which we consider here, we have eq. (2.4.7).

The entropy S^s can be given as

$$S^s = -\left(\frac{\partial F^s}{\partial T}\right)_{A,N^s} = -A\left(\frac{\partial\gamma}{\partial T}\right)_{A,N^s} - N^s\left(\frac{\partial\mu_0{}^s}{\partial T}\right)_{A,N^s} - N^s k\ln\frac{\Gamma}{\Gamma_0}. \tag{2.4.15}$$

Since $\gamma = \gamma_0 - \Gamma kT$ we have

$$-A\left(\frac{\partial\gamma}{\partial T}\right)_{A,N^s} = -A\left(\frac{\partial\gamma_0}{\partial T}\right)_{A,N^s} + kN^s. \tag{2.4.16}$$

The differential entropies $s_{A,T}{}^s$ and $s_{x,T}{}^s$ have been given some consideration in Subsection 2.2.8 (the Clausius–Clapeyron equation). The three entropies become respectively:

$$s^s = -\frac{1}{\Gamma}\left(\frac{\partial \gamma_0}{\partial T}\right)_{A,N^s} - \left(\frac{\partial \mu_0}{\partial T}\right)_{A,N^s} + k - k \ln \frac{\Gamma}{\Gamma_0}, \qquad (2.4.17)$$

$$s_{A,T}{}^s = -\left(\frac{\partial \mu_0{}^s}{\partial T}\right)_{A,N^s} - k \ln \frac{\Gamma}{\Gamma_0}. \qquad (2.4.18)$$

$$s_{\gamma,T}{}^s = -\left(\frac{\partial \mu_0{}^s}{\partial T}\right)_{\gamma,N^s} - k \ln \frac{\Gamma}{\Gamma_0} + \frac{kT}{A}\left(\frac{\partial A}{\partial T}\right)_{\gamma,N^s}. \qquad (2.4.19)$$

Just as in the case of the Clausius–Clapeyron equation we prefer to use in the following the differential entropy $s_{A,T}{}^s$. Writing

$$-\left(\frac{\partial \mu_0{}^s}{\partial T}\right)_{A,N^s} = s_0{}^s$$

we have for the differential energy u^s:

$$u^s = \mu^s + T s_{A,T}{}^s = u_0{}^s = \mu_0{}^s + T s_0{}^s. \qquad (2.4.20)$$

The energy is not affected by the surface concentration Γ and is equal to the standard value $u_0{}^s$. The heat of adsorption is defined as

$$\Delta h = u^s - h^b. \qquad (2.4.21)$$

This is the same definition as the one given in connection with the Clausius–Clapeyron equation. We have explained there too, that our use of u^s is in conformity with the adapted scheme of thermodynamic potentials of interfacial phases.

The heat of adsorption is essentially a negative quantity, adsorption being an exothermic process.

(Ib) The non-zero surface area of each adsorbed particle is taken into account. A fraction $b\Gamma$ of the area is unavailable to the particles. Here b is a constant which is of the order of the area occupied by an adsorbed particle. Only a fraction $(1 - b\Gamma)$ is available. This leads to an equation of state, due to Volmer[46]:

$$\pi(1 - b\Gamma) = \Gamma kT. \qquad (2.4.22)$$

The chemical potential is now

$$\mu = \mu_0{}^s + kT \ln\left(\frac{\Gamma}{\Gamma_0} \cdot \frac{1 - b\Gamma_0}{1 - b\Gamma}\right) + kT\left(\frac{b\Gamma}{1 - b\Gamma} - \frac{b\Gamma_0}{1 - b\Gamma_0}\right). \qquad (2.4.23)$$

Comparison of the expressions for μ, as given by eq. (2.4.5) and by eq. (2.4.23) immediately leads to the pertinent adsorption equation. The heat of adsorption is, just as in the previous case, independent of Γ. The entropy $s^s{}_{A,N^s}$ becomes

$$s^s{}_{A,N^s} = s_0{}^s - k \ln\left(\frac{\Gamma}{\Gamma_0}\frac{1-b\Gamma_0}{1-b\Gamma}\right) - k\left(\frac{b\Gamma}{1-b\Gamma} - \frac{b\Gamma_0}{1-b\Gamma_0}\right). \quad (2.4.24)$$

This becomes minus infinity if $b\Gamma \to 1$, but of course the model breaks down in that case and a new layer builds up on top of the first.

(Ic) If not only the decrease of the available area, but also the inter-particle interaction is taken into account, one arrives at the two-dimensional Van der Waals equation:

$$(\pi + a\Gamma^2)(1 - b\Gamma) = \Gamma kT. \quad (2.4.25)$$

In this equation a† is a parameter accounting for the forces of the interaction between the adsorbed particles.

The chemical potential is

$$\mu = \mu_0{}^s + kT \ln\left(\frac{\Gamma}{\Gamma_0}\frac{1-b\Gamma_0}{1-b\Gamma}\right) + kT\left(\frac{b\Gamma}{1-b\Gamma} - \frac{b\Gamma_0}{1-b\Gamma_0}\right) - 2a(\Gamma - \Gamma_0) \quad (2.4.26)$$

$$\to \mu_0{}^s + kT \ln\frac{\Gamma}{\Gamma_0} + 2b(\Gamma - \Gamma_0)kT - 2a(\Gamma - \Gamma_0). \quad (2.4.26a)$$

Equation (2.4.26a) is the approximation for $b \ll 1$ of eq. (2.4.26). (Of course the small b approximation of eq. (2.4.23) leads to an analogous equation).

Equation (2.4.25), written in virial form up to the second virial coefficients becomes

$$\pi = \Gamma kT + \left(b - \frac{a}{kT}\right)\Gamma^2 kT. \quad (2.4.27)$$

The pertinent chemical potential is given by eq. (2.4.26a). The second virial coefficient is denoted as B:

$$B = b - \frac{a}{kT} = \tfrac{1}{2}\int_0^\infty (1 - e^{-E(r)/kT})2\pi r\, dr. \quad (2.4.28)$$

This is the two-dimensional equivalent of the expression for the second virial coefficient for imperfect gases as, for instance, used by Fowler

---

† There should be no confusion between the use of a here and its use elsewhere for the activity in a one-component system.

and Guggenheim[19b] (their eq. (714.1)). For the hard-sphere model (radius r_0) we write for $E(r)$

$$E(r) = \infty \qquad\qquad\qquad ; \quad r \sim 2r_0 \quad \text{(a)}$$

$$E(r) = -\lambda r^{-n}(n \geqslant 3) \quad \text{and} \quad |E(r)| \ll kT; \quad r > 2r_0 \quad \text{(b)} \quad (2.4.29)$$

where λ is a constant, positive or negative. For B we now write

$$B = 2\pi r_0{}^2 - \frac{\pi\lambda}{(n-2)(2r_0)^{n-2}kT}. \qquad (2.4.30)$$

We readily identify

$$b = 2\pi r_0{}^2 \quad \text{and} \quad a = +\frac{\pi\lambda}{(n-2)(2r_0)^{n-2}}. \qquad (2.4.31)$$

If a (or λ) is positive, the adsorbed particles attract each other. This corresponds to the "usual" Van der Waals case. The possibility of a negative Van der Waals a (or negative λ) will be discussed in Section 3.5. Some theories underlying $E(r)$ will be sketched in Section 4.3. Here we discuss the consequence of a temperature-dependent a. If the temperature dependence can be neglected as is the case for the usual London–Van der Waals forces, then

$$\left.\begin{aligned} s^s{}_{A,N^s} &= s_0{}^s - k \ln \frac{\Gamma}{\Gamma_0} - 2kb(\Gamma - \Gamma_0), \\ u^s &= u_0{}^s - 2a(\Gamma - \Gamma_0). \end{aligned}\right\} \qquad (2.4.32)$$

If a is a linear function of T (this is the case when the adsorbed particles can be represented as *classical* harmonic oscillators, see Section 4.3) then, writing $a = a'kT$, we get

$$\left.\begin{aligned} s^s{}_{A,N^s} &= s_0{}^s - k \ln \frac{\Gamma}{\Gamma_0} - 2k(b-a')(\Gamma - \Gamma_0) \\ u^s &= u_0{}^s \end{aligned}\right\} a = a'kT. \qquad (2.4.33)$$

Thus, although the adsorbed particles show up interaction, this is only revealed by the entropy of adsorption and not by the heat of adsorption.

Finally, if a is inversely proportional to T (this is the case for adsorbed permanent dipoles which behave according to Keesom, see Section 4.3), then

$$\left.\begin{aligned} s_{A,N^s}{}^s &= s_0{}^s - k \ln \frac{\Gamma}{\Gamma_0} - 2k(\Gamma - \Gamma_0)b - \frac{a''}{kT^2}\, 2(\Gamma - \Gamma_0) \\ u^s &= u_0{}^s - 4\frac{a''}{kT}(\Gamma - \Gamma_0). \end{aligned}\right\}\left(a = \frac{a''}{kT}\right). \qquad (2.4.34)$$

A Van der Waals type isotherm may also be obtained when the adsorbed particles are ionized. This is shown as follows:

The (electro-) chemical potential of an adsorbed particle, which carries a charge ze is

$$\tilde{\mu} = \mu_0{}^s + kT \ln \frac{\Gamma}{\Gamma_0} + z\, e\psi_0 \qquad (2.4.35)$$

where we have assumed that the chemical part of $\tilde{\mu}$ is given by only a $\ln \Gamma$-term. As before, the process of the adsorption is assumed to be carried out by means of the addition to the considered system of a neutral component. This means that also a charge $-z\, e\Gamma$ must be accounted for in the system. We have seen, however, that it is often justified to ignore the change of the electrochemical potential of the component carrying the charge $-z\, e$ (Subsection 2.2.7) and therefore that it is justified to use eq. (2.4.3). A Van der Waals-type equation obtains if ψ_0 is proportional to Γ. This is the case if a Debye–Hückel approximation applies (see preceding section):

$$\psi_0 = z\, e(\Gamma - \Gamma_0)\, \frac{4\pi L}{\varepsilon} \qquad (2.4.36)$$

L being the Debye–Hückel length of the material of the adsorbent. Equation (2.4.35) becomes

$$\tilde{\mu} = \mu_0{}^s + kT \ln \frac{\Gamma}{\Gamma_0} + 4\,\frac{\pi}{\varepsilon}\,(z\, e)^2 L(\Gamma - \Gamma_0). \qquad (2.4.37)$$

Defining a "second virial coefficient" B_{el} as

$$B_{el} = \frac{2\pi(z\, e)^2 L}{\varepsilon kT} \qquad (2.4.38)$$

it is seen that eq. (2.4.37) and eq. (2.4.26a) have much in common. In many cases

$$B_{el} \gg B. \qquad (2.4.39)$$

Thus, the absolute value of B is usually of the order of 10^{-14} cm^2. For B_{el} we obtain, with $L = 10^{-4}$ cm, $(z\, e)^2 = 22 \cdot 75 \times 10^{-20}$ e.s.u., $kT = 4 \cdot 1 \times 10^{-14}$ erg and $2\pi \times 22 \cdot 75/\varepsilon \times 41 = 1$, a value of 10^{-9} cm^2, which we consider as a typical value. We have assumed here complete ionization. If the ionization would have been incomplete, we would have had two adsorbed components. Denoting the degree of ionization as α, the surface concentration of the ionized component would have been $\alpha\Gamma$ and the surface concentration of the other component would have been $(1-\alpha)\Gamma$. Even for $\alpha = 10^{-6}$ the influence of the ionized component upon the "true" second virial coefficient would have been noticeable.

Finally we note that, neglecting the dependence of ε and of n_i upon temperature, the entropy contribution arising from the existence of B_{el} is $-\Gamma k B_{el}$ and the contribution to u^s is $B_{el}\Gamma kT$.

(Id) The Freundlich[47] isotherm equation is

$$p = K\Gamma^n \qquad (2.4.40)$$

where the pressure p may be replaced by the activity a, and where K is a constant at a given temperature. Finally n is a number > 1. The equation of state is

$$\pi = n\Gamma kT. \qquad (2.4.41)$$

The physical background of the number n may be sought in two directions. First, each particle may dissociate in n radicals after adsorption, the radicals obeying the ideal law. Second, there may be a repulsive force between the particles, for example because they are ionized. As a general rule, a Freundlich-type isotherm is obtained from an equation of state which can be written as

$$(\pi + \alpha\Gamma) = \Gamma kT \qquad (2.4.42)$$

in which α is a constant. Equations (2.4.41) and (2.4.42) are identical if $\alpha = (1-n)kT$. If α is independent of temperature then

$$\mu = \mu_0{}^s + kT \ln \frac{\Gamma}{\Gamma_0} - \alpha \ln \frac{\Gamma}{\Gamma_0} , \qquad (2.4.43)$$

$$s^s{}_{A,N^s} = s_0{}^s - k \ln \frac{\Gamma}{\Gamma_0} , \qquad (2.4.44)$$

$$u^s = u_0{}^s - \alpha \ln \frac{\Gamma}{\Gamma_0} . \qquad (2.4.45)$$

Thus, if α is temperature-independent, there is a logarithmic fall of the heat of adsorption with increasing coverage.

For the adsorption of ionized particles we use eq. (2.4.35). A Freundlich-type isotherm is obtained for the high-potential approximation of the ψ_0–Γ relationship. This relationship can be written as (see preceding section)

$$z\,e\psi_0 = 2kT \ln \frac{\Gamma}{\Gamma_0} \qquad (2.4.46)$$

leading to $\alpha = -2kT$ or $n = 3$. This equation will be discussed in Chapter 5.

Now localized adsorption, class II, will be considered. We distinguish two cases, adsorption on a homogeneous, and on a heterogeneous surface.

(IIa) Homogeneous surface, Langmuir adsorption. Langmuir[13] assumed the surface to be divided into sites each of which is able to adsorb one particle. For a homogeneous surface all the sites are identical. Their number will be denoted by N_0 or by $\Gamma_{10}A$. A fraction $\theta = N^s/N_0 = \Gamma/\Gamma_{10}$ is assumed to be occupied and a fraction $(1-\theta)$ is empty. According to well-known concepts in statistical thermodynamics[19] the partition function or sum-over-states, Z, can be written as:

$$Z = \frac{N_0!}{N^s!(N_0-N^s)!}\, Q_{N^s}Q_0^{(N_0-N^s)} \qquad (2.4.47)$$

in which Q is the partition function of a single particle in the adsorbed state together with the site, Q_0 is the partition function of this same subsystem without the particle. Furthermore $N_0!/N^s!(N_0-N^s)!$ is the number of distinguishable ways in which N^s particles can be distributed over N_0 sites with the same energy. The Helmholtz free energy F^s of the adsorbed particles is

$$F^s = -kT \ln Z. \qquad (2.4.48)$$

The chemical potential is

$$\mu^s = \left(\frac{\partial F^s}{\partial N^s}\right)_{T,N_0} = -kT \ln \frac{Q}{Q_0} + kT \ln \frac{N^s}{N_0-N^s} \qquad (2.4.49)$$

in which $-kT \ln Q/Q_0$ can be written as μ_0^s. It is the chemical potential at $\theta = N^s/N_0 = \frac{1}{2}$.

Combination of eq. (2.4.49) with eqs. (2.4.4) and (2.4.5) gives the Langmuir isotherm equation

$$\theta = \frac{N^s}{N_0} = \frac{a}{a + a_0 \exp \dfrac{\Delta\mu_0}{kT}} \qquad (2.4.50)$$

in which a and a_0 can often be identified with concentrations or pressures and in which

$$\Delta\mu_0 = \mu_0^s - \mu_0^b. \qquad (2.4.51)$$

This quantity is determined by such properties as the energy with which the particle is kept to the site, by the modes of vibration of the adsorbed particle compared to those of a particle in the bulk and by the modifications which take place in the site upon adsorption. For a homogeneous interface the equation of state becomes

$$\pi = -\Gamma_{10}kT \ln (1-\theta). \tag{2.4.52}$$

For small θ, such that $\ln (1-\theta) \rightarrow -\theta/(1-\theta)$, an equation of state of the Volmer-type obtains. If in eq. (2.4.52) θ is taken proportional to c, the Szyskowski[48] equation of state is obtained. Frumkin[49] proposed the use of:

$$\pi = -\Gamma_{10}kT \ln (1-\theta) - a\Gamma^2. \tag{2.4.53}$$

This is a combination of the Langmuir and of the Van der Waals equation of state. Applications are given in Section 3.5. Finally, we note that Stern[50] used essentially the same model as Langmuir and applied it to ionic adsorption.

(IIb) Assume now the interface to be inhomogeneous.[51-53] Instead of N_0 identical adsorption sites we introduce c kinds of different sites and different partition functions Q_k and Q_{k0} ($k = 1 \ldots c$). Furthermore, the number of sites labelled k is N_{k0} and those occupied is $N_k{}^s$, the degree of coverage of these sites being $\theta_k = N_k{}^s/N_{k0}$. Assuming equilibrium one has for the chemical potential of the adsorbed atoms:[53]

$$\mu_1{}^s = \ldots = \mu_k{}^s = \ldots = \mu_c{}^s = \mu^b. \tag{2.4.54}$$

When in each kind of sites the adsorption takes place according to Langmuir's law, then, comparing the sites labelled 1 and k we get

$$\mu_{10}{}^s - \mu_{k0}{}^s = -kT\left(\ln \frac{N_0{}^s}{N_{10} - N_1{}^s} - \ln \frac{N_k{}^s}{N_{k0} - N_k{}^s}\right) = A_k kT \tag{2.4.55}$$

where we have introduced the parameter $A_k = (\mu_{10}{}^s - \mu_{k0}{}^s)/kT$. This is a constant with respect to variations of the $N_k{}^s$. We arrange the sites such as to make $A_k > A_{k-1}$ ($A_1 = 0$). For the θ_k we have, introducing the notation $x_k = \exp(-A_k)$ (where $1 \geqq x_k \geqq 0$),

$$\theta_k = \frac{a}{a + \left(a_0 \exp \dfrac{\Delta\mu_{10}}{kT}\right)x_k} \tag{2.4.56}$$

where $\Delta\mu_{10} = \mu_{10}{}^s - \mu_0{}^b$. We replace the exponential in eq. (2.4.56) by the symbol u in the following. The total coverage, θ, becomes

$$\theta = \frac{a}{N_0} \sum_{k=1}^{k=c} \frac{N_k{}^s}{a + a_0 u x_k}. \tag{2.4.57}$$

The slope of the isotherm curve, obtained by differentiation is

$$\left(\frac{\partial \ln \theta}{\partial \ln a}\right)_T = 1 - \frac{\sum N_k \, \theta_k{}^2}{\sum N_k \, \theta_k}. \tag{2.4.58}$$

Owing to the fact that $0 < \theta_k < 1$, the slope never exceeds unity, whatever the distribution of the N_k-values.

We now assume that there exists a relation between the N_{k0} and the x_k, to be given as follows:

$$N_{k0} = N_0 b f(x)_k \Delta x_k. \tag{2.4.59}$$

Here N_{k0} is the number of adsorption sites with x_k-values between x_k and $x_k + \Delta x_k$ and b is a normalization factor such that $b \sum f(x_k)\Delta x_k = 1$. For small Δx_k ($\to dx$; we omit the subscript "k") eq. (2.4.57) becomes

$$\theta = ab \int_g^1 \frac{f(x)}{a + a_0 u x} \, dx \tag{2.4.60}$$

the integration limits being $x = 1$ ($A = 0$) and $x = g$ where g which we can write as $g = \exp(-A_c)$ may be close to zero. For $f(x) = x^{-1}$ the well-known Temkin[54] case is obtained

$$\theta = b \ln \frac{a_0 u + a \exp(A_c)}{a_0 u + a} \tag{2.4.61}$$

and b becomes

$$b = -\frac{1}{\ln g} = A_c{}^{-1} = \frac{kT}{\mu_{10}{}^s - \mu_{c0}{}^s}. \tag{2.4.62}$$

For $a \ll a_0 u \ll a \exp(A_c)$ eq. (2.4.61) can be written as

$$\theta = 1 + A_c{}^{-1} \ln \frac{a}{a_0 u} \tag{2.4.63}$$

or, introducing as a new constant $u_c = u \exp(-A_c)$,

$$\theta = A_c{}^{-1} \ln \frac{a}{a_0 u_c} \tag{2.4.64}$$

which is the logarithmic relationship between θ and activity (pressure, concentration) originally envisaged by Temkin and also derived by Brunauer et al.[55]

Other forms of $f(x)$ can also be inserted in eq. (2.4.60). Thus, $f(x) = x^{-\frac{1}{2}}$ gives

$$\theta = \sqrt{\left(\frac{a}{a_0 u}\right)} \tan^{-1} \sqrt{\left(\frac{a}{a_0 u}\right)}. \tag{2.4.65}$$

Here $g = 0$ is taken and b has become equal to 2. The reason that we can here take $A_c = \infty$ is that the sites with a very high A-value are relatively rare. The limiting slope, derived from eq. (2.4.65), is just $\frac{1}{2}$.

The differential entropy of adsorption is found as follows[53]:

Consider the differential dF^s for the adsorption on a heterogeneous surface as defined above:

$$dF^s = -S^s \, dT + \gamma \, dA + \sum_{k=1}^{k=c} \mu_k{}^s \, dN^s. \qquad (2.4.66)$$

Cross-differentiation gives

$$\left(\frac{\partial S^s}{\partial N_k{}^s}\right)_{N^s_{j \neq k}, A, T} = -\left(\frac{\partial \mu_k{}^s}{\partial T}\right)_{N^s_{k}, A}. \qquad (2.4.67)$$

The partial differential entropy appearing in eq. (2.4.67) will be denoted as $s_k{}^s$ and it is easily seen that we can write

$$s_k{}^s = s_{k0}{}^s - k \ln \frac{\theta_k}{1 - \theta_k}. \qquad (2.4.68)$$

We are still interested in the differential entropy $s_{A,T}{}^s$. This is now written as

$$s_{A,T}{}^s = \left(\frac{\partial S^s}{\partial N^s}\right)_{A,T} = \sum_{k=1}^{k=c} \left(\frac{\partial S^s}{\partial N_k{}^s}\right)_{N_{j \neq k}, A, T^s} \left(\frac{\partial N_k{}^s}{\partial N^s}\right)_{A,T}. \qquad (2.4.69)$$

This can be rewritten as

$$s_{A,T}{}^s = s_1{}^s + \sum_{k=2}^{k=c} (s_k{}^s - s_1{}^s) \left(\frac{\partial N_k{}^s}{\partial N^s}\right)_{A,T} \qquad (2.4.70)$$

where we have, for heuristic reasons only, the summation run from $k = 2$ to $k = c$. After working out the differential quotient eq. (2.4.70) can be rewritten as

$$s_{A,T}{}^s - s_0{}^s = -k \ln \frac{\theta_1}{1 - \theta_1} - k \frac{\sum A_k F_k}{\sum F_k} = I + II \qquad (2.4.71)$$

where

$$F_k = \frac{x_k a_0 u N_k{}^s}{(a + a_0 u)^2}.$$

For all $A_k = 0$ (homogeneous surface) eq. (2.4.71) reduces to the familiar Langmuir expression (with $\theta_1 = \theta$). For $A_k \neq 0$ (in fact $A_k > 0$) we have $\theta_1 < \theta$ and contribution I in eq. (2.4.71) is increased compared with $-k \ln \theta/(1 - \theta)$. On the other hand, the contribution II is always negative. A closer inspection of eq. (2.4.71) shows that at $\theta \ll 1$ the entropy $s_{A,T}{}^s - s_0{}^s$ is lower than the familiar Langmuir entropy and at θ-values not too far from unity it is higher. This qualitative result, which is in agreement with numerous experimental data for the physisorption of gases, may also be obtained on the basis of a more general insight into the nature of entropy. At low coverage the effective choice of the adsorbing atoms is limited to only those sites which have a high A_k-value. This leads to an entropy decrease compared with the homogeneous case. Once these sites occupied (at somewhat higher θ) the remaining adsorbed atoms have a wide choice because the remaining sites are relatively empty. Therefore at higher θ-values the adsorption entropy is higher than in the homogeneous case.

Finally, using eq. (2.4.59) we can write

$$s_{A,T}^s - s_0^s = -k \ln \frac{\theta_1}{1-\theta_1} + k \frac{\int G(x) \ln x \, dx}{\int G(x) \, dx} \tag{2.4.72}$$

where

$$G(x) = \frac{xf(x)}{(a+a_0u)^2} \, .$$

For $f(x) = x^{-1}$ (the Temkin case) one obtains (for low θ_1)

$$s_{A,T}^s - s_0^s = -k \ln \theta_1 + \frac{k}{\theta_1} \left[\frac{-g}{1-g} \ln g + \frac{\theta_1+g}{1+g} \ln (g+\theta_1) \right] \tag{2.4.73}$$

where, as before, the integration limits were $x = 1$ and $x = g$.

For $f(x) = x^{-\frac{1}{2}}$ we have (the limits of the integration are $x = 1$ and $x = 0$), again for low θ_1-values,

$$s_{A,T}^s - s_0^s = \frac{4k}{\pi} \sqrt{\theta_1} \ln \theta_1 \tag{2.4.74}$$

Both eq. (2.4.73) and eq. (2.4.74) show only a weak dependence of the entropy upon θ_1. We note that for gas adsorption usually θ_1 is almost proportional to the pressure p. Therefore the entropy–pressure curve is almost parallel to the entropy axis, in conformity with experiments.

In Fig. 2.6 we have given an illustration for a particular choice of A_k-values and N_{k0}-values of the properties of a heterogeneous surface as far as physisorption is concerned.

In conclusion it may be stated that the heterogeneities may affect the interpretation in terms of one of the foregoing equations of state. Thus, the fact that the slope of the adsorption isotherms is reduced below the ideal Henry-value may point to a repulsion between the adsorbed particles, but also to heterogeneities of the surface.

We now turn to the effect of the electrified state of an (ideally polarizable) interface upon the adsorption properties of neutral molecules. The question has arisen in the discussion of the adsorption isotherm, which of the two quantities, the interfacial charge σ or the potential ψ_0 should be held constant when the adsorbed amount is varied. Parsons[56],[57] has suggested that just as for ionic adsorption (see

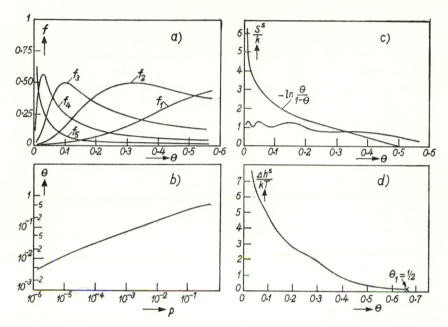

FIG. 2.6. Adsorption on heterogeneous surface. Five different adsorption sites.

$$A_2 - A_1 = A_3 - A_2 = A_4 - A_3 =_5 A - A_4 = 3 \cdot 3\, kT; A_1 = 0$$

For a definition of the A see eq. (2.4).

$$N_1 = 3N_2 = 9N_3 = 27N_4 = 81N_5$$

(a) Fractions f_i adsorbed in different adsorption sites. For any θ-value the sum of all the fractions is unity.

(b) Adsorption isotherm. The normalization is such that $p = 1$ at $\theta_1 = \frac{1}{2}$.

(c) Differential adsorption entropy s_s (eq. 2.4) as a function of θ, for heterogeneous and for homogeneous surface.

(d) Differential heat of adsorption Δh^s, as a function of θ. The normalization is such that $\Delta h^s = 0$ at $\theta_1 = \frac{1}{2}$.

Reprinted from ref. 53b, p. 155, by courtesy of Marcel Dekker Inc.

Chapter 3) the most appropriate quantity to be held constant is the charge. Damaskin[58] and Frumkin[59] argued that the adsorption properties of neutral molecules are easier to understand if the potential is held constant. Frumkin[49] found in 1926 that his experiments concerning the adsorption of 2-methyl-2-butanol (tert.-amylalcohol) at the Hg/l n-NaCl aq. soln. interface could be explained by assuming a $\sigma - \psi_0$ relation of the sort

$$\sigma = (C_0 - \Delta C\theta)\psi_0, \tag{2.4.75}$$

$$\left(\frac{\partial \sigma}{\partial \theta}\right)\psi_0 = -\Delta C\psi_0$$

where C_0 is the capacity at zero degree of coverage of the neutral adsorbate and where ΔC is a constant, positive or negative. A similar relation was assumed by Damaskin for the adsorption of other polar aliphatic compounds from concentrated electrolytes. Cross-differentiation of $d(F^s - \sigma AE)$ leads, under the assumption that $d\psi_0 = dE$, to

$$\left(\frac{\partial \mu^s}{\partial \psi_0}\right)_{T,A,N^s} = -\frac{1}{\Gamma_0}\left(\frac{\partial \sigma}{\partial \theta}\right)_{T,A,\psi_0} \tag{2.4.76}$$

where μ^s and N^s are the chemical potential and the adsorbed number of the neutral component. Integration gives

$$\mu^s = \mu^s(\theta) + \frac{1}{2\Gamma_0}\Delta C\psi_0{}^2 \tag{2.4.77}$$

where $\mu^s(\theta)$ represents that part of the chemical potential which was used so far for the derivation of an adsorption isotherm or an equation of state. For instance, the Langmuir behaviour would lead to an expression for $\mu^s(\theta)$ of the type $\mu_0{}^s + kT \ln \theta/(1+\theta)$. It is seen that the newly added term, $(1/2\Gamma_0)\Delta C\psi_0{}^2$, merely modifies the standard chemical potential $\mu_0{}^s$. This is not the case if σ is held constant. Use of eq. (2.4.75) leads to the addition of the term $(1/2\Gamma_0)\Delta C\sigma^2(C_0 - \Delta C\theta)^{-2}$.

If the potential-charge relationship is not given by eq. (2.4.75) but by

$$\psi_0 = \frac{1}{C_0}\left(1 + \frac{\Delta C}{C_0}\theta\right)\sigma, \tag{2.4.78}$$

$$\left(\frac{\partial \psi_0}{\partial \theta}\right)_\sigma = \frac{\Delta C}{C_0{}^2}\sigma,$$

then the charge enters into the standard chemical potential. Cross-differentiation of dF^s gives

$$\left(\frac{\partial \mu^s}{\partial \sigma}\right)_{T,A,N^s} = -\frac{1}{\Gamma_0}\left(\frac{\partial \psi_0}{\partial \theta}\right)_{T,A,\sigma}. \tag{2.4.79}$$

Integration now leads to

$$\mu^s = \mu^s(\theta) + \frac{1}{2\Gamma_0}\frac{\Delta C}{C_0{}^2}\sigma^2. \tag{2.4.80}$$

For $\Delta C \ll C_0$ the choice of the most appropriate electrical variable is irrelevant. This is probably the case for most non-polar adsorbates. We shall assume (with Parsons[57]) in the following that the standard free energy of adsorption can be written either as a power series in ψ_0

$$\Delta\mu_0 = \Delta\bar{\mu}_0 + a\psi_0 + b\psi_0^2 + \ldots \qquad (2.4.81)$$

or as a power series in σ

$$\Delta\mu_0 = \Delta\bar{\bar{\mu}}_0 + a'\sigma + b'\sigma^2 + \ldots \qquad (2.4.82)$$

in which a, b, b', a' are constants. The linear terms arise from adsorbed ions, dipolar molecules with fixed orientation, or non-polar molecules with an induced permanent dipole. The square terms are discussed above and arise from non-polar molecules or from polar molecules whose dipoles have no fixed orientation. The polarisation terms should not be confused with contributions $z_i e\psi_0$ to the adsorption energy of ions. It may be noted that also $\Delta\bar{\mu}_0$ or $\Delta\bar{\bar{\mu}}_0$ may depend on the electric state of the interface.[60] Thus, if the adsorbent is a semiconductor, charged surface states are usually present compensated by a space charge inside, the total charge being zero. However, the surface charges are discrete entities and local fields are therefore present. If the local field strength is f, then an adsorbed atom with polarizability α is subject to an attractive energy $-\frac{1}{2}\alpha f^2$ and this contribution is contained in the "purely" chemical part of the adsorption energy.

For an ideal polarizable interface we have from eq. (2.4.2)

$$\left(\frac{\partial\Gamma}{\partial\psi_0}\right)_{T,\text{all }\mu} = \left(\frac{\partial(\sigma-\sigma^0)}{\partial\mu}\right)_{T,\mu_j} \qquad (2.4.83)$$

in which it is assumed that $d\psi_0 = dE$; the subscript "i" is omitted. If, instead of $d\gamma$, the differential $d(\gamma + \sigma E)$ is considered,[56] then

$$\left(\frac{\partial\psi_0}{\partial\mu}\right)_{\sigma-\sigma^0,T,\mu_j} = -\left(\frac{\partial\Gamma}{\partial(\sigma-\sigma^0)}\right)_{T,\text{all }\mu_j} \qquad (2.4.84)$$

Parsons derived very useful equations by integrating eqs. (2.4.83) and (2.4.84). Assuming $d\mu = kT \, d\ln c$ eq. (2.4.83) becomes

$$\sigma - \sigma^0 = kT \int \frac{\partial\Gamma}{\partial\psi_0} \, d\ln c. \qquad (2.4.85)$$

As Parsons pointed out, Γ can usually be considered as a function of the product $c \exp -\Delta\mu_0/kT$ which we denote as $c\beta$. Then as a general rule $1/\beta(\partial\Gamma/\partial c)_\beta = 1/c(\partial\Gamma/\partial\beta)_c$ and

$$\sigma - \sigma^0 = kT \frac{\partial\beta}{\partial\psi_0} \int \frac{\partial\Gamma}{\partial\beta} \, d\ln c = kT/\beta \frac{\partial\beta}{\partial\psi_0} \int c\left(\frac{\partial\Gamma}{\partial c}\right) d\ln c. \qquad (2.4.86)$$

The integration of eq. (2.4.86) and the application of eq. (2.4.83) then gives

$$\sigma - \sigma^0 = \Gamma \frac{\partial \Delta \mu_0}{\partial \psi_0} = -(a + 2b\psi_0 + \ldots)\Gamma. \qquad (2.4.87)$$

The capacity contribution $(C - C_0)$ of the adsorbed component at potential E can easily be derived by differentiation:

$$(C - C_0)_E = \frac{\partial}{\partial \psi_0}\left(\Gamma \frac{\partial \Delta \mu_0}{\partial \psi_0}\right)_{T, E} = -(a + 2b\psi_0 + \ldots)\frac{\partial \Gamma}{\partial \psi_0} - (2b + \ldots)\Gamma. \qquad (2.4.88)$$

In an analogous way we obtain from eqs. (2.4.84) and (2.4.82)

$$(E - E^0) = -(a' + 2b'\sigma + \ldots), \qquad (2.4.89)$$

$$(C^{-1} - C_0^{-1}) = -(a' + 2b'\sigma + \ldots)\frac{\partial \Gamma}{\partial \sigma} - (2b' + \ldots)\Gamma. \qquad (2.4.90)$$

2.5.   On the origin of the interfacial tension

There are essentially two ways to arrive at an understanding of the origin of the interfacial tension. In the first, the oldest, approach the transition from one phase to another is considered as continuous, the transition region being many atomic layers thick. In the other approach the transition is assumed to be confined to a single layer. Van der Waals and Kohnstamm[61] and Bakker[62] used the first approach and applied it to a one-component liquid in equilibrium with its vapour near the critical point. Their conclusion was that the seat of the interfacial tension was at the liquid side in the transition region. They used the following definition of the interfacial tension

$$\gamma = -\int_0^\tau (p(x) - p)\, dx \qquad (2.5.1)$$

where τ is the thickness of the transition region, p is the pressure; in a bulk phase it has the same value in any direction, but in the transition region this value is maintained only in a direction normal to the plane of the interface. Finally $p(x)$ is the tangential pressure in a sheet situated between x and $x + dx$ where x marks a position in the transition region, $0 \leqslant x \leqslant \tau$. The boundaries $x = 0$ and $x = \tau$ are outside the transition region in the homogeneous bulk of the adjoining phases. The situation is schematically given in Fig. 2.7. The use of eq. (2.5.1) has been critically examined by Ono and Kondo.[63] The equation is valid for mechanical equilibrium, but Van der Waals and Kohnstamm required explicitly that their treatment be consistent with thermo-dynamic equilibrium. Special models must be used to find $p(x)$. Van der Waals and Kohnstamm applied the Van der Waals equation of state for the adjoining phases and for the transition region

$$\left(p + a\rho^2\right)\left(\frac{1}{\rho} - b\right) = RT \qquad (2.5.2)$$

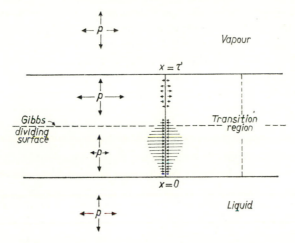

FIG. 2.7. Schematic representation of pressure distribution in transition region after Van der Waals and Kohnstamm,[61] Tolman[64] and Kirkwood and Buff.[65] The pressure p is isotropic in the two bulk phases and aniso-tropic in the transition region.

A negative contribution to the interfacial tension is indicated at the vapour side of the transition region. Ono and Kondo[63] discussed the effect of mechanical non-equilibrium across the transition region. In that case the pressure component normal to the plane of the interface is dependent upon the location in the interfacial region.

where b is the co-volume, ρ the density, and a the parameter represent-ing the influence of the intermolecular forces. It is especially the "a-correction" which is modified in the transition region, in the first place because there each molecule is in non-isotropic surroundings, i.e. each molecule is found in a region where

$$\frac{d\rho}{dx} \neq 0 \quad \text{and} \quad \frac{d^2\rho}{dx^2} \neq 0. \qquad (2.5.3)$$

Van der Waals and Kohnstamm arrived at their conclusion after lengthy, though elementary calculations. Many years later Tolman,[64] in a less precise analysis, followed essentially the same method and arrived at the same result by stating that the seat of the interfacial tension is at the liquid side of the Gibbs dividing surface. For many liquids and among them water, Tolman found that far from the critical temperature the distance between the seat of the interfacial tension and the Gibbs dividing surface, given by $\Gamma/\Delta\rho$, was of the order of Ångstroms. For water at room temperature the distance was 1 Å. Tolman realized that at low temperatures the validity of the theory was in doubt. It was fortunate, therefore, that Kirkwood and Buff,[65] by using the more appropriate approach of Kirkwood, and Born and Green[66] for the intermolecular interactions in liquids, and applying it to eq. (2.5.1) arrived at the same conclusion as Tolman. For argon at 80°K they found that $\Gamma/\Delta\rho \approx 3$ Å. They confirmed in their treatment a result which Fowler[67] obtained in 1937 for the interfacial energy of the liquid–vapour interface.

Concerning the second approach, in which the transition is confined to essentially the outermost layer, the following idealized procedure is carried out. Separate a solid, or a liquid, in a reversible, isothermal way into two parts, thereby creating two new surfaces. The work performed during this procedure leads to a free energy increase of the system. This is written as $2\gamma A$. It is fairly obvious to assume that this work bears some relation to the cohesive energy of the solid or the liquid. A first estimate of γ in terms of the cohesive energy can be given as follows:

We write the cohesive energy per unit volume as $-\frac{1}{2}z_0NE$, where E is the attractive energy between two neighbouring atoms or molecules, z_0 is the coordination number and N is the number of atoms (molecules) per unit volume. Thus $\frac{1}{2}z_0N$ is the number of pairs per unit volume. The nature of the attractive energy E widely diverges amongst various solids and liquids. We have strong covalent bonds in crystals such as those of germanium and silicon, weak Van der Waals-type bonds in liquified and solidified rare gases, and hydrogen bonds of intermediate strength in water and ice. Values of E can be estimated from the heats of vaporization. At the surface we assume that each atom or molecule has z_s, not z_0, nearest neighbours. Naturally $z_s < z_0$, $(z_0 - z_s)$ bonds being disjoined per pair of nearest neighbours. An important contribution to the work of separation is then given by $2(z_0 - z_s)/z_0 N_s E A$, where N_s is the number of atoms or molecules per unit area. After formation of the surface usually a rearrangement of the surface atoms or molecules takes place. Experimental evidence for such a rearrangement is given in Chapter 6 for germanium and silicon surfaces. The rearrangement leads to a free energy contribution $2F_{\mathrm{rearr}}{}^s N_s A$ which is negative because it is the result of a spontaneous process. We include in this contribution any change ΔE of the attractive energy, such changes being an almost natural consequence of a rearrangement. We expect the contribution $2F_{\mathrm{rearr}}{}^s N_s A$ to be relatively small.

A contribution which we want to consider separately is given by the thermal vibrations of the atoms or molecules. Modes of vibration at the surface will be different from those in the bulk. This leads to a contribution $2F_v{}^s N_s A$. We expect $F_v{}^s$ to be negative because surface vibrations will be characterized by lower frequencies (and larger amplitudes) than those in the bulk.

Combination of these three contributions would lead to

$$\gamma = \left(\frac{z_0 - z_s}{z_0} E + F_{\text{rearr}}{}^s + F^s \right) N^s \qquad (2.5.4)$$

as an expression for the interfacial tension. Lack of experimental data prevents an adequate check on the validity of this expression for solids. However, according to Wolf[68] the equation works quite well for liquids. Here an abundance of data is available. Most of these can be represented in the following form:

$$\gamma = [\alpha k T_0 - \beta k (T - T_0)] N^s \qquad (2.5.5)$$

where α and β are constants and where T_0 is a chosen temperature which should be much lower than the critical temperature.

Eötvös[69] predicted that for many non-polar liquids β has the same value. Indeed it has been found that

$$\beta \cong 2. \qquad (2.5.6)$$

For associating, usually polar, liquids β is lower than 2. For water one has $\beta = 1 \cdot 1$ ($T_0 = 300°K$). Concerning α, let us assume that eq. (2.5.4) is also applicable for liquids, and equate that part in [ ] of eq. (2.5.5) that is independent of temperature, to $(z_0 - z_s)/z_0$ times E. Then, ignoring $F_{\text{rearr}}{}^s$,

$$(\alpha + \beta) k T_0 = \frac{z_0 - z_s}{z_0} E. \qquad (2.5.7)$$

For a large number of liquids it has been found, according to Wolf, that $(z_0 - z_s)/z_0 \cong \frac{1}{3}$. For liquids, in which association can be expected, this ratio was lower. For water at $T_0 = 300°K$ one has a value of about $0 \cdot 2$.

A justification for the use of eq. (2.5.7) in the case of liquids may be found in the fact that for a wide variety of compounds the latent heat of melting, L, is about ten times smaller than the heat of vaporization, H_v. For instance, for water $L = 6000$ J mol^{-1} and $H_v = 45,000$ J mol^{-1} (0°C), and for a substance as widely different from water as argon these values are $L = 1140$ J/mole and $H_v = 6300$ J/mole (77°K). In these two cases the ratio L/H_v is somewhat larger than 1/10. For many metals (Frenkel[70] mentions sodium, zinc, lead and mercury) the ratio is about 1/30–1/40. Therefore, rather than as a compressed gas, a liquid may be considered as a distorted solid (although it has lost its anisotropic properties) and this view has been worked out by Frenkel. One of the consequences is that in liquids the concept of a coordination number is still meaningful. Thus, X-ray data have indicated that for liquid water $z_0 \sim 4\cdot2$,[71] i.e. that the tetrahedral structure of ice is still largely preserved. Assuming the validity of eq. (2.5.7) for the water surface, one has in this case $z_0 - z_s$ 0·85. This means that, in creating a new water surface, less than one bond per water molecule must be broken.

A low value of $(z_0 - z_s)/z_0$ seems to be correlated to a low β-value. This can be understood in the following way:

According to eqs. (2.2.31) and (2.2.35) the quantity βk is the difference of the entropies of a molecule in the surface layer and a molecule in the bulk. One has

$$S^s = S^s - S^{\mathrm{I}} = \beta k. \tag{2.5.8}$$

This entropy difference is a positive quantity, about $1\cdot1k$ $(T_0 = 300°\mathrm{K})$ for water and about $2k$ for other, non-associating, liquids. We interpret a positive value as a larger amount of disorder at the surface than in the bulk. When relatively few bonds must be broken for the creation of a new surface (a low value of $(z_0 - z_s)/z_0$) the amount of disorder at the surface will be relatively small, in agreement with experimental data.

Frenkel[70] wrote for β

$$\beta = \ln \frac{\nu^{\mathrm{I}}}{\nu^s} \tag{2.5.9}$$

where ν^{I} and ν^s are average frequencies which are characteristic of bulk and surface waves respectively. Frenkel estimated that $\nu^{\mathrm{I}} \cong 10\nu^s$, in agreement with the experimental value of β, of Eötvös, and in qualitative agreement with what has been said upon the introduction of the contribution $F_\nu{}^s$ in eq. (2.5.4).

Turnbull et al.[72] have given values of the interfacial tension, γ_{SL}, of the interface between a solid and its melt, these values being roughly ten times smaller than the γ-values of the melts. These γ_{SL}-values were obtained by measuring nucleation rates and applying a nucleation rate theory, the nuclei being small crystallites embedded in the liquid phase. Since the obtained γ_{SL}-values were not markedly dependent on the size of the nuclei, the results were considered to be valid for large, flat interfaces. Turnbull remarked that the ratio γ_{SL}/L was usually very close to 0·45 (especially metals), or, for some other cases, close to 0·3 (for instance, water and germanium). This suggests, that a model in which the ratio $(z_0 - z_s)/z_0$ is an essential notion, is widely applicable. For the surface tension of liquids varying from liquified rare gases to liquid water, liquid metals and many organic liquids, a newly created surface requires the breaking of $(z_0 - z_s)$ bonds per atom or molecule at the surface; for the interfacial tension γ_{SL} of many solids and their melt it means that about $(z_0 - z_s)$ bonds per surface atom or molecule must be "weakened" to a degree indicated by the ratio L/H_v. Table 2.1

TABLE 2.1

Liquid	$U^s/(H_v N^s A)$ (Wolf)	Solid/melt	$\gamma_{SL}/(LN^s)$ (Turnbull)
Argon (85°K)	0·37	Ag	0·45
Oxygen (70°K)	0·40	Cu	0·44
SiCl$_4$	0·33	Fe	0·45
Hexane	0·34	Ga	0·44
Benzene	0·33	Hg	0·53
Fatty acids	0·17–0·25	Ni	0·44
Water (300°K)	0·15	Sn (tetrag.)	0·42
Na (100°C)	0·30	Water	0·32
Hg (20°C)	0·49	Bi	0·33
Ag and Au (1000°C)	0·17	Ge†	0·35

† According to Jaccodine (ref. 73) the surface free energy of solid Ge ({111} planes) is 1060 mJm^{-2} ($=$1060 erg cm^{-2}) and that of solid Si {111} is 1230 mJm^{-2} ($=$1230 erg cm^{-2}). Estimates for the {100} planes led to values which were 1·8 times higher. For the {110} planes estimates led to 1·3 times higher values. These different values are indicative of the fact that our second approach is only one of a qualitative character. The ratio $U^s/(H_v N^s A)$ is sometimes called Stefan's number (U^s is defined on page 35).

collects a number of examples obtained from Wolf's book and from Turnbull's papers. The success of this approach suggests that the neglect of contributions such as $F_{rearr}{}^s$ is justified. However, we want to reconsider here this contribution because it has a bearing on the double layer that may be formed when a new surface is created. The formation of a double layer in this case is the result of a redistribution of electric charge carriers. As such it can be considered as participating in the rearrangement leading to the term $F_{rearr}{}^s$ and we may write at once for our one-component system:

$$F_{rearr}{}^s N^s = F_{rearr}{}^s (\text{chem}) N^s - \int_0^{\psi_0} \sigma \, d\psi_0. \qquad (2.5.10)$$

Values of the integral amount typically to a few percent of γ and, although one may expect that it may rise to a considerable fraction of the value of γ_{SL}, the fact that their neglect does not prevent the ratio γ_{SL}/L from being constant, suggests that the influence of the double layer is small.

In summary, both when the liquid is considered as a compressed vapour and when it is considered as a distorted solid, a semi-quantitative understanding of the nature of the interfacial tension can be obtained. The electrical double layer plays here usually only a subordinate role.

REFERENCES

1. J. W. GIBBS, *The Collected Works* I, Yale U.P., 1948, p. 314.
2. R. DEFAY and I. PRIGOGINE, *Tension superficielle et adsorption,* Liège, 1951.
3. Ref. 2, p. 1.
4. G. WULFF, *Z. Kristallog.*, 1901, **34,** 449; see for a generalization C. A. JOHNSON, *Surface Sci.*, 1965, **3,** 429.
5. T. YOUNG, *Phil. Trans.,* 1805, **95,** 65, 82.
6. P. S. LAPLACE, *La Théorie de l'action capillaire,* Paris, 1806.
7. J. D. VAN DER WAALS and PH. KOHNSTAMM, *Lehrbuch der Thermostatik* I, Leipzig, 1927.
8. K. L. WOLF, *Physik und Chemie der Grenzflächen* I, Springer, 1957.
9. G. LIPPMANN, *Ann. Chim. Phys.*, 1875, (5), **5,** 494; *J. Phys. Radium*, 1883, (2), **2,** 116.
10. H. G. BUNGENBERG DE JONG in *Colloid Science* (ed. H. R. KRUYT), Elsevier, Vol. II, p. **433** (1949).
11. J. TH. G. OVERBEEK in *Colloid Science* (ed. H. R. KRUYT), Elsevier, Vol. I, p. 146 (1952).
12. J. T. DAVIES and E. K. RIDEAL, *Interfacial Phenomena*, Academic Press, 1961, 2nd ed., 1963.
13. I. LANGMUIR, *J. Am. Chem. Soc.,* 1916, **38,** 2221; 1917, **39,** 1848.
14. J. W. GIBBS, *The Collected Works* I, Yale U.P., 1948.
15. E. A. GUGGENHEIM, *Thermodynamics*, North Holland Publ. Co., 1959, 4th ed.
15a. A. SANFELD, *Thermodynamics of Charged and Polarized Layers*, Wiley–Interscience, 1968.
16. E. H. MacDOUGALL, *Thermodynamics and Chemistry*, 3rd ed., Wiley, 1945.
17. K. G. DENBIGH, *The Principles of Chemical Equilibrium*, Cambridge U.P., 1957, in Dover Publications, 1963.
18. D. H. EVERETT, *Chemical Thermodynamics*, 3rd impression, Longmans, 1962.
19. For statistical aspects see, for example:
    (a) J. W. GIBBS, *Collected Works* II, Yale U.P., 1948.
    (b) R. H. FOWLER and E. A. GUGGENHEIM, *Statistical Thermodynamics*, Cambridge U.P., 1938.
    (c) R. C. TOLMAN, *Statistical Mechanics*, McGraw-Hill, Oxford U.P., 1938.
    (d) TERRELL HILL, *Statistical Mechanics*, McGraw-Hill, 1956.
    (e) A. KHINCHIN, *Mathematical Foundations of Statistical Mechanics*, Dover Publications, 1949.
    (f) H. L. FRIEDMAN, *Ionic Solution Theory*, Interscience, 1962.
20. Ref. 14, p. 219.
21. Ref. 15, p. 326 and E. A. GUGGENHEIM, *Trans. Faraday Soc.,* 1940, **46,** 453.
22. Ref. 7, p. 283.
23. D. C. GRAHAME and R. PARSONS, *J. Am. Chem. Soc.,* 1961, **83,** 1291.

24. S. R. DE GROOT and H. A. TOLHOEK, *Proc. Kon. Ned. Akad. Wetenschap.*, 1951, **B54**, 42.
25. Ref. 14, p. 429.
26. Ref. 15, p. 374 and E. A. GUGGENHEIM, *J. Phys. Chem.*, 1929, **33**, 842.
27. Ref. 7, p. 286.
28. L. ONSAGER, *Chem. Revs.*, 1933, **13**, 73.
29. H. D. HURWITZ, A. SANFELD and A. STEINCHEN-SANFELD, *Electrochimica Acta*, 1964, **9**, 929.
30. Ref. 7, p. 302.
31. D. C. GRAHAME, *Chem. Revs.*, 1947, **41**, 441.
32. R. PARSONS and M. A. V. DEVANATHAN, *Trans. Faraday Soc.*, 1953, **49**, 404.
33. R. PARSONS, *Can. J. Chemistry*, 1959, **37**, 308.
34. Ref. 7, p. 204.
35. D. STIGTER and TERRELL L. HILL, *J. Phys. Chem.*, 1959, **63**, 551; D. STIGTER, *ibid.*, 1960, **64**, 838.
36. I. LANGMUIR, *J. Chem. Phys.*, 1938, **6**, 893.
37. Ref. 17, p. 197.
38. D. H. EVERETT, *Trans. Faraday Soc.*, 1950, **46**, 942.
39. E. MATIJEVIĆ and P. A. PETHICA, *Trans. Faraday Soc.*, 1958, **54**, 589 and D. G. HALL and B. A. PETHICA in *Non-ionic Surfactants* I, Chap. 6 (ed. M. J. Schick), Marcel Dekker, 1967.
40. G. GOUY, *Ann. Chim. Physique*, 1908, (8), **8**, 294; a review in *Ann. Phys.* (Paris), 1917, (9), **9**, 129.
41. P. DEBYE and E. HÜCKEL, *Physik. Z.*, 1923, **24**, 185.
42. E. J. W. VERWEY and J. TH. G. OVERBEEK, *Theory of the Stability of Lyophobic Colloids*, Elsevier, 1948 (chap. I, p. 37).
43. M. J. SPARNAAY, *Rec. Trav. Chim. Pays-Bas*, 1959, **78**, 680.
44. B. M. W. TRAPNELL, *Chemisorption*, Butterworth, 1955; 2nd ed. 1963 by HAYWARD and TRAPNELL.
45. D. H. EVERETT, *Proc. Chem. Soc.*, Feb. 1957, p. 38.
46. M. VOLMER, *Z. physik. Chemie*, 1925, A**115**, 253.
47. H. FREUNDLICH, *Kapillarchemie* I and II, Leipzig, 1930 and 1932.
48. B. V. SZYSKOWSKI, *Z. physik. Chemie*, 1908, **64**, 385.
49. A. N. FRUMKIN, *Z. Physik*, 1926, **35**, 792.
50. O. STERN, *Z. Elektrochemie*, 1924, **30**, 508.
51. TERRELL L. HILL, *J. Chem. Phys.*, 1949, **17**, 762.
52. J. M. DRAIN and J. A. MORRISON, *Trans. Faraday Soc.*, 1952, **48**, 316.
53a. M. J. SPARNAAY, *Surface Sci.*, 1968, **8**, 100.
53b. M. I. SPARNAAY, in *Clean Surfaces* (ed. G. GOLDFINGER), Marcel Dekker, 1970, p. 153.
54. M. TEMKIN and V. PYZHEV, *Acta Phisicochim. U.S.S.R.*, 1940, **12**, 327; M. TEMKIN, *Zhur. Fiz. Khim.*, 1941, **15**, 296.
55. S. BRUNAUER, K. S. LOVE and R. G. KEENAN, *J. Am. Chem. Soc.*, 1942, **64**, 751.
56. R. PARSONS, *Trans. Faraday Soc.*, 1959, **55**, 999.
57. R. PARSONS, *J. Electroan. Chem.*, 1963, **5**, 397.
58. B. B. DAMASKIN, *J. Electroan. Chem.*, 1964, **7**, 155.
59. A. N. FRUMKIN, *J. Electroan. Chem.*, 1964, **7**, 152.
60. M. J. SPARNAAY, *J. Phys. Chem. Solids*, 1960, **14**, 111.
61. Ref. 7, Ch. 8.
62. G. BAKKER, *Handbuch der Experimentalphysik*, VI, Leipzig, 1928.

D

63. S. ONO and S. KONDO, *Encyclopedia of Physics*, 1960, **X**, 134, Springer.
64. R. C. TOLMAN, *J. Chem. Phys.*, 1949, **17**, 118.
65. J. G. KIRKWOOD and F. P. BUFF, *J. Chem. Phys.*, 1949, **17, 338**.
66. See textbooks such as T. L. HILL, *Statistical Mechanics*, McGraw-Hill, 1956.
67. R. H. FOWLER, *Proc. Roy. Soc.*, 1939, **A159**, 229 and ref. 19(b).
68. Ref. 8, Sect. 7.
69. E. VON EÖTVÖS, *Ann. d. Physik*, 1886, **27**, 448.
70. J. FRENKEL, *Kinetic Theory of Liquids*, first Russian ed. 1943, Dover, 1946, chap. VI.
71. D. EISENBERG and W. KAUZMANN, *The Structure and Properties of Water*, Oxford U.P., 1969.
72. D. TURNBULL, *J. Appl. Phys.*, 1950, **21**, 1022.
73. R. J. JACCODINE, *J. Electrochem. Soc.*, 1963, **110**, 524.

CHAPTER 3

THE Hg/AQUEOUS SOLUTION INTERFACE

3.1. Electrocapillary curves. Double-layer capacities

If, using the equipment schematically represented in Fig. 2.3, the interfacial tension γ of the interface between mercury and an aqueous salt solution is measured as a function of the applied potential difference (potentiometer reading E), an electrocapillary curve can be drawn. To obtain such a curve in a reliable and reproducible manner the equipment and the chemicals used should be rigorously cleaned and the solution should be air-free. It should, as a general rule, be free of reducible substances. If not, electrochemical reactions may take place at the interface and an electric current may pass. In that case the interface is no longer ideally polarizable and E can no longer be regarded as an independent variable. Unwanted effects might also be expected to occur if in the cell the calomel electrode is used as the reversible electrode. Then mercurous ions are present which show the reversible reaction $Hg^+ + e^- \leftrightharpoons Hg$ with the mercury electrode. However, the Hg^+-concentration is only 10^{-18} M and its effects can be neglected. Under favourable circumstances E can be varied over more than 2 V without an appreciable current passing across the interface. This limit is set by the reduction reaction at cathodic potentials, surpassing the value of the hydrogen overvoltage and by oxidation reactions at high anodic potentials.

Examples of electrocapillary curves are shown in Fig. 3.1. They show a maximum, the electrocapillary maximum (ecm) which depending on the salt and its concentration may be found at γ-values of 0·40–0·42 N m^{-1} ($= 400$–420 dyn cm^{-1}). At either side of the ecm γ may fall to 0·3 N m^{-1} ($= 300$ dyn cm^{-})[1], or thereabouts. At the ecm the charge σ on the mercury is zero as may be seen from the Lippmann equation:

$$\left(\frac{\partial \gamma}{\partial E}\right)_{T,p,\mu} = -\sigma. \tag{3.1.1}$$

At the anodic side $\sigma > 0$ and this positive charge is compensated by an excess of anions near the interface. At the cathodic side $\sigma < 0$ and there is an excess of cations in the bulk phase. Figure 3.1 shows that for each solution the ecm is situated at a different value of E. This means, that for each case at $\sigma = 0$ the potential has a different value. There are good reasons to believe that part of this potential jump can be explained by the specific adsorption, at $\sigma = 0$, or even at $\sigma < 0$, of anions, the cations being merely present to fulfil the requirement of electroneutrality. Specifically adsorbed anions lead to a potential jump across the interface, which shifts the E-value at the ecm in the cathodic direction. There is one important exception, provided by fluoride ions.

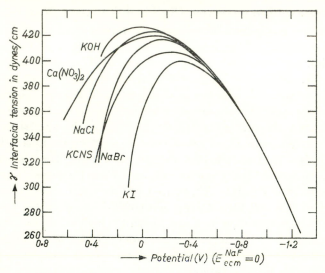

FIG. 3.1. Electrocapillary curves (D. C. Grahame[1]); 1 M aqueous salt solutions in contact with mercury. $E_{ecm}^{NaF} = 0$ means: the origin of the potential scale is chosen at the electrocapillary maximum of NaF solutions (1 dyn/cm = 10^{-3} Nm^{-1}).

No simple salt solution gives an ecm which is at a more positive potential than that of a fluoride solution. Therefore it is probable that in this case neither cations nor anions are specifically adsorbed.[1,2] This is confirmed by further experimental and theoretical investigations (see next section). The non-specificity of the cations is borne out by the fact that salts containing different cations but the same anions have the same electro-capillary curve at sufficiently negative σ-values (where anions are desorbed). However, exceptions to this rule were reported by Frumkin[3] and his school. Tl$^+$-ions[4] and Cs$^+$-ions[4b] were found to be specifically adsorbed. On the other hand, Parsons and Zobel[5] found that fluoride solutions could sometimes be replaced by solutions containing as the anion dihydrogen phosphate ($H_2PO_4^-$). This anion is specifically adsorbed only at positive σ-values.

The concentration of solutions containing anions which may be specifically adsorbed has a noteworthy effect upon the E-value at the ecm or rather upon the position of the whole electrocapillary curve. Esin and Markov[6] found that the rate of change of the potential at the ecm (E_{ecm}) with changing concentration was often 100–200 mV per decade of concentration. Frumkin[6] attributed this rather large value to the discrete nature of the charges of the adsorbed anions. Esin and Shikov,[7] Ershler,[8] Grahame,[9] Levich $et\ al.$,[10] Barlow and Ross Macdonald[11] and Levine $et\ al.$[12] worked out this concept theoretically. Parsons[13] gave a thermodynamic analysis. The addition of non-ionic substances to the solution often leads to profound change of the electro-capillary curve.[14,15] There is a maximum, but this is often at a much lower value than that obtained in the absence of organic material. According to the original Gibbs adsorption equation this means that at $\sigma = 0$ (and at small absolute values of σ) there is a strong adsorption of the non-ionic substance. The lowering of γ is often absent at high anodic and high cathodic potentials. Then, due to the high field strength, the strongly polar water molecules displace the usually less polar non-ionic molecules. This was first pointed out by Gouy.[15,16] Later developments are discussed in Section 3.4.

At a first glance the electrocapillary curves of Fig. 3.1 seem to have a parabolic shape. This would mean that the second derivative of γ with respect to E, which is the differential capacity, C, is constant if E is varied. This is only true to a very rough approximation. Figure 3.2 shows curves of C versus E, where

$$C = \left(\frac{\partial \sigma}{\partial E}\right)_{T,p,\mu} = -\left(\frac{\partial^2 \gamma}{\partial E^2}\right)_{T,p,\mu}. \tag{3.1.2}$$

The curves show, especially at low concentrations, a pronounced minimum and there is a sharp rise if the potential is made highly anodic. There are, at 20–40 μF/cm^2, indeed more or less flat portions in a number of cases. A first interpretation of these flat portions can be given in terms of the simple Helmholtz picture of the double layer. The value of $\varepsilon_1 \delta^{-1}$ (δ is the thickness, ε_1 the relative dielectric constant of the Helmholtz layer) then is of the order of $2 \cdot 5 \times 10^8$–5×10^8 cm^{-1}. Since δ will be 3–4 Å at most, the value of ε_1 must be of the order of 10–15, which is much lower than that of bulk water, but still higher than the limit which characterizes complete saturation. We shall see that the estimate of ε_1 so obtained is an upper limit.

The simple Helmholtz picture leaves most of the curve unexplained and it has proved necessary to reconsider the whole problem both from an experimental and from a theoretical point of view. Thus, usually the

$C–E$ curves are not found from the electrocapillary curve, but C is directly measured at various potentials, for instance by using a dropping mercury electrode. Figure 3.3 shows schematically an equipment as used by Grahame[17,18] (see also Damaskin[19]). The mercury droplet emerges from a thin glass capillary (outer diameter about 0·1 mm; inner diameter about 0·04 mm). It is surrounded by a platinum gauze

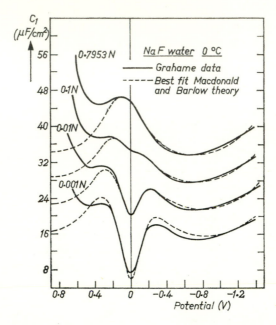

FIG. 3.2. Differential capacity C versus applied potential. Origin at $E_{ecm}^{NaF} = 0$ (no specific adsorption). Experimental data (solid lines): D. C. Grahame, *J. Am. Chem. Soc.* 1954, **76**, 4819. Dashed lines: theory of J. Ross Macdonald and C. A. Barlow, *J. Chem. Phys.*, 1962, **36**, 3062 discussed in Section 3.2. There is a poor fit at the cathodic side of the ecm ($\sigma \gg 0$) (1 dyne/cm = 10^{-3} N m^{-1}).

which is an auxiliary electrode and which has a fixed potential difference with the reversible (often calomel) electrode M. The capacity is measured between the Hg-droplet and the auxiliary Pt-electrode by means of a bridge circuit. Since the surface area of the Pt-electrode is large compared with that of the Hg-droplet, the latter determines the total capacity. This capacity varies in time because the size of the Hg-droplet increases in time until it breaks off. Then a new droplet starts to form. The arrangement of the bridge circuit is chosen in such a way (suitable values for C_1, R_1, R_2 and R_3) that the bridge has a balance point once during the lifetime of a droplet. This balance point gives the capacity of the droplet/solution interface. It is now essential to have the exact value of the surface area of the droplet at the moment of the balance point. This can be calculated from the rate of flow of the mercury and the time τ elapsed after the birth of the droplet. The rate of flow can be given a constant value (in Grahame's case between 0·5 and 1·5 mg/sec). The time τ is of the order of seconds. Its exact determination is all-important and this problem has been given much attention.

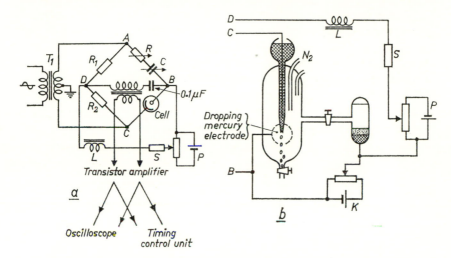

FIG. 3.3. Principle of equipment used by Grahame[17,18] as modernized by Hills and Payne.[20] (a) Bridge circuit (schematically).[20] A, B, C, D, corners of bridge. T_1, T_2, transformers. R_1, R_2, resistors, up to 10^4 ohms. C, variable capacitor, small steps (e.g. 10^{-4} μF), up to ~ 1 μF. R, variable resistor, small steps (e.g. 10^{-1} ohm), up to $\sim 10^3$ ohms. P, potential divider, 2-V battery. S, 10-kohm resistor; L inductor block the a.c. path. The 0·1 μF capacitor placed between T_2 and B blocks d.c. path. (b) Dropping mercury electrode (schematically). B, C, D; L, P, S, same meaning as in (a). K auxiliary potential divider, fixes potential difference between calomel electrode and Pt gauze surrounding mercury electrode. The gauze may be spherical or cylindrical.

Grahame used an electromechanical device, Hills and Payne[20] used an electronic chronometer system. Bockris et al.[21] determined the capacity at a predetermined time after the birth of the droplet, usually 9·35 seconds. All these authors (see also ref. 19) discussed the electronic problems pertinent to the use of this particular electrochemical cell. A result was that reliable measurements with an accuracy of less than 0·1% can be carried out only at frequencies roughly between $\frac{1}{2} \times 10^3$ c/sec and $\frac{1}{2} \times 10^5$ c/sec. In this region the obtained capacity values were independent of frequency. For organic adsorbates, which are the subject of Section 3.4, difficulties may arise, however.

Concerning the theory, low C-values were explained by Gouy[23,24] through the introduction of the concept of diffuse ionic layer, the Gouy layer. We have given Gouy's theory in Section 2.3. The capacity C_G attributable to the Gouy layer appeared to be (see eq. (2.3.19))

$$C_G = \frac{\varepsilon\kappa}{4\pi} \cosh \frac{z}{2} . \qquad (3.1.3)$$

To obtain the capacity C of the double-layer system, assume that the ions of the Gouy layer cannot come closer to the interface than a finite distance $x = \delta$. Grahame called[1] the plane $x = \delta$ the outer Helmholtz plane (OHP). For an ideally polarizable interface any change dE can, at constant temperature, be identified with $d\psi_0$, where ψ_0 is the potential difference between the plane $x = 0$ and the bulk of the solution where $x = \infty$. Likewise we define for the OHP a potential difference ψ_δ. In the absence of any specific adsorption we have for the capacity C

$$\frac{1}{C} = \frac{1}{C_1} + \frac{1}{C_G} \qquad (3.1.4)$$

where

$$\frac{1}{C} = \left(\frac{\partial\psi_0}{\partial\sigma}\right)_T; \quad \frac{1}{C_1} = \left(\frac{\partial(\psi_\delta - \psi_0)}{\partial\sigma}\right)_T; \quad \frac{1}{C_G} = \left(\frac{\partial\psi_\delta}{\partial\sigma}\right)_T. \qquad (3.1.5)$$

Equation (3.1.4) is due to Grahame. For C_1, the capacity of the inner region, one has the old Helmholtz expression

$$C_1 = \frac{\varepsilon_1}{4\pi\delta} . \qquad (3.1.6)$$

The dimensionless quantity z in eq. (3.1.3) must be identified with $e\psi_\delta/kT$ (we assume 1–1 electrolytes). For a 10^{-3} N 1–1 electrolyte the Debye–Hückel length $\kappa^{-1} = 100$ Å and since δ is of the order of 5 Å, C_G^{-1} in eq. (3.1.4) is of the same order of magnitude as C_1^{-1} and it even dominates around the capacity minimum at $z = 0$. Thus, eq. (3.1.4), in combination with eq. (3.1.3), explains the experimentally observed capacity minimum. According to Grahame[2] it does this with great precision. Thus, if at one NaF concentration the C–E curve was measured around the minimum, the curve at another concentration could be predicted with a high degree of precision by using eqs. (3.1.3) and (3.1.4) and assuming the same $C_1 - E$ curve for each NaF concentration. This suggests that the Poisson–Boltzmann equation, as it is used in Gouy's theory, is based on sound assumptions, in particular for low ionic concentrations when C_G^{-1} dominates most.

In many cases there is specific adsorption of ions and eq. (3.1.4) is invalid. Denoting the surface charge of these ions by σ_A, it is convenient to write:

$$\frac{\partial \psi \delta}{\partial \sigma} = \frac{\partial \psi \delta}{\partial (\sigma + \sigma_A)} \cdot \frac{\partial (\sigma + \sigma_A)}{\partial \sigma} = C_G^{-1}\left(1 + \frac{\partial \sigma_A}{\partial \sigma}\right). \qquad (3.1.7)$$

The charge in the Gouy layer is $-(\sigma + \sigma_A)$. Instead of eq. (3.1.4) one has

$$\frac{1}{C} = \frac{1}{C_1} + \frac{1}{C_G}\left(1 + \frac{\partial \sigma_A}{\partial \sigma}\right). \qquad (3.1.8)$$

From a physical point of view it is unlikely that specific adsorption takes place at the outer Helmholtz plane, this plane being the limiting plane for ions in the Gouy layer. Grahame denoted the plane where specifically adsorbed ions are located as the inner Helmholtz plane (IHP). Ions attached to the surface by specific forces (which may have a chemical origin) will have lost part of their hydration shell. Therefore we expect the distance, β, between the IHP and the wall to be smaller than the distance δ between the OHP and the wall. The physical structure of the inner layer and therefore the values of ε_1 and C_1 are (probably) strongly affected by specifically adsorbed ions.

3.2. The inner region in the absence of specific adsorption

Curves of C_1 versus σ are given in Fig. 3.4 for the Hg/NaF solution interface at various temperatures. The curves are rather non-symmetric around the point $\sigma = 0$ and there is a characteristic "hump" at the anodic side ($\sigma \approx 1$ μCoulomb $\simeq 6 \times 10^{12}$ e cm^{-2}) at low temperatures. This hump is not only present in the case of aqueous solutions but also in a number (not all[2]) of non-aqueous systems containing polar liquids (e.g. formamide solutions[25]). Since the measurements have been carried out at a more than one temperature, the entropy of the system can be calculated as a function of σ. The entropy of the inner region, as given by Hills and Payne,[20] shows a maximum at a surface charge of -10μC/cm^2, i.e. at the cathodic side of the ecm. Hills and Payne noted that the excess entropy of the inner region is negative, the entropy loss at the ecm being about $1 \cdot 2$ k (k is Boltzmann's constant) per molecule transported from the bulk to the inner region. This is about half the entropy of freezing of bulk water ($-2 \cdot 64$ k). Therefore the entropy calculations by Hills and Payne support the "ice-like" layer model of the inner region suggested by Grahame,[26] although, as also noted by Grahame, the structure of liquid water will not be "forgotten" in the inner region.

It is difficult to give an unambiguous interpretation of the existing data. Thus, although a "natural orientation" of the water molecules at the metal surface (the orientation at the ecm) has often been proposed there is no agreement as to which side of the dipole is turned to the metal surface. Ross Macdonald and Barlow[27] concluded that the positive H-atoms were turned to the metal surface, whereas Grahame, Hills and Payne arrived at the opposite conclusion. In our attempt we shall find agreement with Ross Macdonald and Barlow but we shall also show that our conclusion must be considered as tentative.

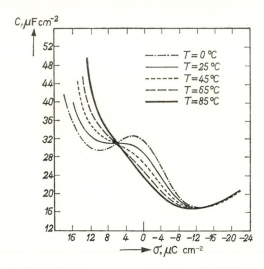

FIG. 3.4. Inner layer differential capacity versus applied potential at various temperatures (D. C. Grahame, *J. Am. Chem. Soc.* 1957, **79**, 2093).

Furthermore, one factor which determines the capacity of the inner region is the thickness of this region and its dependence upon the field strength of the applied field. Any prediction as to the value of the compressibility, based on a comparison with data obtained in macroscopic experiments, is dangerous because we can hardly predict anything about the relationship between the structure of the inner region and the structure of existing liquids or solids. For these reasons our attempt is tentative and does not go beyond the presentation, in conjunction with a review of existing theories of a qualitative interpretation. Our line of thought resembles somewhat that of Ross Macdonald and Barlow. In contrast to these authors we shall stress the possibility that the structure of liquid water has an important effect upon the properties of the inner region. Such structure effects are implicit in the treatment of the problem by Watts-Tobin.[28]

The "inner region" is assumed to consist here of a monolayer of about 10^{15} water molecules per cm^2. Regarding the orientation of these water molecules and their movements we distinguish four influences:

1. The external field with field strength E_1 in a direction normal to the plane of the interface (the x-direction). A water dipole is subject to the influence of an energy $-pE_1 \cos \theta$ where p is the dipole moment† and θ is the angle between the direction of the field and the direction of the dipole. For positive E_1-values the field has the tendency to orient a dipole such that its positive side is turned toward the mercury surface. We note that if at room temperature $pE_1 = kT$, we have $E_1 = 6.7 \times 10^6$ volt/cm.

2. The "natural field" with field strength ΔE tending to place the dipoles in the x-direction. This field leads to an energy contribution $-p\Delta E \cos \theta$. We assume ΔE to be independent of σ and, therefore, of E_1. The total field is given by $E_1 + \Delta E$, which we denote by E.

3. The force of attraction between a dipole and a metal. Considering the metal surface as an ideal mirror this (image) force leads to an energy contribution W_1

$$W_1 = -\frac{p^2}{2(2r)^3}(a_1 + a_2 \cos^2 \theta) \tag{3.2.1}$$

where r is the distance between the dipole and the mirror and where a_1 and a_2 are constants. For a single dipole and a mirror $a_1 = a_2 = 1$. There are energy minima for $\theta = 0$ and for $\theta = \pi$. There are maxima for $\theta = \frac{1}{2}\pi$ and for $\theta = -\frac{1}{2}\pi$. This means that the image force tends to place the dipole in the x-direction.

Also the outer Helmholtz plane can act as a mirror because the dielectric constant of the water phase is much higher than that of the inner region. In that case both a_1 and a_2 deviate from unity. For a dipole placed between two mirrors Ross Macdonald and Barlow found for a_2, the only constant that interests us, a value close to 3.

4. The molecular structure of water. While we have considered the applied field, the natural field and the image effect, there is no reason to ignore the fields arising from the presence of the water molecules in the adjacent bulk and of other water molecules in the inner region. It is difficult to present their contribution in a more than speculative way. Only if we can make a detailed calculation of the thermodynamic quantities of a subsystem consisting of a water molecule in the inner

---

† $p_{water} \simeq 1.8 \times 10^{-19}$ e.s.u. cm. By way of comparison: a charge $-e$ and a charge $+e$, placed 1 Å apart, would form a dipole with a moment of 4.8×10^{-18} e.s.u. cm, or 4.8 Debye. Since $e = 4.8 \times 10^{-10}$ e.s.u. $= 1.6 \times 10^{-19}$ Coulomb, this moment is equal to 1.6×10^{-29} Cm; 1 Debye $= \frac{1}{3} \times 10^{-29}$ Cm.

region and its neighbours in the inner region and in the bulk phase can we hope to be able to provide for a quantitative and reliable prediction. In anticipation of such a calculation we assume that the potential energy of this subsystem at a given θ-value (this potential energy is defined as the difference between the free energies of the subsystem with $\theta = 0$ and of the same subsystem with the given non-zero θ-value) has the following form:

$$W_2 = -(apE + W_0) \sin^2 \theta \qquad (3.2.2)$$

where a is a dimensionless constant and where W_0 is a positive energy constant. We assume that eq. (3.2.2) is also applicable, when a relatively weak specific adsorption, for instance of F^- ions (Schiffrin[26]) has taken place. In that case W_0 may be increased, because the F^- ion is a structure former, the structure being alien to that of the inner region.

A physical justification of eq. (3.2.2) can be given as follows:

The quantity $(apE + W_0)$ can be positive or negative. Assume first that it is positive. In that case there is a tendency to counteract the alignment in the x-direction promoted by the influences mentioned under (1), (2) and (3). This tendency is ascribed to the thermal movements of the neighbouring water molecules. The orientation of the water molecules in the inner region can be of importance here, i.e. it can be of importance whether these molecules have turned their H-atoms or their O-atoms toward the bulk. If, for instance, the "H-atom" position fulfils better the requirements of the bulk water structure, then the thermal movements of the bulk water molecules have a larger disorienting effect than in the case of a "O-atom" position. The "H-atom" position requires a positive a. For positive a we see that increasing E-values lead to an enhancement of the disorienting effect.

When $(apE + W_0)$ is negative the bulk water structure helps to align the water molecules in the inner region in the x-direction.

It is evident that both a and W_0 may depend on temperature, but so far we do not want to make predictions as to the values of $\partial a/\partial T$ and $\partial W_0/\partial T$. We merely note that non-zero values of all those quantities are necessary to explain an appreciable separation of the capacity and the entropy maxima.

Collecting all contributions (we are only interested in the θ-dependent part) we have

$$W = -p(E_1 + \Delta E) \cos \theta - (apE + W_0 + a_1 p^2/16 r^3) \sin^2 \theta \qquad (3.2.3)$$

We introduce the following notation:

$$x = p(E_1 + \Delta E)/kT = pE/kT, \qquad (3.2.4)$$

$$z = (apE + W_0 + a_2 p^2/16 r^3)/kT = ax + z_0, \qquad (3.2.5)$$

$$z_0 = (W_0 + a_2 p^2/16 r^3)/kT. \qquad (3.2.6)$$

We now calculate the thermodynamic quantities of a water dipole which is subject to the four influences which we have enumerated. First we note that for $z = 0$ we have the well-known Langevin case. For this case the partition function Z_L (the subscript "L" pertains to the Langevin case) is

$$Z_L = \tfrac{1}{2} Z_{L0} \int_0^\pi e^{+x\cos\theta} \sin\theta \, d\theta \qquad (3.2.7)$$

where Z_{L0} is that part of the partition function which is independent of θ. For $x = 0$ one has $Z_L = Z_{L0}$.

The Helmholtz free energy becomes

$$\Delta F_L = -kT \ln \frac{\sinh x}{x} \qquad (3.2.8)$$

which really is the free energy difference between the case where the dipole is placed in a field with non-zero field strength E and the case where $E = 0$. Likewise the entropy difference becomes

$$T\Delta S_L = kT\left(\ln \frac{\sinh x}{x} - x \, \overline{\cos\theta_L} \right) \qquad (3.2.9)$$

where

$$\overline{\cos\theta_L} = \coth x - \frac{1}{x} . \qquad (3.2.10)$$

This quantity, the average value of $\cos\theta$, is calculated by means of:

$$\overline{\cos\theta_L} = (Z_{L0}/2Z_L) \int_0^\pi \cos\theta \, e^{+x\cos\theta} \sin\theta \, d\theta. \qquad (3.2.11)$$

In Fig. 3.5 the entropy of eq. (3.2.9) is plotted against x. Also drawn is the entropy curve given by Hills and Payne who plotted the excess entropy of the inner region in cal deg^{-1} cm^{-2} (1 cal deg^{-1} cm^{-2} = $4 \cdot 186 \times 10^4$ J deg^{-1} m^{-2}; instead of "deg^{-1}" one often writes "°K^{-1}") against σ in μC cm^{-2}. Scale conversion was carried out with the assumption that 10^{15} water molecules per cm^2 were present in the inner region, whereas σ and x were assumed related through the relation

$$\sigma = -\frac{\varepsilon_1}{4\pi}(E - \Delta E) = -\frac{\varepsilon_1}{4\pi} kT/p(x - \Delta x), \qquad (3.2.12)$$

$$\approx -10^4(x - \Delta x) \quad \text{(e.s.u.)},$$

$$\approx -3\tfrac{1}{2}(x - \Delta x) \quad (\mu\text{C cm}^{-2}),$$

$$\approx -2 \times 10^{13}(x - \Delta x) \quad \text{(electron charges per cm}^2\text{)}.$$

Hills and Payne ascribed the variation of the entropy as a function of varying σ to a variation of the density of the water molecules in the inner region. They found that, when each water molecule suffered the same entropy loss on its transport from the bulk to the inner region, an increase by $10\mu C$ cm^{-2} of the surface charge (starting from the ecm) led to an increase of about 6×10^{14} water molecules per cm^2 in the inner region. This number seems unlikely because $-10\mu C$ cm^{-2} represents the charge of about 6×10^{13} electrons. So far optical measurements

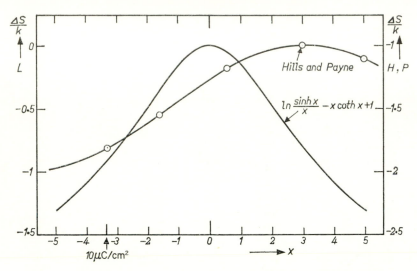

Fig. 3.5. Entropy curves. Left curve: theoretical, Langevin model (eq. (3.2.9)); $x = pE/kT$, at $x = 0$ the Langevin entropy is maximal and zero (left abscissa). Right curve: experiments by Hills and Payne,[20] giving the excess entropy of the inner region. Same entropy scale but the entropy maximum is at $x = 3$, its value being $-k$.

by M. Stedman,[20c] who used the method of ellipsometry; (more information concerning this method will be given in Chapter 6) do not confirm such a strong variation of the concentration of water molecules in the inner layer. Therefore the removal of one electron from the mercury surface would lead to the adsorption of ten water molecules which we consider as improbable. Hills and Payne assumed the entropy of a water molecule in the inner region to be independent of σ. We see that already the simple Langevin model leads to an entropy–surface charge relationship which is comparable to the one found by Hills and Payne. Therefore we consider the electric field as the main cause of this relationship. A comparison of the positions of the entropy maxima in Fig. 3.5 would lead to the conclusion that the (negatively charged)

O-atoms of the water molecules in the inner region are pointing toward the mercury surface but this conclusion may be incorrect owing to the crudeness of the Langevin model. The Langevin model does show that the maxima of the capacity and of the entropy are not found at the same σ-value. This is implicit in the calculations of Ross Macdonald and Barlow. These authors wrote:

$$C_1 = \frac{\partial \sigma}{(\psi_0 - \psi_x)} = \frac{\partial \sigma}{\partial E_1} \left(\frac{\partial (\psi_0 - \psi_x)}{\partial E_1} \right)^{-1} \quad (T \text{ const.}) \quad (3.2.13)$$

in which $\psi_0 - \psi_x = -E_1 \delta$. With the use of the relation $4\pi\sigma = -\varepsilon_1 E_1$ eq. (3.2.13) becomes

$$C_1 = \frac{1}{4\pi\delta} \left(\varepsilon_1 + E_1 \frac{\partial \varepsilon_1}{\partial E_1} \right) \left(1 - \frac{\partial \ln \delta}{\partial \ln E_1} \right)^{-1}. \quad (3.2.14)$$

The term $(1 - (\partial \ln \delta / \partial \ln E_1))$ describes the compression of the inner layer after application of the field. Its value may be estimated from the experiments by Hills and Payne. These authors not only measured the temperature coefficient, but also the pressure coefficient of the double layer capacity, their pressures ranging from 1 atmosphere to as high as 3000 atmospheres. Therefore they were able not only to obtain the excess entropy, but also the excess volume of the interface as a function of σ. They found a decrease of the excess volume of about 0.35×10^{-8} cm^3/cm^2 when σ was increased from zero (the ecm) to about $10 \mu C$/cm^2. We interpret this result as a compression of 0.35 Å of the inner layer. Assuming that $\delta \approx 10^{-5}$ Å we see that the compression leads to a correction of C_1 of 7% or less.

The compression is related to other properties of the inner layer as follows. Consider the differential of (G_1) of the inner layer:

$$d(G_1) = -S_1 \, dT + V_1 \, dp + \mu_1 \, dN_1 - \frac{D_1 V_1}{4\pi} \, dE_1 \quad (3.2.15)$$

where the symbols have their usual meaning and where the subscript "1" indicates quantities related to the inner layer. There is only one component, water, present in the inner layer. With $D_1 = \varepsilon_1 E_1$ one relation, obtained by cross-differentiation, is

$$\left(\frac{\partial V_1}{\partial E_1} \right)_{T,p,N_1} = -\frac{V_1 E_1}{4\pi} \left(\frac{\partial \varepsilon_1}{\partial p} \right)_{T,E_1,N_1} - \frac{E_1 \varepsilon_1}{4\pi} \left(\frac{\partial V_1}{\partial p} \right)_{T,E_1,N_1}. \quad (3.2.16)$$

Since $V_1 = A_1 \delta$ where A_1 is the surface area of the inner layer, this equation can be used to correlate the pressure dependence of the dielectric constant and the thickness of the inner layer to the compression term in eq. (3.2.14). Estimates were given by Ross Macdonald and Barlow, who followed a somewhat different route.

For ε_1 one has (see, for example, Böttcher[29])

$$\varepsilon_1 = \varepsilon_\infty + 4\pi n_1 \frac{p}{E} \tag{3.2.17}$$

where n_1 is the density of water molecules $(N_1 = n_1 V_1)$ taken as 3×10^{22} cm^{-3}, $\bar{p}$ is an average value of the dipole moment in the direction of the field and ε_∞ is the high-frequency limit of the dielectric constant, where molecular association is wholly absent. Measurements at frequencies of about 10^{11} sec^{-1} lead to $\varepsilon_\infty \approx 6$[30] but this value should be considered with care.[31]

The Langevin model gives for $\bar{p}$

$$\bar{p} = p \cos \theta_L \rightarrow p \left(\frac{x}{3} - \frac{x^2}{45} \right) \quad \text{for} \quad |x| = p|E|/kT \ll 1. \tag{3.2.18}$$

Ignoring the compression term in eq. (3.2.14) the expression for C_1 can be written as

$$C_1 = \frac{1}{4\pi \, \delta} \left(\varepsilon_1 + E \, \frac{\partial \varepsilon_1}{\partial E} - \Delta E \, \frac{\partial \varepsilon_1}{\partial E} \right). \tag{3.2.19}$$

The Langevin model gives

$$E \, \frac{\partial \varepsilon_1}{\partial E} = -4\pi n_1 \frac{p^2}{kT} \left(\frac{1}{x} \coth x + \frac{1}{\sinh^2 x} - \frac{2}{x^2} \right). \tag{3.2.20}$$

For small $|x|$ the bracketed part becomes equal to $2x^2/45$. The factor $4\pi n_1 \, (p^2/kT)$ is, at room temperature, about equal to 30.

The temperature derivative of C_1 is calculated by means of

$$T \, \frac{\partial \varepsilon_1}{\partial T} = -4\pi n_1 \frac{p^2}{kT} \left(\frac{1}{x^2} - \frac{1}{\sinh^2 x} \right) \tag{3.2.21}$$

in which the bracketed part becomes, for small $|x|$, equal to $\frac{1}{3} - x^2/15$. Furthermore,

$$ET \, \frac{\partial^2 \varepsilon_1}{\partial E \, \partial T} = 4\pi n_1 \frac{p^2}{kT} \left(\frac{2}{x^2} - 2 \frac{x \coth x}{\sinh^2 x} \right) \tag{3.2.22}$$

in which the bracketed part becomes, for small $|x|$, equal to $2x^2/15$. Curves representing eqs. (3.2.17) and (3.2.20)–(3.2.22) are given in Fig. 3.6 for $\Delta x = p\Delta E/kT = 1$. Graphical construction leads, for each desired value of Δx, to a C_1–x curve and to a $(\partial C_1/\partial T) - x$ curve. It is seen that the capacity maximum is not found at $x = 0$ but, as inferred from the small $|x|$-approximation, at $x = \frac{1}{3}\Delta x$. The maximum value of C_1 is

$$(C_1)_{max} = \frac{\varepsilon_\infty}{4\pi\,\delta} + \frac{n_1p^2}{3kT\,\delta}\left(1 + \frac{2}{45}(\Delta x)^2\right). \tag{3.2.23}$$

A minimum of $\partial C_1/\partial T$ is found at the same value $x = \frac{1}{3}\Delta x$. It has a value of

$$\left(\frac{\partial C_1}{\partial T}\right)_{min} = -\frac{n_1p^2}{3kT^2\,\delta}\left(1 + \frac{1}{15}(\Delta x)^2\right). \tag{3.2.24}$$

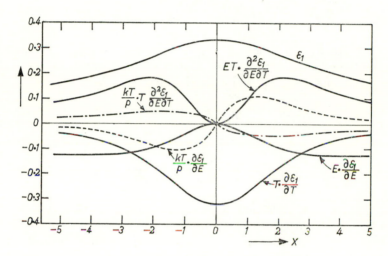

FIG. 3.6. Curves necessary to obtain C_1 and $\partial C_1/\partial T$ (Langevin model). See eqs. (3.2.17)–(3.2.22). Dashed lines: E replaced by kT/p. These lines do not have a maximum or a minimum at $x = 0$.

We note that this is a negative value. Graphical construction shows that apart from this minimum a maximum may be found, depending on the value of Δx, which is positive. Thus, for $\Delta x = 2$ it is found that at $x = -2$ there is a maximum of about $0\cdot1$ $(n_1p^2/kT^2\,\delta)$, which is positive. Such a behaviour is characteristic for the "hump"-region.

To gain some insight into the effect of the structure of water upon all these curves we replace the partition function Z_L of eq. (3.2.7) by

$$Z = \tfrac{1}{2}Z_0 \int_0^\pi e^{x\cos\theta}(1 + z\sin^2\theta)\sin\theta\,d\theta. \tag{3.2.25}$$

We have assumed $|z| \ll 1$, i.e. the structure effect is considered as a correction. Again $Z = Z_0$ when $x = z = 0$. Instead of eq. (3.2.9) we get

$$T\Delta S = T\Delta S_L - 2T\frac{\partial W_s}{\partial T}\frac{\overline{\cos\theta_L}}{x} - 2W_s\frac{T\,\partial\overline{\frac{\cos\theta_L}{x}}}{\partial T} \qquad (3.2.26)$$

where we have expressed the result in terms of the uncorrected quantities. We remember that $W_s = zkT$ (see eq. (3.2.5)).

Instead of eq. (3.2.10) we get

$$\overline{\cos\theta} = \overline{\cos\theta_L} - 2zx\left[\left(\frac{\overline{\cos\theta_L}}{x}\right)^2 - \frac{1}{x^2} + \frac{3}{x^2}\frac{\overline{\cos\theta_L}}{x}\right]. \qquad (3.2.27)$$

Equations for $E(\partial\varepsilon_1/\partial E)$ and $T(\partial\varepsilon_1/\partial T)$ can also be given. We restrict ourselves here to a determination of the positions of the capacity and entropy maxima. Approximations for small $|x|$ lead to:

$$x_{\text{cap max}} \cong \tfrac{1}{3}\Delta x + \tfrac{4}{3}a\frac{1 + \tfrac{2}{3}x\Delta x - x^2}{1 + \tfrac{4}{3}z} \qquad z = ax + z_0, \qquad (3.2.28)$$

$$x_{\text{ent max}} \cong -\frac{2T(\partial a/\partial T)}{1 + (4/15kT)(2W_0 - T(\partial W_0/\partial T))}(W_0 = z_0kT). \quad (3.2.29)$$

These results were obtained upon considering $z = W_s/kT = ax + z_0$. The addition of contributions of the type bx^2 would not have affected the values of $x_{\text{cap max}}$ and $x_{\text{ent max}}$. The experiments indicate that $x_{\text{cap max}} - x_{\text{ent max}} = -4$. They also indicate that $x_{1\text{cap max}}$, which is obtained by use of the relation $x_1 = x - \Delta x$, is negative, whereas $x_{1\text{ent max}}$ is positive. In view of the eqs. (3.2.28) and (3.2.29) we arrive at the result that $T(\partial a/\partial T)$ must be fairly large and negative, whereas a must be fairly small. This means that it is difficult to say which side of a water molecule is attached to the metal surface.

If eq. (3.2.15) is valid for the inner layer, then by cross-differentiation we get

$$\left(\frac{\partial S_1}{\partial(E_1{}^2)}\right)_{T,p,N_1} = \frac{V_1}{8\pi}\left(\frac{\partial\varepsilon_1}{\partial T}\right)_{p,E_1,N_1} + \frac{\varepsilon_1}{8\pi}\left(\frac{\partial V_1}{\partial T}\right)_{p,E_1,N_1}. \qquad (3.2.30)$$

In our calculations we have left out the thermal expansion term. This is justified as long as (see eq. (3.2.21))

$$\frac{n_1p^2}{6kT} \gg \frac{\varepsilon_1T}{8\pi V_1}\left(\frac{\partial V_1}{\partial T}\right).$$

Since, at room temperature $n_1 p^2/6kT \approx 0.4$ we expect that this is the case.

As noted, the cathodic side of the capacity curves usually is independent of the specific properties of the cations involved. This is surprising, because if the charge $\sigma = -10\mu C$ cm^{-2}, there should be about 6×10^{13} cations per cm^2 present close to the interface. Mott *et al.*[32] suggested that the non-specificity is due to the fact that the cations can not penetrate the region of low dielectric constant ε_1, characteristic for the inner layer, but that they remain a few angstroms

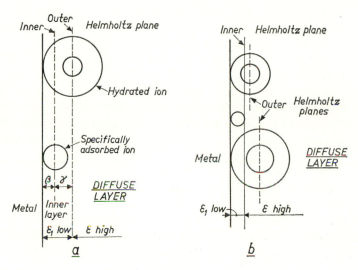

FIG. 3.7. Model of the Hg/aq. soln. interface according to Mott *et al.*,[32] (b) compared with the old model (a).

more inside the solution, where the dielectric constant is already high. Most of the potential drop over the chargeless region between the mercury surface and the other Helmholtz plane (the plane of nearest approach of the cations) then falls inside the region of low dielectric constant, and the specificity of the cations as reflected in their size, is no longer of importance. Figure 3.7 gives an illustration.

It is improbable that the high capacity values at the far anodic side can satisfactorily be explained by invoking electrostriction. Watts-Tobin[32] worked out (Gerischer's[32]) idea of "ad-atoms", which are Hg-atoms containing (positive) charges protruding from the metal surface. The formation of ad-atoms would be the first stage of the formation of an anodic reaction product. It is also possible that the anodic capacity rise is due to specifically adsorbed OH$^-$ ions. Parsons,[33] however, found no effect of the OH^--concentration upon the capacity value in the anodic region, unless pH $\geqslant$ 11. Bockris *et al.*[34] showed, by applying an optical method (ellipsometry), that a calomel film may form after anodic polarization. Such a film (its formation may be introduced by "ad-atoms", see above) may be held responsible for the anomalous capacity rise in the anodic region.

3.3.   The specific adsorption of the anions

Stern[35] completed Gouy's theory by incorporating the concept of the specific adsorption of ions. He essentially used the same model as Langmuir. The surface was assumed to be divided in Γ_0 sites per cm². Each site could be occupied by one ion (of the kind i). The plane of location of the adsorbed ions is the inner Helmholtz plane. If Γ_i' sites per cm² were occupied, $(\Gamma_0 - \Gamma'_i)$ sites were empty. The total amount of ions i present per cm² in the sites and in the Gouy layer is $(\Gamma_i - \Gamma_i')$. There will be an adsorption energy $\varphi_i + z_i\,e\psi_A$ in which ψ_A is the electrostatic potential at the empty site. The bondstrength of an absorbed ion will be affected by the charge on the metal and therefore φ_i will be a function of σ, but we assume it to be no explicit function of ψ_A. As Grahame[1] pointed out, it is not justified to identify ψ_A with ψ_δ, the limiting potential of the Gouy layer (at the outer Helmholtz plane) because the underlying physical concepts are different. The ions in the Stern layer have probably lost part of their hydration shell and will share some water molecules with the metal surface. The ions participating in the Gouy layer are all subject to the same thermal agitation as those in the bulk of the solution and have retained their full hydration shell. As we shall see the distinction between ψ_A and ψ_δ is particularly important in connection with the "discreteness-of-charge" (or "discrete-ion") effect.

The adsorption equilibrium of the ions i in the Stern-layer requires:

$$Kc_i(\Gamma_0 - \Gamma_i') = \Gamma_i'\, e^{(\varphi_i + z_i e\psi_A)/kT} \tag{3.3.1}$$

in which c_i is the bulk concentration of the ions i and K is a constant. Stern put $Kc_i = n_i M^{-1}$ where M is the number of available sites in 1 cm³ of the solution for the ions i and n_i is the number of ions i per cm³ in the solution. Mott and Watts-Tobin[28] wrote $Kc_i\Gamma_0 = n_i a$ where a is a length of the order of the diameter of a molecule. Since Stern's model is essentially the same as that of Langmuir (Section 2.4), Kc_i can be replaced by exp μ_i/kT or by Ka_i, and φ_i by $\Delta\mu_{i0}$. As pointed out later in this section, the Stern–Langmuir theory should only be considered as a first approach. Relations which may exist between φ_i and ψ_A on one side, and Γ_i' on the other side, should be clarified. On the other hand, the relation between Γ_i' and c_i (or a_i), which is given by the adsorption isotherm, should be known. These two problems can hardly be disentangled.

The electroneutrality condition is

$$\sigma + \sigma_1 + \sigma_G = 0. \tag{3.3.2}$$

If only ions i are adsorbed in the Stern layer, we get

$$\sigma_1 = z_i\, e\Gamma_i' = \frac{z_i\, e\Gamma_0}{1 + \dfrac{1}{Kc_i}\, e^{(\varphi_i + z_i e\psi_A)/kT}} \tag{3.3.3}$$

and for the ions in the Gouy layer one has

$$\sigma_G = \sqrt{\left(\frac{1}{\pi}\, 2\varepsilon nkT\right)} \sinh \tfrac{1}{2} \frac{z\, e\psi_\delta}{kT} \tag{3.3.4}$$

where we have assumed that we have a z–z electrolyte with n cations and n anions per cm^3. In many cases we shall have $z = z_i$ and $n = n_i$. In their treatment of the AgI/aq. soln. interface in 1948, Verwey and Overbeek identified ψ_A and ψ_δ and constructed, with the aid of the three preceding equations, and of eq. (3.3.5)

$$4\pi\sigma = -\varepsilon_1(\psi_0 - \psi_\delta) \tag{3.3.5}$$

a figure which still provides an insight in the numerical relations between the various quantities appearing in eqs. (3.3.2)–(3.3.5). This is essentially our Fig. 3.8.

For $\Gamma_i' \ll \Gamma_0$ eq. (3.3.1) or eq. (3.3.3) simplifies to a Henry-type equation:

$$\Gamma_i' = \Gamma_0 K c_i\, e^{-(\varphi_i + z_i e\psi A)/kT} = \Gamma_0 K' c_i\, e^{-(z_i e\psi A)/kT} \tag{3.3.6}$$

in which $K' = K\, e^{-(\varphi_i/kT)}$.

The quantity Γ_i' is experimentally accessible in the following way: for simplicity we specify the case of the adsorption in the Stern layer of anions A whereas the solution is assumed to contain only monovalent cations C and monovalent anions A. Equation (2.2.109) provides the way to obtain Γ_C' the total amount per unit area of cations C, adsorbed in the Gouy layer and in the Stern layer at a given value of σ. An equation, analogous to eq. (2.2.109) but now derived for anions, would have led to the determination of Γ_A, again at a given value of σ. Curves are given in Fig. 3.9. The electroneutrality condition can also be written as

$$\sigma + e(\Gamma_C - \Gamma_A) = 0. \tag{3.3.7}$$

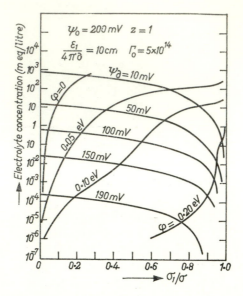

Fig. 3.8. Corresponding values of σ_1/σ, ψ_∂ and electrolyte concentration for different values of adsorption potential ϕ. Wall potential $\psi_0 = 200\,\mathrm{mV}$; $\varepsilon_1/4\pi\delta = 10^7$ cm; density of adsorption sites $\Gamma_0 = 5 \times 10^{14}$ per cm^2; 1–1 valent electrolytes. From E. J. W. Verwey and J. Th. G. Overbeek, *Theory of the Stability of Lyophobic Colloids*, Elsevier, 1948, p. 45.

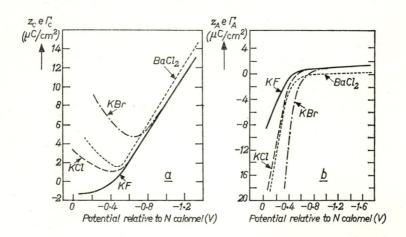

Fig. 3.9. Surface charge attributable to cations ($z_C\,e\Gamma_C$) and to anions ($z_A\,e\Gamma_A$) versus applied potential for 0·1M aqueous solutions of KF, KCl, KBr and BaCl$_2$ at 25°C (D. C. Grahame and B. A. Soderberg[1a]).

Now the Gouy theory allows the calculation of $(\Gamma_C - \Gamma_C')$ and of $(\Gamma_A - \Gamma_A')$ (Fig. 3.10). (As observed in Section 3.1, there usually is no specific cationic adsorption. This makes $\Gamma_C' = 0$.)

The excess of anions in the Gouy layer can be defined as

$$(\Gamma_A - \Gamma_A') = \int_\delta^\infty (n_A - n)\, dx \qquad (3.3.8)$$

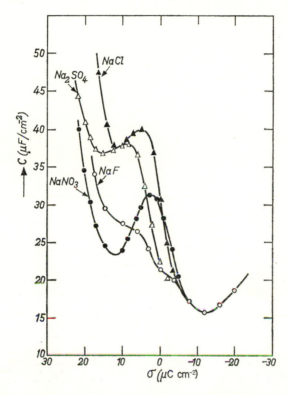

FIG. 3.10. Charge dependence of the differential capacity for 0·1 M aqueous solutions of NaCl, Na₂SO₄, NaNO₃, and NaF, at 20°C and 1 atm (Hills and Payne[20a]).

where $n_A = n\, e^{-y}$ $(y = e\psi/kT)$ is the number per cm³ of anions A in the Gouy layer at given position x. With the aid of eq. (2.3.11) this equation can be transformed into

$$(\Gamma_A - \Gamma_A') = \frac{n}{2\kappa} \int_z^0 \frac{1 - e^{-y}}{\sinh \frac{1}{2}y}\, dy = \frac{2n}{\kappa}(e^{\frac{1}{2}z} - 1)[z = e\psi_\delta/kT]. \qquad (3.3.9)$$

For cations one finds in an analogous manner

$$(\Gamma_C - \Gamma_{C'}) = \frac{2n}{\kappa}(e^{-\frac{1}{2}z} - 1) \quad [z = e\psi_\delta / kT]. \quad (3.3.10)$$

Once $(\Gamma_C - \Gamma_{C'})$ is known, and consequently ψ_δ is known, $(\Gamma_A - \Gamma_{A'})$ can be calculated and vice versa. Assuming that $\Gamma_{C'} = 0$, values of $\Gamma_{A'}$ can be obtained. These values can be plotted as a function of the activity of the anions a_A (or rather: as a function of the mean activity of

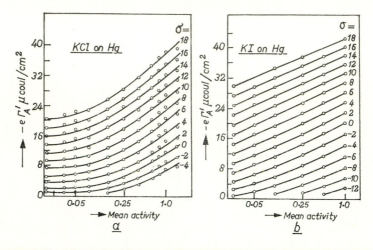

FIG. 3.11. Adsorption isotherms of I^- and Cl^- ions at 25°C for various values of the surface charge (in $\mu C/cm^2$); 10 $\mu C/cm^2$ represents 6×10^{13} ions/cm²; the curves are plotted against the mean ionic activity, not the salt activity. (a) Grahame and Parsons,[37] (b) Grahame[36]).

the dissolved ions). Then, provided the measurements were carried out at one temperature, an adsorption isotherm is obtained. Figure 3.11, taken from the work of Grahame[36] and of Grahame and Parsons,[37] gives isotherms at various values of σ for the adsorption of I^- ions and Cl^- ions. The data also allows to plot the potential difference across the inner layer: $V_1 = \psi_0 - \psi_\delta$ as a function of $\Gamma_{A'}$. Such plots are given in Fig. 3.12 taken from work of the same authors. The value of V_1 at $\sigma = 0$, extrapolated to $\Gamma_{A'} \to 0$, was taken as zero. This extrapolation was facilitated by the fact that the data almost fell on straight lines, although the systematic deviations at small $\Gamma_{A'}$ for KCl should then be disregarded. For all salts taken, this extrapolation at $\sigma = 0$ for $\Gamma_{A'} \to 0$ provided the same value of the E_{ecm} to within 15 mV. The linearity of the curves in Fig. 3.12 and their approximate parallelism may seem somewhat surprising[36] but, accepting these facts, they can be used to estimate the location of the inner Helmholtz plane in a simple way.[37] The curves of Fig. 3.12 can be represented by

$$V_1 = e\Gamma_A' K_\sigma^{-1} + \sigma K_\Gamma^{-1} \tag{3.3.11}$$

where K_σ is the capacity pertinent to the specific adsorption of anions and K_Γ is the capacity pertinent to the surface charge σ. The measure-

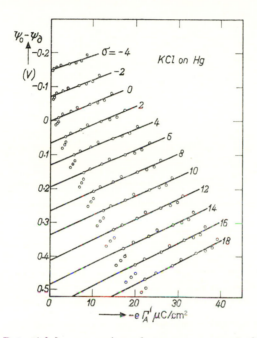

FIG. 3.12. Potential drop across inner layer versus amount of specifically adsorbed anions (25°C) $-e\Gamma_A$ (Grahame and Parsons[37]).

ments indicate that K_σ is only a weak function of σ and that K_Γ is only a weak function of Γ_A'. One may put

$$K_\sigma = \frac{\varepsilon_1}{4\pi\gamma} \qquad K_\Gamma = \frac{\varepsilon_1}{4\pi(\beta+\gamma)} \tag{3.3.12}$$

in which γ is the distance between inner and outer Helmholtz plane and β is the distance between the metal surface and the inner Helmholtz plane. We note that $\beta+\gamma = \delta$. It follows from eq. (3.3.12) that

$$\frac{K_\Gamma}{K_\sigma} = \frac{\gamma}{\beta+\gamma} . \tag{3.3.13}$$

The thickness ratio, obtained by this method (the capacity method), is sometimes denoted as $(x_2 - x_1)/x_2$. Figure 3.13 provides this value through eq. (3.3.11). It appeared that $(x_2 - x_1) < x_1$ which would only mean that the positions of inner and outer Helmholtz layer, are relatively close together. This is somewhat improbable from a physical point of view. However, the eqs. (3.3.12), used for this estimate, are certainly over-simplifications. This is borne out by the discussions of the previous sections. Thus, at $\Gamma_A' = 0$, K_Γ should be identical with the integral

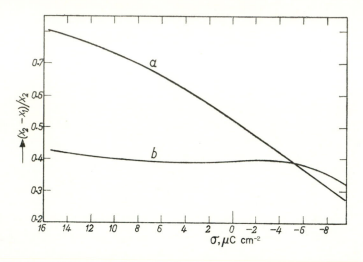

FIG. 3.13. Thickness ratio $(x_2 - x_1)/x_2$, obtained by the capacity method (curve b) and by the potential method (curve a) for KI, versus interfacial charge σ. Capacity method: eq. (3.3.11). Potential method: eq. (3.3.19). The thickness ratio obtained via the capacity method is sometimes denoted by $(x_2 - x_1)/x_2$ and that obtained via the potential method is sometimes denoted by $\gamma/(\beta + \gamma)$.

capacity of the inner region found for fluoride solutions. This means that

$$K_{\Gamma=0} = \frac{1}{\psi^{02}} \int_0^{\psi^{02}} C_1 \, d\psi^{02} \tag{3.3.14}$$

in which ψ^{02} (Grahame's notation) is used to indicate σK_Γ^{-1} and C_1 is the differential capacity of the inner region defined in Section 3.2. Figure 3.14 shows that this requirement of identity is fairly well met for KCl solutions. This means that the suggestion made by Mott et al.[32] of a sharply increasing dielectric constant a few angstroms away from the metal surface (i.e. beyond the inner Helmholtz plane) is pertinent here. These authors obtain an estimate of γ which is too low because they take a value of the dielectric constant of the matter between inner and outer Helmholtz layer which is too low.

Hills and Payne find a large increase of entropy and volume accompanying the adsorption of Cl^-, NO_3^- and $SO_4^=$ ions and explain this result by assuming a desorption of about ten water molecules per adsorbed anion, the hydration shells remaining intact. A desorption ratio of ten water molecules per anion seems rather high. We believe that the specific anionic adsorption leads to an important change of the

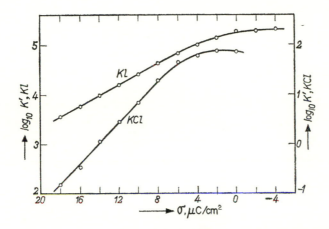

Fig. 3.14. Integral capacities of the inner region at zero amount of specifically adsorbed ions (25°C) versus interfacial charge σ (Grahame and Parsons[37]). For $K_{\Gamma = 0}$ see eq. (3.3.14).

water structure of the inner layer and that this change has an important effect upon the entropy and the volume change of the inner layer.

Returning to the adsorption isotherms of Fig. 3.11 they should provide information concerning the adsorption free energy. For the interpretation of ψ_A, it is necessary to refer to the Esin and Markov effect[6] and to Frumkin's explanation of the effect in terms of the discreteness of the charges of the specifically adsorbed anions. Figure 3.15 may be illuminating. It shows first (15a) the simple case of a flat condenser with uniform charge densities $+e\Gamma$ and $-e\Gamma$, the condenser plates being a distance δ apart. If the condenser is placed in a medium with relative dielectric constant ε, the potential difference ψ developed by this system is well known to be $(4\pi \delta)^{-1}\varepsilon e\Gamma$. The field outside the plates is zero and inside the plates it is ψ/δ. If for the moment the positive charges are assumed to be placed at zero potential and are identified with the charges of the Gouy layer located at the outer Helmholtz plane and if the negative charges are assumed to be the anions at the inner Helmholtz plane, then the potential ψ_A of the anions is identical with ψ. The situation is less simple when the discreteness of the charges is taken into account. Then the potential gradient outside the plates is not zero, although it generally approaches the zero value if x becomes large compared with $\Gamma^{-\frac{1}{2}}$, and its value and sign depend on the position considered with respect to the location of the discrete charges. Thus, close to a positive charge the potential gradient

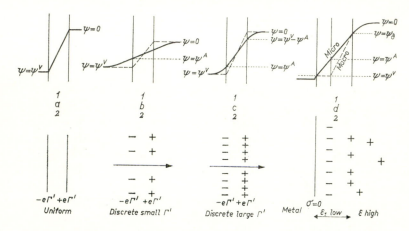

FIG. 3.15. Influence of the discreteness of the charges upon potential–distance relationship. Flat condensers. Curved continuous lines: micropotential (term introduced by Ershler[1]) versus distance. Slotted lines: macropotential versus distance. a 1, 2, uniform surface charge; b 1, 2 and c 1, 2, discrete charges; $\psi = \psi^A$ and $\psi = \psi^V - \psi^A$: potential at (adsorption) site; arrow: path of unit charge leading through empty (adsorption) sites d 1, 2, metal/solution interface, discrete (negative) charges at localized adsorption sites.

outside the layers is positive, whereas at a position just in between two positive charges it is negative. To discuss potential–distance relationships near layers of discrete charges, a model should be made of the charge distributions and the positions should be fixed, where the potential must be considered. Following Ershler[8] this potential is called micropotential (ψ^a). Potentials pertaining to smoothed charge distributions are macropotentials. Esin and Shikov,[7] Ershler[8] and Grahame[9] assumed a charge distribution according to a two-dimensional hexagonal lattice, each lattice point being occupied by a charge. Two opposite charges, a positive one and a negative one, were assumed to be removed and the micropotential was determined on the line connecting the empty lattice points. There is a natural tendency of the system to rearrange itself in such a way that the "holes" with the empty lattice points as their centres are eliminated. According to Grahame there is only a negligible energy gain involved in such a rearrangement but this statement has been challenged.[11,38] The situation is schematically drawn in Fig. 3.15b, c for two values of Γ. The full lines in the upper drawings indicate the micropotential vs. distance relationship, the dotted lines indicate macropotentials. For small Γ the potential gradient, or the field in between the layers and to a good distance beyond the layers, is a constant. In contrast, such a linear relationship does not or hardly exist if Γ is large. Then, especially near the layers, there may be a marked curvature as shown in Fig. 3.15c. The curves of Figs. 3.15b and c indicate the general trend by the calculations of Esin and Shikov and by Grahame. For low $\Gamma(\delta\sqrt{(\pi\Gamma)} \ll 1)$ Grahame found

$$\psi^a/\psi = 0.766\,\delta\sqrt{(\pi\Gamma)} - 0.697(\delta\sqrt{(\pi\Gamma)})^3. \qquad (3.3.15)$$

If now the positive charges are again identified with the charges in the Gouy layer (assumed localized at the outer Helmholtz plane) and the negative charges with the anions, the potential ψ_A would not have been given by ψ, but by $\psi - \frac{1}{2}(\psi - \psi^a)$. However, when this result is directly applied to the Hg/aq. soln. interface, the Esin and Markov effect is over-explained. This was shown by Esin and Shikov. Ershler then pointed out that the effect of the discreteness of the charges is weakened when multiple reflections of the charges within the double layer are taken into account. The specifically adsorbed anions are located in a region with comparatively low relative dielectric constant. Assume that the relative dielectric constant in the whole inner region ($0 \leqslant x \leqslant \delta$) is ε_1. The distribution of the lines of force around a charge e, located in this dielectric a given distance β away from the plane of the interface with another dielectric (ε_2), is the same as the distribution of the lines of force around the charge e when placed in the bulk of a dielectric (ε_1) at a distance 2β of a charge $[(\varepsilon_1 - \varepsilon_2)/(\varepsilon_1 + \varepsilon_2)]\,e$. The plane of the interface considered here is called a mirror plane. In the actual situation two mirror planes are present, one at $x = 0$ and one at $x = \delta$. Ershler assumed that both these planes were perfect mirrors (i.e. $\varepsilon_2 \gg \varepsilon_1$) for the specifically adsorbed ions. The effect of this situation upon the value of the micropotential can be estimated if a system of images is constructed. First the charges of the anions are reflected in the outer Helmholtz plane. Then the "dipole" layer of these charges and their images, a distance 2δ away, is reflected in the metal. The "image dipole" layer in the metal is in turn reflected by the outer Helmholtz plane and so on. The important point of the result is, that if prior to considering the images, the potential-distance relationship was linear, this linearity is preserved after considering the images. The condition for linearity is $\gamma \sqrt{(\pi \Gamma_A)} \ll 1$. This means that, with $\gamma = 1\frac{1}{2} - 3$ Å, $\Gamma_A' \ll 5 \times 10^{11}$ cm^{-2}, or that $e\Gamma_A' \ll 80\mu$C cm^{-2}. The slope of the straight potential–distance curve is different, however, and is no longer proportional to $\Gamma^3_{/2}$. The situation is sketched in Fig. 3.15d, where $\sigma = 0$ is chosen and where the potential at the outer Helmholtz plane is ψ_δ. The field in the inner region is constant and equal to $V_1/(\beta + \gamma)$ (constant field approximation). The potential of the anions at the inner Helmholtz plane is a fraction $\gamma/(\beta + \gamma)$ of the whole potential drop V_1 across the inner layer, plus the potential at the outer Helmholtz plane:

$$\psi_A = \frac{\gamma}{\beta + \gamma} V_1 + \psi_\delta. \qquad (3.3.16)$$

This result was improved upon by Levich et al.[10] who carried out a mathematical analysis and added a term to the right-hand side of eq. (3.3.16) which is approximately equal to $e/\gamma \varepsilon_1 \ln 2$. For sufficiently low Γ_A', where the constant-field approximation applies, the potential ratio $\gamma/(\beta + \gamma)$ should equal the distance ratio. For large large Γ_A' where the constant-field approximation no longer applies, a larger fraction of the potential drop than given by the distance ratio must take place between inner and outer Helmholtz planes. We approach the smeared-out case and eventually, for $\sigma = 0$, the potential ratio approaches unity (see also Fig. 3.15d). Furthermore, the assumption of perfect imaging is only justified for an inner region placed between a metal and a concentrated electrolyte. For dilute electrolytes imaging is less perfect and this may manifest itself in a more pronounced discrete-ion effect. A third reason why eq. (3.3.16) must be considered with some reserve is the assumption of a uniform relative dielectric constant ε_1 in the whole inner region. This assumption has already been challenged.[31] Much theoretical work has been done in a recent past as to the validity of the (hexagonal-lattice) model and of the three assumptions just mentioned. Some of this work will be discussed below. Other aspects will be discussed in Chapters 4, 5 and 7. Here we proceed with a simple analysis first given by Ershler, by Grahame and by Grahame and Parsons.

Esin and Markov measured the E_{ecm} as a function of the salt concentration c in the solution. Starting from

$$e\Gamma_A' = e\Gamma_0 K' c_A \, e^{xe\psi_A/kT} = K_\sigma V_1, \qquad (3.3.17)$$

which is a combination of eq. (3.3.6) and eq. (3.3.12), Ershler obtained, upon neglecting the potential drop in the diffuse region,

$$\frac{\partial V_1}{\partial \ln c} = \frac{kT}{2e} \frac{1}{(kT/eV_1) - (\partial \psi_A/\partial V_1)} = \frac{\partial E_{\mathrm{ecm}}}{\partial \ln c} \qquad (3.3.18)$$

in which $d \ln c_A = \frac{1}{2} d \ln c$ is taken. The change of V_1 at $\sigma = 0$ is equal to a change of the E_{ecm}. Values of β and of γ were determined from data on ionic radii and quite close agreement was reported[10] between predictions based on eqs. (3.3.18) and (3.3.16) as corrected by Levich *et al.* Grahame and Parsons had also data for non-zero values σ. Therefore they were able to obtain the potential ratio (which is denoted as

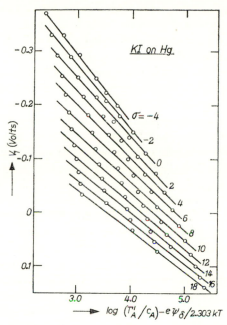

FIG. 3.16. Potential drop $(\psi_0 - \psi_0)$ across inner region versus log $\Gamma'_A/c_A - e\psi_\delta/2 \cdot 303 kT$ (see eq. (3.3.19)). (Grahame and Parsons[37]).

$\gamma/(\beta + \gamma)$ although it may not coincide with the distance ratio), and also the capacity ratio, denoted as $(x_2 - x_1)/x_2$ and also $\ln K'$ as a function of σ. They used a combination of eq. (3.3.6) and eq. (3.3.16)

$$\ln \frac{\Gamma_A'}{c_A} - \frac{e\psi_\delta}{kT} = \ln \Gamma_0 K' + \frac{eV_1}{kT} \frac{\gamma}{\beta + \gamma} . \qquad (3.3.19)$$

Plots of $\ln \Gamma_A'/c_A - e\psi_\delta/kT$ against V_1 appeared to be straight lines for each σ, at least for KI solutions (Fig. 3.16). The slope of these lines provides $\gamma/(\beta + \gamma)$ and the intercept gives $\ln K'\Gamma_0$. The values of the potential ratio for each σ are also given in Fig. 3.13 and can be compared with the capacity ratio $(x_2 - x_1)/x_2$. Both $(x_2 - x_1)/x_2$ and $\gamma/(\beta + \gamma)$ are larger for KI than for KCl but, especially at large σ-values, there is an important discrepancy between the two ratios. Capacity measurements indicated that the value of x_2 ($\equiv \delta$) is approximately the same in both cases and equal to $0.3\varepsilon_1$ Å. Therefore, according to Grahame and Parsons, it is probable that the adsorbed Cl$^-$ ions are located closer to the outer Helmholtz layer than are the I$^-$ ions. The values found for $\gamma/(\beta + \gamma)$ provide a linear relationship between ln K' and σ, especially at high σ-values. Here an increase of 5μC cm^{-2} (for instance from 10μC cm^{-2} to 15μC cm^{-2}) corresponds to a tenfold decrease of K'. A more constant value of K' would obtain if, instead of $\gamma/(\beta + \gamma)$, the capacity ratio $(x_2 - x_1)/x_2$ would have been taken. This was pointed out by Parry and Parsons.[39] Lawrence et al.[37a] used Br$^-$ ions instead of Cl$^-$ or I$^-$ ions and compared also electrocapillary measurements with capacity measurements. This work largely confirmed the results given in this section. An important problem in the analysis leading to knowledge of $\gamma/(\beta + \gamma)$ is an appropriate choice of the adsorption equation of the adsorbed anions. The eqs. (3.3.18) and (3.3.19) are derived by using eq. (3.3.6) which itself is based on an ideal equation of state of the adsorbed ions. This already does not seem to be compatible with the hexagonal-lattice model. Depending on the size and the charge of the anions, saturation values for $\sigma_1(= e\Gamma_A')$ were found to be roughly of the order of 30–50μC cm^{-2}. Many experiments have been carried out at a degree of coverage of 0.1 or more. Under these circumstances an important condition for the validity of an ideal equation of state will be violated. Therefore it can be expected that the introduction of excluded-area correction, or a positive second virial coefficient in the two-dimensional equation of state of the adsorbed species leads to an improved interpretation of the results. Such an introduction is implicit in the treatment by Parry and Parsons who returned to the original Stern (Langmuir) equation and who also used the Temkin equation. It is also implicit in the adsorption equation suggested by Levine et al.[40] who used a Flory–Huggins[41] approach. These authors obtained values for $\gamma/(\beta + \gamma)$ which were independent of σ. This must be considered as an improvement over the old interpretation because the locations of inner and outer Helmholtz planes are now less dependent upon the value of σ. In this respect it may be noted that it is also probable that the location of the outer Helmholtz plane is independent of the size of the cations (and of anions such as I$^-$ and Cl$^-$, see above[36,37] and of the electrolyte concentration. As argued by Payne[42] and by Levine et al.[12] the assumption that $(\beta + \gamma)$ and independently ε_1 do not depend very much on σ offers on the whole less difficulties than the assumption that only the ratio $\varepsilon_1/(\beta + \gamma)$, obtained via capacity measurements, depends little on σ.

Another way to account for a variable $\gamma/(\beta + \gamma)$ is a reconsideration of the multiple reflection theory of the charges in the inner region.

Ross Macdonald and Barlow[43,11] have shown this. They considered the outer Helmholtz plane as an imperfect mirror and obtained qualitative agreement with the (σ-dependent) potential ratio of Grahame and Grahame and Parsons. Imperfect imaging was also invoked by Dutkiewiecz and Parsons[44] for an explanation of the difference in the results obtained with KF + KI mixed solutions which were compared with results obtained with KI solutions at the same I^- concentrations. The electrolyte concentration in the KF + KI solutions being larger, more perfect imaging could be expected here than in the KI case. Imperfect imaging can be accounted for, in a qualitative way, by assuming an effective distance of the outer Helmholtz plane from the wall that is larger than this distance taken in the case of perfect imaging. Dutkiewiecz and Parsons found a larger (effective) distance in the KI case than in the KF + KI case.

A third factor in the analysis which deserves attention is the value of K' and its possible variations with varying σ. Again one feels inclined to consider an interpretation leading to an invariable K' as a more satisfactory result because $\ln K'$ is the chemical part of the free energy of adsorption. However, some care is needed. A sharp distinction between a "chemical" part and an "electrostatic" part is sometimes difficult to give. In this connection we mention the discussion by Dutkiewiecz and Parsons[44] about the question which parameter should be used in a description of ionic adsorption: the activity of the adsorbing ion or the salt activity. In their work, plots of σ_A, the charge due to the adsorbed I^- ions, against $\log a_{I^-}$ ($\approx \log a_{\pm}$) obtained by using KI solutions, were different from those obtained by using the mixed KF + KI solutions. When σ_A was plotted against $\log a_{K^+} a_{I^-}$ ($\approx \log a_{\pm}^2$) instead of against $\log a_{I^-}$, the curves were coincident. Therefore Dutkiewiecz and Parsons replaced the bulk ionic activity in eq. (3.3.6) by $a_{\pm}^2$. Ion-pair adsorption rather than the adsorption of I^- ions alone must be envisaged. The location of the K^+ ions which accompany the I^- ions must be somewhere in the Gouy layer (including positions at the outer Helmholtz plane). Here they modify imaging parameters. Formally the chemical potential of the K^+ ions can be absorbed in the $\ln K'$ term. Then the "new" free energy of adsorption varies in a different way with varying σ than does the "old" free energy of adsorption. *Some* variation, depending on the choice of the adsorption equation, will always be found. This was argued by Levine, Mingins and Bell.

The hexagonal-lattice model itself was subject to discussions by Ross Macdonald and Barlow, by Levine *et al.*, and by Buff and Stillinger.[45] The latter authors assumed only weak interactions between the adsorbed ions; the strong Coulombic forces were assumed almost absent owing to the presence of the conducting body at $x \leqq 0$. In this

E

view the adsorbed ions were in a state of disorder. The two-dimensional density distribution could be calculated by means of cluster theory. The two models, the "lattice" model and the "disorder" model, may be considered as two extremes. The "lattice" model gives the highest possible estimate of the discreteness-of-charge effect, the "disorder" model gives the lowest possible estimate. At low temperatures and (or) high density, thermal motions become unimportant and the "lattice" model will represent the physical situation. At high temperatures and (or) low densities thermal motions will have destroyed any ordered lattice. Estimates at room temperature, made by Bell *et al.*,[46] gave as a lower limit for the existence of an hexagonal lattice a surface area per ion of about 70 Å^{+2}, which is equivalent to a surface charge of monovalent ions of 23 $\mu\text{C cm}^{-2}$. Ross Macdonald and Barlow[47] arrived at a higher critical surface area: 200 Å^{+2} or $8\mu\text{C cm}^{-2}$. They also gave expressions for the temperature dependence. The difference between the two estimates is mainly due to the different values of ε_1 used by these authors[31]. Bell, Mingins and Levine used $\varepsilon_1 \approx 10$, and Ross Macdonald and Barlow used $\varepsilon_1 \approx 6$. In the latter case Coulombic interaction is more important than in the former case. When the temperature rises, ε_1 will rise and the critical surface area per ion decreases. It seems impossible to give any quantitative estimate, for one thing because little is known about the water structure in the inner region occupied with specifically adsorbed ions. It may well be that the anisotropy of this region must be taken into account. Such an anisotropy is almost certainly present because the electric fields in the x-direction and in the y- and z-directions are highly different. What is taken here as ε_1 is the dielectric constant in the x-direction. This will differ from that in the y- and z-directions. However, it is the latter quantity which is pertinent to the Coulombic interaction between the adsorbed ions. Future calculations concerning an inner region where crossed electric fields are present should contain the quadrupole-moment of the water molecule as an important parameter.

The nature of the transition between the "lattice" and the disordered state is unknown. Bell, Mingins and Levine, and Ross Macdonald and Barlow did not consider a real phase transition but used the (Lindemann) criterion that, in order for the transition to take place the root-mean-square of the amplitude of an oscillating ion in the lattice must exceed a certain fraction of the nearest-neighbour distance. A model which is intermediate between the "lattice" model and the "disorder" model is the "cut-off-disc" model, worked out by Levine *et al*. Here each ion was ascribed a charge-free circular zone with radius r_0 ($r_0 = \sqrt{\{1/\pi\Gamma_A'\}}$). Outside this circle the charge was considered smeared out. The radius r_0 was called Grahame radius because Grahame[9] used this parameter in an illustrative way. The physical meaning of a cut-off disc is that, when an ion A is held fixed at a given position, no charge is assumed present within a circle with radius r_0. For low Γ_A' this is unrealistic and it was shown in a thermodynamic treatment that other ions A penetrated to a non-negligible extent in the forbidden region. Instead of the parameter r_0 Levine *et al.* introduced a parameter r_β, where $r_\beta/r_0 \leq 1$. The nature of this ratio was extensively discussed by Levine and co-workers and it was pointed out that it was a sensitive function of the interaction energy between the adsorbed ions.

Krylov[48] calculated the effect of the discreteness of the charges in the inner region upon the potential distribution in the Gouy layer. He found an increase of the Gouy capacity compared with the smeared-out case. This means that the potential decays faster than calculated for the classical smeared-out case. For a 10^{-3} N solution the effect was small, but for a 10^{-1} N solution Krylov found a 7% increase of the Gouy capacity. This is a consequence of the tendency of the ions in the diffuse layer to crowd in hemispheres around the specifically adsorbed ions. For concentrated solutions the Gouy layer has almost disappeared and the early Ershler model becomes valid. In this case the outer Helmholtz plane serves as an ideal mirror.

Krylov also compared the cut-off disc model with the (hexagonal) lattice model and found, at least for the metal/solution interface, only a slight effect upon the micropotential.

3.4.   The adsorption of organic compounds

It was already recognized by Gouy, that the Hg/aq. soln. interface was equally well suited for the study of the adsorption of organic, ionized or non-ionized, substances as for that of the inorganic ions considered so far. Figure 3.17 gives two examples. The first is a set of electro-capillary curves of solutions of tertiary amyl alcohol (2-methyl-2-butanol) in 1 M NaCl.[49] The ecm is shifted to more anodic potentials and, especially at the cathodic side, the interfacial tension can be made appreciably lower than measured in the absence of the organic sub-stance. For extreme values of the applied potential, positive or negative, the curves approach the basis curve indicating desorption of the alcohol molecules under these circumstances. This desorption was explained by Gouy to the replacement of the alcohol molecules by the more polar water molecules (see Section 3.1). The shift of the ecm was considered indicative for the adsorption of the alcohol molecules with the positive end of their dipole oriented toward the surface and consequently with their hydrophobic part directed to the surface. However, some care is needed because the role of the water molecules in this adsorption process may be such as to cause a shift of the ecm in the same direction. The second example is given by a set of electrocapillary curves of thiourea dissolved in 0·1 M NaF solutions.[50] Here the situation is different. The ecm is shifted to more cathodic potentials and the largest lowering of the interfacial tension is found at the anodic side of the curve. An important feature is that even at strong anodic potentials the lowering is not absent, indicating that under these circumstances thioureum is still adsorbed.

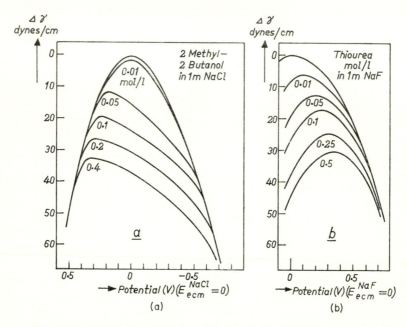

Fig. 3.17. (a) Influence of tertiary amylalcohol(2-methyl-2 butanol) upon the electrocapillary curve of 1 M NaCl aqueous solutions. Lowering $\Delta\gamma$ of the interfacial tension compared with ecm (A. N. Frumkin[49]), (1 dyne/cm = 10^{-3} N m^{-1}). (b) Influence of thiourea upon the electrocapillary curve of a 1 M NaF aqueous solution (Schapink et al.[50]).

Frumkin[49] verified Gouy's measurements and gave, in 1926, a quantitative interpretation in terms of a simple theory. He assumed a Langmuir equation of state in which Van der Waals-type interactions were introduced (see eq. (2.4.53))

$$\pi = -\Gamma_0 kT \ln (1-\theta) - a\Gamma^2. \qquad (3.4.1)$$

The adsorption equation is derived from the condition $\mu_s = \mu_{\text{ads}}$ where μ_s is the chemical potential of the dissolved organic molecules and μ_{ads} the chemical potential of the organic molecules at the interface. For an ideal solution (concentration c) one has

$$\mu_{s0} + kT \ln c = \bar{\mu}_0 + kT \ln \frac{\theta}{1-\theta} - 2a\Gamma. \tag{3.4.2}$$

The influence of the applied potential E upon the standard chemical potential $\bar{\mu}_0$ was given in the form of polarization energies as follows:

$$\bar{\mu}_0 = \mu_0 + \frac{1}{2\Gamma_0}(C_0 - C')E^2 + \frac{1}{\Gamma_0}C'EE_{ads} \tag{3.4.3}$$

in which C_0 and C' are respectively the capacities of the double layer in the absence of organic molecules ($\theta = 0$) and in the saturated state ($\theta = 1$ or $\Gamma = \Gamma_0$). Furthermore, E_{ads} is the potential drop due to the fixed orientation of the dipoles of the alcohol molecules. It is considered a constant. E is taken zero at the ecm; it may be identified with ψ_0 of the preceding sections. Gouy's qualitative explanation of the expulsion of adsorbed molecules is reflected in the term $1/2\Gamma_0\,(C_0 - C')E^2$ in which $C_0 > C'$. For extreme values of E^2 the value of the degree of coverage becomes small and the (negative) contribution to γ, given by eq. (3.4.1) becomes small. For moderate E-values Frumkin obtained good agreement with the experimental data when $E_{ads} \cong -0.5$ volt, $C' \cong 4.4$ μF cm^{-2}, $C_0 \cong 42$ μF/cm^2, $a \cong 1.8$ kT/Γ_0 and $(\mu_0 - \mu_{s0})/kT \cong -2\frac{1}{2}$ (standard concentration 1 M).† The paper of Frumkin may serve here as a basis for further discussion, just as it served as the basis for much further work in this field. We briefly discuss the following points:

(a) The capacity–potential relation.
(b) Multilayer adsorption. "Hemi-micelle" formation.
(c) Polarization forces at the charged interface.
(d) The position of an adsorbed organic molecule. The physical nature and the value of $\Delta\mu_0$.
(e) The Van der Waals a. The equation of state.
(f) The adsorption of weakly ionized acids and of dibasic acids H$_2$A.

This classification is somewhat arbitrary. For example, it is hardly possible to consider the equation of state as a topic which is independent of the points mentioned under (a)–(d).

(a) It is sometimes assumed that the capacity C of the double layer can be written as

$$C = C_0'(1 - \theta) + C'\theta \tag{3.4.4}$$

---

† The adsorption coefficient $\exp -(\mu_0 - \mu_{s0})/kT \cong 12.7$ and was denoted as B_0.

but this is an over-simplification. In this respect Fig. 3.18, from Parsons,[51] is illuminating. It gives schematically various pertinent quantities plotted against the potential. The capacity–potential curve provides two peaks, each at a potential where θ is small. Proskurnin and Frumkin[52] found such peaks experimentally in 1935. To show theoretically the possibility of peaks, consider a simple Henry adsorption equation for Γ and insert this into eq. (2.4.88). We assume, for convenience, that $E_{ads} = 0$ (this means that in eq. (2.4.88) $a = 0$). The

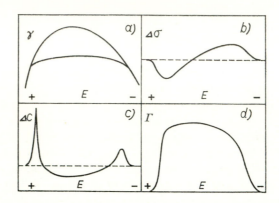

FIG. 3.18. Schematic representation of the effect of adsorbed organic molecules upon electrocapillary properties. (a) Electrocapillary curve and lowering of interfacial tension caused by the adsorption of organic material. (b) Effect upon surface charge. (c) Effect upon capacity. (d) Adsorbed amount. Horizontal: applied potential E (Parsons[2a]).

case $E_{ads} = 0$ leads to somewhat more complicated expressions. With these assumptions eqs. (3.4.2)and (3.4.3) can be combined to give

$$\Gamma = \Gamma_{max}\, e^{-b\psi^2 0/Tk} \qquad (3.4.5)$$

where Γ_{max} is the adsorption at $E = 0$ (we identify E with ψ_0) at a given concentration c. Furthermore, $b = \frac{1}{2}\Gamma^{-1}(C_0 - C')$ is the same constant as that used in eq. (2.4.88). Equation (3.4.5), inserted into eq. (2.4.88), gives

$$(C - C_0)_E = -2b\Gamma\left(1 - \frac{2b\psi_0^2}{kT}\right). \qquad (3.4.6)$$

At $\psi_0 = 0$ the capacity contribution of the adsorbed species has a minimum and at $\psi_0 = \pm(3kT/2b)^{\frac{1}{2}}$ there are maxima. At these maxima $\Gamma = 0\cdot223\Gamma_{max}$. If the absolute value of ψ_0 is increased beyond this value, there is a rapid desorption. Therefore desorption is indicated by a capacity peak. This conclusion will be valid for other equations of state as well provided there is a square term in the $\Delta\bar{\mu}_0 - \psi_0$ relationship. However, not every capacity peak indicates desorption, because peaks are sometimes found at a potential where γ was not yet equal to $\gamma°$. Thus, Eda[53] and Damaskin *et al.*[54] found, for the case of alkylsulphates (C_{10}, C_{12}) in ≈ 0.25 M K_2SO_4 or Na_2SO_4 solutions, not only the usual

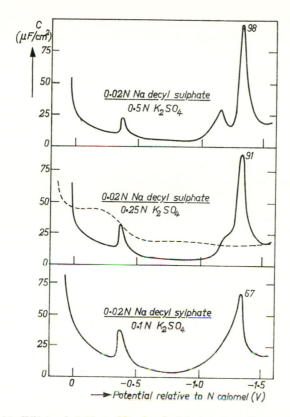

Fig. 3.19. Effect of 0.02 M Na decylsulphate upon the capacity of Hg/aq.soln. interface; K_2SO_4 solutions (Eda and Tamamushi[53]).

desorption peaks, but also two small peaks at ψ_0-values (-1.1 V and -0.3 V, referred to normal calomel) where the adsorption must be appreciable (see Fig. 3.19). These small peaks were attributed to a reorientation of the adsorbed molecules. The nature of this reorientation may be connected with a first stage of micelle formation on the surface as suggested by Damaskin *et al.*[54]

(b) Multilayer adsorption was first detected by Melik-Gaikazyan[55] for 1-octanol and 1-hexanol, who found a minimum capacity of

2 $\mu F/cm^2$ or lower. For the sulphate case it was about 4 $\mu F/cm^2$. Multilayer formation should be connected with a slow diffusion process of the organic molecules (ions) toward the surface. Measurements of the frequency dependence in the sulphate case did not point to such a slow diffusion process[54] and, together with the relatively high capacity value, it was concluded that no multilayers were formed in that case. Good evidence for multilayer formation was found by Barradas and Kimmerle[56] for the adsorption of the surfactants "Triton-X-100" or "Triton-X-305" (non-ionic iso-octylphenoxypolyethoxy ethanols). When the surfactant concentration was increased beyond about 2×10^{-5} M, a sharp drop of the interfacial tension was found, this concentration being independent of the applied potential over a wide range of values (of the order of 1 V) at the cathodic side of the ecm. At concentrations beyond $2 \cdot 45 \times 10^{-4}$ M the interfacial tension tended to a constant value, $0 \cdot 03 - 0 \cdot 04$ N m^{-1} ($= 30 - 40$ dyn cm^{-1}) lower than γ^0, the value of the interfacial tension at the same potential value but in the absence of the surfactant. The concentration of $2 \cdot 45 \times 10^{-4}$ M is the critical micelle concentration (cmc). At this concentration the surfactant molecules in the solution start to "cluster" leading to the formation of micelles (see Chapter 5) and the concentration of the individual molecules is only slightly affected by the addition of more surfactant to the solution. This explains why the drop of the interfacial tension is only slightly affected at concentrations beyond the cmc. The sharp drop of γ, starting at the concentration (c_h) of 2×10^{-5} M, points to an enhanced adsorption which was interpreted by Barradas and Kimmerle in terms of multilayer formation on the surface. This multilayer was called "hemi-micelles" by Somasunduran et al.[57] who studied similar phenomena in other systems (see Chapter 4). The concentration c_h was called the hemi-micelle concentration (hmc).

(c) Hansen et al.[58] studied the adsorption–desorption properties of a number of organic molecules and ions. An interesting example was provided by acetylacetone (2,4-pentanedione). Here the desorption peaks were anomalously small. Now acetylacetone occurs in the keto form and in the enol form and Hansen et al. offered as a possible explanation of the impaired desorption at extreme potential values, a shift of the keto–enol equilibrium in the direction of the more polarizable enol form, when the field strength was increased.

There has been some discussion as to which parameter is the more meaningful from a physical point of view in the analysis of adsorption data at charged interfaces, the potential ψ_0 or the surface charge σ which is proportional to the field strength in the inner region (see Section 2.4). Some support for the use of σ is provided by the explanation of the behaviour of adsorbed acetylacetone. Parsons[59] preferred to use, instead of eq. (3.4.3), the following equation:

$$\Delta\bar{\mu}_0 = \Delta\mu_0 + \frac{1}{2\Gamma_0}(C_0^{-1} - C'^{-1})\sigma^2 + \frac{4\pi}{\varepsilon_1}p\sigma \qquad (3.4.7)$$

in which the dipole moment $p = (4\pi\Gamma_0)^{-1}\varepsilon_1 E_{ads}$ of an adsorbed molecule is introduced. Then, as pointed out in ref. 59 and in Section 2.4, the function $(\gamma + \sigma E)$ should be used for the derivation of thermodynamic relations instead of γ. Thus, the surface pressure is

$$\pi = -(\gamma - \gamma^0) - \sigma(E - E^0). \qquad (3.4.8)$$

(d) Starting from eq. (3.4.8), Parsons[60] analysed data obtained by Schapink et al.[50] for the adsorption of thiourea. It was known that there is a strong specific interaction between the Hg and the S atoms of thiourea. The standard† free energy of adsorption is -96 kJ per mole $(= -38 \, kT$ per molecule at room temp.) and that desorption was taking place at strongly cathodic potentials. Parsons found that one single (virial) equation of state fitted the data for values of σ between zero and $-12 \, \mu C \, cm^{-2}$. At each σ another value of $\Delta\bar{\mu}_0$ was found. From the slope of the (linear) $\Delta\bar{\mu}_0$–σ curve the value of $(4\pi/\varepsilon_1)p$ was obtained. With $p = 4.89 \times 10^{-18}$ e.s.u. cm values of ε_1 were obtained ranging from 15 $(\sigma = 0)$ to 8 $(\sigma = -12 \, \mu C \, cm^{-2})$ which is in qualitative agreement with the ε_1-values of the inner region given in the previous sections. On the whole the same behaviour was found for thiourea as for specifically adsorbed anions. Just as for these anions the potential drop across the inner region had two components and could be represented by an equation, similar to eq. (3.3.11)

$$\psi^0 = \frac{\sigma}{K_{\Gamma=0}} + \Gamma\left(\frac{4\pi p}{\varepsilon_1}\right)_{\sigma=0}. \qquad (3.4.9)$$

The ε_1-values found through eq. (3.4.9) were in agreement with those given above and independent of Γ. The inner layer differential capacity $(\partial\sigma/\partial\psi^0)_\Gamma$ appeared to decrease with increasing Γ. This pointed to an increase of the inner layer thickness from about $3\frac{1}{2}$ Å $(\sigma = 0; \Gamma = 0)$ to about $4\frac{1}{2}$ Å $(\sigma = 0; \Gamma = 1\frac{1}{2} \times 10^{14} \, cm^{-2})$ which is understandable because the length in the C–S direction of the molecule is about 5.7 Å.

A similar analysis was carried out by Parry and Parsons[39] for the adsorption of the (fully ionized) Na-benzene m-disulphonate. The adsorption isotherms could be fitted to a Temkin isotherm. A rather small saturation value of the adsorbed amount existed, the area per

---

† Standard states: 1 M and 1 molecule cm^{-2}.

ion being 125 Å². This large value suggested that the plane of the benzene ring was lying flat on the metal surface. Such a position of the benzene ring was also inferred from the value of the inner layer capacity (about 20 μF/cm²) indicating a thickness of about 4 Å. The analysis outlined in the previous section provided a distance from the metal surface to the inner Helmholtz plane of about 3 Å. The measurements were carried out at various temperatures and it was found that the heat of adsorption increased markedly with increasing temperature. According to Parry and Parsons this seems characteristic of this type of adsorption processes. At room temperature the heat of adsorption is $-19\frac{1}{2}$ kJ mol⁻¹ ($=4\frac{1}{2}$ kcal mol⁻¹).

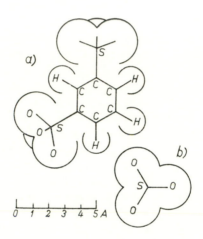

FIG. 3.20. Structure of the benzene m-disulphonate ion (Parry and Parsons[39]).

There is now much evidence that the principal adsorption forces between flat-lying aromatic rings and the mercury are due to π-electron interaction. Blomgren et al.[61] compared the adsorption properties of butyl-, phenyl- and naphthyl compounds dissolved in a 0·1 N HCl aqueous solution. The adsorption coefficient k where k was defined by

$$k = \frac{\theta}{c(1-\theta)} \qquad (3.4.10)$$

was found for the butyl compounds between 10³ (n-butylsulphonic acid) and 10⁷ (n-dibutylketone): for the phenyl compounds k-values were found between 10⁴ (aniline) and 3×10^7 (diphenylsulphide) whereas for the naphthyl compounds the k-values for all adsorbates investigated were found near 10^7. (The bulk adsorbate concentration c was expressed in moles per litre.) For the aliphatic compounds there was a shift of the ecm in the positive direction and for the aromatic compounds there was a shift in the negative direction. The aliphatic compounds contained polar groups with their negative end pointing away from the hydrocarbon radical. Therefore it is likely that the aliphatic molecules are adsorbed with their hydrocarbon radical at the surface and the substituent group in the solution. The shift in the negative direction of the

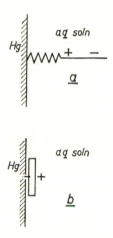

FIG. 3.21. Orientation of adsorbed organic molecules. (a) Aliphatic molecule, polar head tending into solution. (b) Aromatic molecule, benzene ring lying flat on metal surface and held there through π-electron interaction.

potential of the ecm for the aromatic compounds is consistent with a π-electron interaction between the metal and the aromatic radical. Since the k-values show less spread for the aromatic compounds than for the aliphatic compounds, adsorption forces arising from π-electron interaction will dominate.

Blomgren *et al.* used a Langmuir model (as is already clear from eq. (3.4.10)), a fraction θ covered by the adsorbate and a fraction $(1 - \theta)$ covered by water molecules. The following expression was given for $\Delta\mu_0$:

$$\Delta\mu_0 = (\mu_{0,A}{}^s - \mu_{0,A}{}^b(- (\mu_{0,H_2O}{}^s - \mu_{0,H_2O}{}^b) \qquad (3.4.11)$$

$$= \Delta\mu_{0,A} - \Delta\mu_{0,H_2O}$$

where $\mu_{0,A}{}^s$ and $\mu_{0,H_2O}{}^s$ are the standard electrochemical potentials of adsorbate (A) and water in the adsorbed state where the standard state was a unit mole fraction on the surface. Furthermore, $\mu_{0,A}{}^b$ and $\mu_{0,H_2O}{}^b$

are the standard electrochemical potentials of the adsorbate and of water in solution referred to unit mole fractions in the solution. An "intrinsic" free energy of adsorption was defined as follows:

For each given adsorbate the $\Delta\mu_0$-value was plotted against the molar free energy of dissolution $\Delta\mu_0^*$:

$$\Delta\mu_0^* = \mu_{0,A}^b - \mu_{0,A}^* \qquad (3.4.12)$$

where μ_0^* is the standard electrochemical potential of the pure adsorbate in liquid or solid form at the temperature of the adsorption experiment. The quantities $\mu_{0,A}^b$ and $\mu_{0,A}^*$ are discussed in Section 2.2 (e.g. eqs. (2.2.15), (2.2.16) and (2.2.17)). Three curves relating $\Delta\mu_0$ to $\Delta\mu_0^*$ were obtained, for the butyl-, the phenyl- and the naphthyl compounds. Extrapolation to $\Delta\mu_0^* = 0$ provided values of $\Delta\mu_0$ which we denote as $(\Delta\mu_0)_0$:

$$(\Delta\mu_0)_0 = -6\tfrac{1}{2}\,\text{kJ mol}^{-1}\,(=1\tfrac{1}{2}\,\text{kcal mol}^{-1})\,\text{(butyl)},$$

$$(\Delta\mu_0)_0 = -19\tfrac{1}{2}\,\text{kJ mol}^{-1}\,(=4\tfrac{1}{2}\,\text{kcal mol}^{-1})\,\text{(phenyl)}, \qquad (3.4.13)$$

$$(\Delta\mu_0)_0 = -29\,\text{kJ mol}^{-1}\,(=6{\cdot}9\,\text{kcal mol}^{-1})\,\text{(naphthyl)}.$$

The "intrinsic" free energies of adsorption were defined as

$$(\Delta\mu_{0,A})_0 = (\Delta\mu_0)_0 + \Delta\mu_{0,\text{H}_2\text{O}}. \qquad (3.4.14)$$

The name "intrinsic" was chosen because the role of the solvent is eliminated. To obtain numerical values for the intrinsic free energy of adsorption, it is necessary to know the value of $\Delta\mu_{0,\text{H}_2\text{O}}$. This was estimated by Blomgren et al.[61] at about $-22\tfrac{1}{2}$ kJ mol^{-1} ($=5{\cdot}4$ kcal mol^{-1}). Use was made here of the results of old measurements by Kemball[62] concerning the adsorption of water from the vapour phase on the mercury surface.

The importance of π-electron interaction was also demonstrated by Frumkin et al.[63] who compared the adsorptive properties of aromatic compounds with the corresponding hydroaromatic compounds, both with respect to the Hg/aq. soln. interface, for instance aniline ($\text{C}_6\text{H}_5\text{NH}_2$) and cyclohexylamine ($\text{CH}_6\text{H}_1\text{NH}_2$). (The case of aniline has received wide attention, see refs. 64 and 65.) They also determined the bulk concentrations of aromatic and the corresponding hydroaromatic compounds which were needed to obtain a certain decrease $\Delta\gamma$ of the interfacial tension between Hg and a solution, and compared it with the bulk concentrations which were needed to obtain the same decrease $\Delta\gamma$ for the solution/air interface. The ratio of the concentrations for the hydroaromatic compounds was closer to unity than that for the aromatic compounds. This pointed to a π-bond contribution which was estimated at 4 kJ mol^{-1} ($\cong 1$ kcal mol^{-1}). The orientation of an adsorbed molecule often is a function of the surface charge and (or) of the bulk concentration of the adsorbate. Aniline and pyridine molecules only lie flat on the surface, if the potential is more cathodic than about -0.5 V (with respect to the normal calomel electrode). Then $\sigma > 0$ and the π-electrons are able to interact with the positive charges in the metal. At less cathodic potentials the adsorbed molecules assume a vertical position. This was inferred by Damaskin[66] from a pronounced capacity peak at about -0.5 V which was ascribed to a reorientation process and not to a desorption process. The vertical position, occupying a relatively small surface area, was also favoured by a relatively large bulk concentration. For the p-toluenesulphonate ion Parry and Parsons[67] found below a concentration of 0.557 M only the "flat" position for adsorbed ions whereas at concentrations of 0.567 M and higher two positions were found depending on the charge on the mercury. For $\sigma > 0$ only the "flat" position was found. At negative surface charges there was a rise in the amount adsorbed. This rise (at $\sigma = -4$ $\mu C/cm^2$ and $c = 2.268$ M an adsorption peak corresponding to an area per sulphonate ion of 32 Å^2 was found) was explained by assuming a reorientation of the sulphonate ion in a direction perpendicular to the plane of the interface. In the flat position the area per ion was 53 Å^2, in the perpendicular position it was 32 Å^2. A Langmuir model was used for the latter case. This was modified when the former case was pertinent and a Flory–Huggins approach was used. Here the number of solvent molecules replaced per adsorbate molecule is greater than one, r say, and the Langmuir expression, eq. (3.4.10), must be replaced by

$$k = \frac{\theta}{r(1-\theta)r_c} \quad \text{(Flory–Huggins)}. \tag{3.4.15}$$

Parry and Parsons found $r \cong 1$ for the perpendicular orientation and $r \cong 2$ for the flat orientation. This would involve an area of roughly 30 Å^2 for a water molecule at the mercury surface. A possible explanation for this large surface area was that water molecules at the metal surface exist in the form of clusters.[68] Concentration-dependent reorientation phenomena have been found for aliphatic compounds as well. For octanoic acid Hansen et al.[58] found that at small coverage the molecules were lying flat on the surface and were standing right up at higher coverages. Frumkin's theory was used and, for octanoic acid, it was found that the best fit of the experimental data was obtained when Γ_0 was allowed to decrease at increasing concentration. It should be noted that Frumkin already came to the same conclusion (although tentatively) in 1926 for the adsorption of tertiary amylalcohol. The phenomenon of a variable Γ_0 and its interpretation is quite common for the adsorption at the air/water interface. Hansen et al. observed that all their compounds had, at the ecm, the positive end of their dipoles turned to the mercury surface, in conformity with the results of Gouy, Frumkin and Blomgren, Bockris and Jesch in 1961. The value of the free energy of adsorption was, for aliphatic compounds as used by Hansen *et al.*, proportional to the value of the difference ΔP of the optical polarizabilities per cm^3 between adsorbate and solvent. This suggests that dispersion forces were largely responsible for the free energy of adsorption because both in ΔP and in the expressions of the dispersion forces the same natural frequencies of adsorbate and solvent molecules appear. Specific functional groups, being turned to the solution side, play a minor role here. This should be contrasted with the findings of Blomgren *et al.*[61] who found, for butyl compounds, a wide variety of values of the adsorption coefficient k.

(e) Just as in the case of the adsorption of the inorganic anions it is here impossible to characterize unambiguously the behaviour of the adsorbed molecules by a certain equation of state. After a rough characterization of the adsorption forces almost all one can hope for is to arrive at a statement concerning the sign of the second virial coefficient which may be obtained from a Frumkin, a Temkin, or a Van der Waals equation of state. Hansen *et al.* and Damaskin *et al.* found the Frumkin equation (eq. (3.4.1)) applicable for a fairly wide variety of adsorbates. Parry and Parsons considered a variety of equations of state (Temkin, Frumkin, virial). Lorenz *et al.*[69] found, for the adsorption of the large cations of triethyl, tetraethyl- and tetrapropylammonium-chloride, a best fit with a Langmuir isotherm. For the adsorption of aliphatic amines, for instance triethylamine, they found a best fit with a virial isotherm with $a > 0$ (attraction between adsorbed molecules). These isotherms were obtained by plotting the bulk concentrations against $(C - C_0)$ at the capacity minimum where according to eq. (3.4.6) the surface coverage is maximal under the given circumstances.

The second virial coefficient $B = b - a/kT$ has been introduced in Section 2.4. The contribution b may be derived from the Volmer equation of state, or from the Langmuir equation of state although the meaning of b differs somewhat in the two cases. As pointed out in Chapter 2, section 4, the value of B depends on the energy $E(r)$ of the interaction between two adsorbed particles, a distance r apart, where r may assume each value between 0 and infinity. There are several contributions to $E(r)$. For ionized particles Coulombic interactions may dominate. Dispersion interaction certainly contributes to (Er) whether the adsorbed molecules are ionized or non-ionized, polar or non-polar. It leads to a positive contribution to a. This will be shown in Chapter 4 where the intermolecular interaction, applied to the interactions between colloidal particles is dealt with. Here we confine ourselves to the polar–polar interaction because polar organic molecules have often been used in the experiments. The dipole contribution to a may be either positive or negative. It is positive when the adsorbed dipoles assume all directions in space. Then the value of λ in eq. (2.4.31) can be calculated to be $+p^4/3\varepsilon_1^2r^6$ where p is an "effective" dipole moment of an adsorbate molecule. As explained in the next chapter, p may be written as $(p_A^{\frac{1}{2}} - p_{H_2O}^{\frac{1}{2}})^2$ in which p_A is the dipole moment of an isolated organic molecule. When the adsorbed dipoles are placed in one layer and oriented in a direction normal to the plane of this layer, there is a negative contribution to a. The dipoles repel each other, the energy of the repulsion being given by p_A^2/ε_1r^3. Since this repulsion energy is inversely proportional to r^3 it will dominate other contributions to a, such as dispersion contributions, because the decay of these contributions is much faster, i.e. proportional with r^{-6}. For thiourea molecules it is probable that a dipole–dipole repulsion such as the one just described, exists. Parsons found that here $a = -60$ Å^2.

If there is a multilayer adsorption of polar molecules, the Van der Waals a may have an anisotropic character. Dipoles in the second layer attract those in the first with an energy $-2p_A^2/\varepsilon_1r^3$ if all the dipoles in the two layers are aligned in a direction normal to the plane of adsorption. Dipoles in one layer repel each other. The repulsion in the second layer will be weaker than in the first, or turned into an attraction because the orientation in the second layer will be less pronounced than in the first.

(f) If the adsorbate is a substance, which is weakly ionized, the simultaneous and competitive adsorption should be considered of molecules and the corresponding ions. The ions are usually more hydrophilic than the molecules and their adsorption is relatively low. This is reflected in the lowering of γ, resulting from the addition to the base solution of such a substance. For weak mono- and dibasic fatty acids the lowering was shown to be smaller at high pH than it was at low pH.[70] Similar phenomena have been observed at the air/water interface (Chapter 5). Assuming a weak acid H_2A, dissolved in an aqueous HCl solution, the capillary equation becomes

$$d\gamma = -\sigma \, dE - \Gamma_{H_2A} \, d\mu_{H_2A} - \Gamma_{HA} \, d\mu_{HA} - \Gamma_A \, d\mu_A \qquad (3.4.16)$$
$$- \Gamma_{Cl} \, d\mu_{Cl} \Gamma_H \, d\mu_H.$$

After noting that

$$\sigma = e(\Gamma_{HA} + 2\Gamma_A + \Gamma_{Cl} - \Gamma_H),$$
$$d\mu_{HCl} = d\mu_H + d\mu_{Cl}$$
$$d\mu_{HA} = d\mu_{H_2A} - d\mu_H,$$
$$d\mu_A = d\mu_{H_2O} - 2d\mu_H,$$

eq. (3.4.16) can be rewritten as

$$d\gamma = -\sigma\, d\left(E - \frac{\mu_H}{e}\right) - (\Gamma_{H_2A} + \Gamma_{HA} + \Gamma_A)\, d\mu_{H_2A} - \Gamma_{Cl}\, d\mu_{HCl} \quad (3.4.17)$$

in which $d(E - \mu_H/e)$ is written as dE_H, which is the change of the applied potential of a cell in which a H_2 electrode is the reversible electrode. As noted by Blomgren and Bockris,[65] the information concerning the adsorption is pertinent to a sum, $(\Gamma_{H_2A} + \Gamma_{HA} + \Gamma_A)$ in our case, and not to the separate values. Cross-differentiation leads to

$$\left(\frac{\partial\sigma}{\partial\mu_{H_2A}}\right)_{T,p,E_H,\mu_{HCl}} = \left(\frac{\partial(\Gamma_{H_2A} + \Gamma_{HA} + \Gamma_A)}{\partial E_H}\right)_{T,p,\mu_{H_2A},\mu_{HCl}}. \quad (3.4.18)$$

It is difficult to work out the right-hand side of this equation, for one thing because specific interactions may occur between the various adsorbed molecules and ions. Thus Grand[70] found around pH = $\frac{1}{2}(pK_1 + pK_2)$ in which pK_1 and pK_2 are ionization constants of the dibasic acid, indications of a specific two-dimensional arrangement, formed by HA^- ions. Ostrowski *et al.*[71] measured, in strongly acid solutions (0·1 N HCl and 6 N HCl), the capacity–potential relationships pertaining to the adsorption of sulphoxides R_1R_2SO. There is an equilibrium

$$R_1R_2SO + H^+ \leftrightarrows (R_1R_2SOH)^+. \quad (3.4.19)$$

In 0·1 N HCl solutions only few oxide ions are formed. In 6 N HCl the equilibrium is more at the side of the sulphoxonium ion. This was inferred from u.v. spectroscopy. R_1 stood for a benzyl- or a hexyl-group, or for an aliphatic chain (C_4 to C_{12}), whereas R_2 stood for a benzyl- or a hexyl group. The capacity–potential curve showed a pronounced minimum just as in other cases mentioned above, which will be due to sulphoxide adsorption, but in the very strong (6 N) acid solutions with R_1 being a long chain, there was a characteristic hump in the centre of the low-capacity region. This hump was frequently independent up to 10^4 Hz, and was ascribed to a reorientation of adsorbed pairs of Cl^- ions and sulphoxonium ions. If both R_1 and R_2 were benzyl groups, the hump was not observed probably because π-electrons give rise to a strong attachment of the molecules to the surface.

REFERENCES

1. D. C. Grahame, *Chem. Revs.*, 1947, **41**, 441.
1a. D. C. Grahame and B. A. Soderberg, *J. Chem. Phys.*, 1954, **22**, 449.
2. R. Parsons in *Modern Aspects of Electrochemistry*, 1954, **1** (ed. J. O'M. Bockris and B. E. Conway), Butterworth.
2a. R. Parsons in *Advances in Electrochemistry and Electrochemical Engineering*, 1961, **1**, Interscience (ed. Delahay).
2b. R. Parsons, *Annual Reports of the Chemical Society for* 1964, **61**, 80.
2c. D. J. Schiffrin, *Trans. Faraday Soc.*, 1971, **67**, 331.
3. A. N. Frumkin, *J. Electrochem. Soc.*, 1960, **107**, 461.
4. A. N. Frumkin and A. Titievskaja, *J. Phys. Chem. U.S.S.R.*, 1957, **31**, 485.
4a. A. N. Frumkin and D. Poljanovskaja, *J. Phys. Chem. U.S.S.R.*, 1958, **32**, 157.
4b. B. B. Damaskin, N. Nikolajewa-Fedorovitch and A. N. Frumkin, *Doklady Akad. Nauk U.S.S.R.*, 1958, **121**, 129.
4c. T. N. Andersen and J. O'M. Bockris, *Electrochimica Acta*, 1964, **9**, 347.
5. R. Parsons and F. G. R. Zobel, *J. Electroanal. Chem.*, 1965, **9**, 333.
6. O. A. Esin (Essin) and B. F. Markov, *Acta Physicochim. U.S.S.R.*, 1939, **10**, 353. These authors attribute the idea of the discreteness-of-charge effect to Frumkin, see A. N. Frumkin, *Phys. Z. Sowjetunion*, 1933, **4**, 239; *Usp. Khim.*, 1935, **4**, 987. These references were given by C. A. Barlow and J. Ross Macdonald, ref. 11.
7. O. A. Esin and V. Shikov, *Zhur. Fiz. Khim.*, 1943, **17**, 236.
8. B. V. Ershler, *Zhur. Fiz. Khim.*, 1946, **20**, 679.
9. D. C. Grahame, *Z. Elektrochemie*, 1958, **62**, 264.
10. V. G. Levich, V. A. Kyrianov and V. S. Krylov, *Doklady Akad. Nauk U.S.S.R.*, 1960, **135**, 1425.
11. C. A. Barlow and J. Ross Macdonald in *Advances in Electrochemistry and Electrochemical Engineering*, 1967, **6**, 1, (ed. Delahay and Tobias).
12. S. Levine, J. Mingins and G. M. Bell, *J. Electroanal. Chem.*, 1967, **13**, 280.
13. R. Parsons, *Proc. 2nd Intern. Congr. Surface Act.*, 1958, **III**, 38, Butterworth.
14. G. Gouy, *Ann. Chim. Phys.*, 1908, (8), **8**, 294.
15. G. Gouy, *Ann. Phys.*, 1917, (9), **9**, 129, a review.
16. J. A. V. Butler, *Proc. Roy. Soc. A*, 1929, **122**, 399; also in *Electrica Phenomena at Interfaces*, 1951, chap. II, London.
17. D. C. Grahame, *J. Am. Chem. Soc.*, 1941, **63**, 1207.
18. D. C. Grahame, *J. Am. Chem. Soc.*, 1949, **71**, 2975; *Z. Elektrochem.*, 1955, **59**, 740.
19. B. B. Damaskin, *Zhur. Fiz. Khim.*, 1958, **32**, 2199.
20a. G. J. Hills and R. Payne, *Trans. Faraday Soc.*, 1965, **61**, 316.
20b. G. J. Hills and R. Payne, *Trans. Faraday Soc.*, 1965, **61**, 326.
20c. M. Stedman, *Symp. Faraday Soc.*, 1970, **4**, 64.
21. J. O'M. Bockris, E. Gileadi and K. Mueller, *J. Chem. Phys.*, 1966, **44**, 1445.
22. D. C. Grahame, E. M. Coffin, J. I. Cummings and M. A. Poth, *J. Am. Chem. Soc.*, 1952, **74**, 1207.
22a. A. N. Frumkin and V. I. Melik-Gaikazyan, *Zhur. Fiz. Khim.*, 1950, **26**, 560.
23. G. Gouy, *J. Phys.*, 1910, (4), **9**, 457.
24. See also D. C. Chapman, *Phil. Mag.*, 1913, (6), **25**, 475.

25. S. Minc, J. Jastrzebska and M. Brzostowska, *J. Electrochem. Soc.*, 1961, **108**, 1161.

25a. For a "hump" in methylformamide solutions: B. B. Damaskin and Yu. M. Povarov, *Doklady Akad. Nauk U.S.S.R.*, 1961, **140**, 394; see also ref. 2b.

26. D. C. Grahame, *J. Am. Chem. Soc.*, 1957, **79**, 2093.

27. J. Ross Macdonald and C. A. Barlow, *J. Chem. Phys.*, 1962, **36**, 3062.

28. R. J. Watts-Tobin, *Phil. Mag.*, 1961, **6**, 133; N. F. Mott and R. J. Watts-Tobin, *Electrochim. Acta*, 1962, **4**, 79.

29. C. J. F. Böttcher, *Theory of Electric Polarisation*, Elsevier, 1952.

30. R. W. Rampola, R. C. Miller and C. P. Smyth, *J. Chem. Phys.*, 1959, **30**, 566.

31. J. Ross Macdonald and C. A. Barlow, Jr., *J. Electrochem. Soc.*, 1966, **113**, 978, discussion remark, pp. 991–2.

32. N. F. Mott, R. Parsons and R. J. Watts-Tobin, *Phil. Mag.*, 1962, **7**, 483.

33. M. J. Austen and R. Parsons, *Proc. Chem. Soc.*, July 1961, 239.

34. J. O'M. Bockris, M. A. V. Devanathan and A. K. N. Reddy, *Proc. Roy. Soc.* A, 1964, **279**, 327.

35. O. Stern, *Z. Elektrochemie*, 1924, **30**, 508.

36. D. C. Grahame, *J. Am. Chem. Soc.*, 1958, **80**, 4201.

37. D. C. Grahame and R. Parsons, *J. Am. Chem. Soc.*, 1961, **83**, 1291.

37a. J. Lawrence, R. Parsons and R. Payne, *J. Electroanal. Chem.*, 1970, **16**, 193.

38. C. A. Barlow and J. Ross Macdonald, *J. Chem. Phys.*, 1965, **43**, 2575.

39. J. M. Parry and R. Parsons, *Trans. Faraday Soc.*, 1963, **59**, 241.

40. S. Levine, G. M. Bell and D. Calvert, *Can. J. Chem.*, 1962, **40**, 518.

41. M. L. Huggins, *J. Chem. Phys.*, 1941, **9**, 440; P. J. Flory, *J. Chem. Phys.*, 1941, **9**, 660; 1942, **10**, 51. See textbooks such as H. Morawetz, *Macromolecules in Solution*, Interscience, 1966.

42. R. Payne, *J. Chem. Phys.*, 1965, **42**, 3371.

43. J. Ross Macdonald and C. A. Barlow, *J. Electrochem. Soc.*, 1966, **113**, 978.

44. E. Dutkiewiecz and R. Parsons, *J. Electroanal. Chem.*, 1966, **11**, 100.

45. F. P. Buff and F. H. Stillinger, *J. Chem. Phys.*, 1963, **39**, 1911.

46. G. M. Bell, J. Mingins and S. Levine, *Trans. Faraday Soc.*, 1966, **62**, 949.

47. J. Ross Macdonald and C. A. Barlow, Jr., *Can. J. Chem.*, 1965, **43**, 2985.

48. V. S. Krylov, *Electrochimica Acta*, 1964, **9**, 1247.

49. A. N. Frumkin, *Z. Physik*, 1926, **35**, 792.

50. F. W. Schapink, M. Oudeman, K. W. Leu and J. R. Helle, *Trans. Faraday Soc.*, 1960, **56**, 415.

51. Ref. 2a., fig. 9.

52. M. Proskurnin and A. N. Frumkin, *Trans. Faraday Soc.*, 1935, **31**, 110.

53. K. Eda and B. Tamamushi, *Proc. 3rd Intern. Congr. Surface Act.*, 1960, **II**, 291 (Cologne).

54. B. B. Damaskin, N. V. Nikolaeva-Fedorovitch and R. V. Ivanova, *Zhur. Fiz. Khim.*, 1960, **34**, 894.

55. W. I. Melik-Gaikazyan, *Zhur. Fiz. Khim.*, 1952, **26**, 1184; ref. 22a.

56. R. G. Barradas and F. M. Kimmerle, *J. Electroanal. Chem.*, 1965, **9**, 483 and 1966, **11**, 128.

57. P. Somasundaran, T. W. Healy and D. W. Fuerstenau, *J. Coll. and Interf. Sci.*, 1966, **22**, 599.

58. R. S. Hansen, R. E. Minturn and D. A. Hickson, *J. Phys. Chem.*, 1956, **60**, 1185; 1957, **61**, 953.

59. R. PARSONS, *Trans. Faraday Soc.*, 1955, **51**, 1518.
60. R. PARSONS, *Proc. Roy. Soc.* A, 1961, **261**, 79.
61. E. BLOMGREN, J. O'M. BOCKRIS and C. JESCH, *J. Phys. Chem.*, 1961, **65**, 2000.
62. C. KEMBALL, *Proc. Roy. Soc.* A, 1947, **190**, 117.
63. A. N. FRUMKIN, R. I. KAGANOVICH and E. S. BIT-POPOVA, *Doklady Akad. Nauk U.S.S.R.*, 1961, **141**, 670.
64. M. A. GEROVICH and N. S. POLYANOVSKAYA, *Doklady Nauk Vysshei Shkoly Khim. i Khim. Teknol.*, 1958, 651 (this ref. given by BLOMGREN and BOCKRIS, ref. 65).
65. E. BLOMGREN and J. O'M. BOCKRIS, *J. Phys. Chem.*, 1959, **63**, 1475.
66. B. V. DAMASKIN, *Electrochimica Acta*, 1964, **9**, 231.
67. J. M. PARRY and R. PARSONS, *J. Electrochem. Soc.*, 1966, **113**, 992.
68. R. PARSONS, *J. Electroanal. Chem.*, 1964, 8, 93.
69. W. LORENZ, F. MOCKEL and N. MÜLLER, *Z. Phys. Chem.* N.F., 1960, **25**, 145.
70. R. GRAND, *C.R.Acad. Paris*, 1963, **257**, 1315.
71. Z. OSTROWSKI, H. A. BRUNE and H. FISCHER, *Electrochimica Acta*, 1964, **9**, 175.

CHAPTER 4

THE AgI/AQUEOUS SOLUTION INTERFACE

4.1. Potential determining ions. Introductory remarks

The potential ψ_0 at a reversible interface is determined by the activity of a particular ionic species, called potential determining ions, and also by the properties of the solvent molecules. In the present case, the Ag^+ ions (or the I^- ions) are potential determining. Application of eq. (2.2.45) for the Ag^+ ions gives

$$\tilde{\mu}_{Ag} = \mu_{Ag}{}^s + e\psi_0 + e\chi \quad \text{(interface)}, \qquad (4.1.1)$$

$$= \mu_{Ag}{}^0 + kT \ln c_{Ag} \text{ (bulk)}, \qquad (4.1.2)$$

where we assume ideal behaviour of the solution. For ψ_0 we get

$$\psi_0 = -\chi + \frac{1}{e}(\mu_{Ag}{}^0 - \mu_{Ag}{}^s) + kT/e \ln c_{Ag}. \qquad (4.1.3)$$

As noted in Subsection 2.2.7, a Ag^+ concentration can be found, where $\psi_0 = 0$, corresponding to $\sigma = 0$, the point of zero charge. This concentration is denoted as c_{0Ag} or as c_{pzc}. The point of zero charge (pzc) is equivalent to the ecm as has been pointed out earlier. One of the methods to find c_{0Ag} is to measure the velocity of colloidal AgI particles in a solution under the influence of an electric field as a function of c_{Ag}, and then to plot this electrophoretic velocity (v), against c_{Ag}. The point $v = 0$ corresponds to c_{0Ag}. If only the concentration is a variable and the solvent remains unaltered, then

$$\psi_0 = kT/e \ln \frac{c_{Ag}}{c_{0Ag}}. \qquad (4.1.4)$$

For the AgI/aqueous solution interface Van Laar[2] found $c_{0Ag} = 3 \times 10^{-6}$ M, corresponding to $c_{0I} = 3 \times 10^{-4}$ M, but these values may depend on the way of preparation of the colloid[3] (see below). If the composition of the solvent is altered at constant ψ_0, then the χ-potential may be affected. Indicating two solvent compositions by superscripts "A" and "B" one has

$$\Delta\chi = \chi^A - \chi^B = 1/e(\mu_{Ag}{}^{0A} - \mu_{Ag}{}^{0B}) + kT/e \ln \frac{c_{Ag}{}^A}{c_{Ag}{}^B} \qquad (4.1.5)$$

$$- 1/e(\mu_{Ag}{}^{sA} - \mu_{Ag}{}^{sB}).$$

Mackor[4] found by the electrophoresis method that the addition of only 3–4 mol% acetone shifted c_{0Ag} to 2×10^{-3} M. Attributing this shift only to a shift of the χ-potential, Mackor concluded that $\chi^{H_2O} - \chi^{acetone} = -199$ mV, this change of the χ-potential being due to the expulsion of water molecules adhering to the surface by acetone molecules. Analogous phenomena in the previous chapter were discussed in terms of the inner layer potential drop. This is a matter of terminology. Overbeek et al.[5] repeated Mackor's experiments and extended them to AgBr and

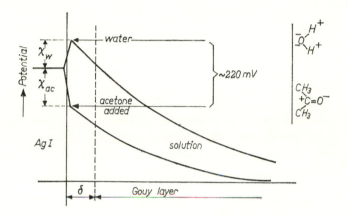

FIG. 4.1. Possible potential distributions at the interface between AgI and water and between AgI and a water (96%)–acetone(4%) mixture. (1 erg $= 10^{-2}$ J).

AgCl. The shift was here respectively -225 mV and -105 mV. From the sign of the effect it can be inferred that the adhering water molecules have a greater tendency to turn their negative side toward the crystal surface than have the acetone molecules. Estimates by Lijklema[6] made probable a preferential orientation of the negative side of the water molecule toward the AgI surface, the "wall anisotropy energy" being of the order of kT.

Indifferent ions (ions other than Ag^+ or I^- ions) have on the whole little effect upon c_{0Ag}. Therefore there is less specific ionic adsorption in the AgI case than in the Hg case.

For oxides, such as SiO_2, Al_2O_3 and TiO_2, Verwey[7] found that H^+ ions were potential determining and this will be the case for other oxides too, such as Fe_2O_3.[8] Also for carbon,[9] H^+ ions are potential determining, the pzc here depending on the sample history. Thus, if prior to contacting an (oxygen-free) solution, the carbon was given a heat treatment (300–400°C) in H_2 atmosphere, the pzc was found at pH = 0. In neutral, oxygen-free solutions, the interface was found electrically negative and this points to a surface reaction $H_2 + OH^- \rightarrow 2H_2O + 2e^-$ (e^- denotes an electron in the carbon surface region). If oxygen is admitted, the pzc shifts to higher pH values. If an oxide layer is formed by heating the carbon to 400°C in an oxygen atmosphere, an oxide dipole layer forms and the pzc shifts to negative pH values. Many other examples originating from the classical colloid chemical literature of the first half of this century can be found in Overbeek's contributions to *Colloid Science*. An interesting, more recent, example is given in Chapter 6: the Ge/aqueous solution interface. This interface is negatively charged and H^+ ions are potential determining. These properties should be ascribed to surface oxidation followed by acid formation and subsequent ionization.

Returning to the AgI/aqueous solution interface, this interface is often available in the form of a colloidal solution, a sol. One way to prepare a AgI sol.[1,10] is the careful addition of a 10^{-3}–10^{-4} M aqueous $AgNO_3$ solution to a 10^{-3}–10^{-4} M aqueous KI solution or vice versa. The pzc should be avoided because no stable sol can be prepared at a concentration c_{0Ag}. The reasons for this behaviour will be pointed out in Section 4.3. It goes without saying that highly purified chemicals, a thoroughly degreased and steamed glass apparatus and high-quality deionized water should be used. The (only) chemical reaction which takes place when this way of preparation is followed is

$$AgNO_3 + KI \rightleftharpoons AgI + KNO_3. \qquad (4.1.6)$$

The insoluble salt, the AgI, is present in the form of particles of about $0 \cdot 1\mu$ diameter although the sol is far from monodisperse. The particles have (very) roughly a spherical shape. The remaining ions, and other ions which may be present, can partly be removed by forcing them through a membrane which is impervious to colloidal particles (dialysis). The properties of the AgI sols so obtained are constant only after about a day or so. Verwey and Kruyt[11] observed during the first minutes and hours after the formation of the sol a marked decrease even up to a factor 10 in special cases, of the adsorption of I^- ions measured in samples taken from the same sol after given time intervals had passed. The surface area decreased only slightly in the case of AgI sols and therefore the particle size remained constant. (This is no longer the case for AgBr sols as pointed out by Jonker.[12]) This "ageing" process is speeded up upon increasing the temperature. The process probably consists, in the AgI case, of a recrystallization of the particles. Just after formation the (polycrystalline) particles probably contain numerous imperfections, cracks, and so on. In particular the surface will be highly heterogeneous and will contain numerous sites which are particularly favourable to the adsorption of I^- ions. The recrystallization process makes the particles, together with its surface, more homogeneous.

Ottewill and Woodbridge[3] prepared monodisperse silver halide sols by a process in which the particle growth was kept under control. This method was similar to that employed by LaMer and Dinegar[14] for the preparation of sulphur sols and uses the fact that water-soluble complexes are formed in concentrated $AgNO_3$ or KI solutions with additions of respectively KI and $AgNO_3$. Such complexes have been studied by Tezak and his school.[15] Thus, 150 cm^3 of a complex solution which was 3×10^{-5} normal with respect to AgI and 10^{-1} M with respect to KI was diluted with 500 cm^3 distilled water. After about 1 minute monodisperse particles were formed, the monodispersity being checked by optical methods (tyndallograms). The excess electrolyte was removed by dialysis or by a treatment with ion exchange resins. The approximately spherical particles had a radius of between 0·38 and 0·40 μm, as determined under the electron microscope.

For the monodisperse sols the pzc was reported to be at a much higher Ag^+ concentration than for the "classical" sols and close to the pzc the electrophoretic velocity was found to depend only weakly (much weaker than for the "classical" sols) upon c_{Ag}. These two facts may be connected with differences of the composition of the particles and differences in the crystal structure which in neither case will be very perfect. Although no direct information is available as to the solid-state properties of the sol particles, it is useful to mention here some facts concerning investigations of AgI in its "dry" solid state: AgI is an ionic conductor as first pointed out by Tubant[16] in 1921. Its conductivity at room temperature is of the order of 10^{-6} ohm^{-1} cm^{-1}. Weiss et al.[11] found that between 50°C and 140°C the conductivity λ was given by

$$\lambda = 5 \times 10^4 \, e^{-26T_0/T} \text{ ohm}^{-1} \text{ cm}^{-1} \qquad (4.1.7)$$

where $T_0 = 300°K$. Measurements below 50°C showed some scatter. The cation lattice is more disordered than the anion lattice[18] and the conduction is attributed to mobile cations and cation vacancies. It is probable that Frenkel defects dominate rather than Schottky defects. Frenkel defects[18] are obtained when an ion leaves its lattice position, thus creating a lattice vacancy, and is transferred to an interstitial position, Schottky defects are obtained when ions of both signs leave their lattice position and find new lattice positions at the surface. Assuming ideal behaviour of interstitial ions and vacancies, the Frenkel model provides the result that the product of the concentrations of interstitial ions and vacancies is a constant. It is here assumed that the defects are electrically charged. Ionization reactions can be introduced to describe the equilibrium between charged and neutral defects.

There are three phases of solid AgI: a cubic centred phase (α) at temperature above 145°C, a wurtzite (hexagonal) phase (β) showing birefringence at temperatures between 110°C and 145°C, and a zincblende (cubic) phase (β') at temperatures lower than about 110°C. At room temperature one often has a mixture of the β and the β' phases. Conductance measurements at room temperature can therefore not be interpreted in an unambiguous way. The given transition temperatures depend upon the purity of the crystal. Thus, the addition of 0·15 mol% Ag_2Se, which is incorporated in the lattice, leads to a decrease of the $\alpha \leftrightarrows \beta$ transition temperature to about 135°C. The conductance of AgI containing additions of divalent ions differs markedly from that of pure AgI. This is not (always) the case when monovalent foreign ions are added. Thus, the addition of 0·5 mol% KI leaves the conductance measured between 50°C and 140°C unaltered.[17] When a new phase is formed this may be different. Some years ago a phase denoted as KAg_4I_5 has been described,[19] having at temperatures as low as 50°C a conductivity of about 10^{-2} ohm^{-1} cm^{-1}.

With this knowledge we compare the equilibrium at the AgI/aqueous solution interface with a Donnan equilibrium. The interface is considered as an semipermeable membrane impervious to I$^-$ ions. At both sides the sum $\mu_{Ag} + \mu_I$ is constant and equal to μ_{AgI}. The "ionic solution" in the AgI phase consists of mobile interstitial cations and mobile negative cation vacancies. When both move out of the crystal this remains electrically neutral but it has a non-stoicheometric composition, excess I atoms being present. Foreign atoms and modifications of the lattice structure can affect the concentrations of interstitials and vacancies.

In this view the charge of a particle depends on the situation at either side of the "membrane". The situation at the liquid side can, of course, be easily modified through the addition of $AgNO_3$ or of KI. The situation at the solid side is rather more fixed but the way of preparation is of importance. Thus, to increase the c_{0Ag} the preparation must be such as to eliminate mobile vacancies or (what is the same thing in the Frenkel model) to promote the formation of interstitial Ag$^+$ ions in the lattice.[3,20]

These views brought Honig[20] to the following experiment: silver bromide (melting point 434°C), obtained by precipitation of a colloidal solution, was melted with small amounts of $PbBr_2$ or Ag_2S. The melt was poured on a cold tile and, after solidification, and tempering during

24 hours, was crushed in an agate mortar. AgBr was chosen instead of AgI to avoid difficulties which may arise in connection with the AgI phase transformations described above. The AgBr particles were immersed in an aqueous solution of known Ag^+ and Br^- concentration and the pzc was determined. Results are given in Table 4.1. The table clearly shows an influence of the divalent cationic and anionic impurities upon c_{0Ag}. Each S^{--} ion needs two Ag^+ ions for charge

TABLE 4.1

(ref. 20, Honig and Hengst)

$-\log x_S$	$-\log x_{Pb}$	$-\log c_{0Ag}$	Temperature of tempering (°C)
$4\cdot36 \pm 0\cdot03$		$3\cdot47$	350
$3\cdot33 \pm 0\cdot01$		$2\cdot63$	350
	$3\cdot87 \pm 0\cdot04$	$9\cdot61$	—
	$3\cdot84 \pm 0\cdot03$	$9\cdot56$	252
	$3\cdot84 \pm 0\cdot03$	$9\cdot62$	348
	$3\cdot30 \pm 0\cdot03$	$10\cdot37$	—
—	—	$5\cdot40$	(25)

$x_S = n_S/n_L$; $n_{Pb} = n_{Pb}/n_L$, n_S and n_{Pb} being the number per unit volume of S^{--} and Pb^{++} ions in the lattice, and n_L being the number per unit volume of lattice sites.

compensation. Thus, the addition of S^{--} ions tends to increase the interstitial Ag^+ ion concentration (and to decrease the vacancy concentration). This leads to an increase of c_{0Ag}. The decrease of c_{0Ag}, observed when Pb^{++} ions are added, can be explained along similar lines.

4.2. The capacity

Mackor[4] devised a method to study the capacity of the (reversible) AgI/aqueous solution interface. Knowledge of the capacity requires knowledge of σ as a function of ψ_0. Since

$$\sigma = e(\Gamma_{Ag} - \Gamma_I) \qquad (4.2.1)$$

and since ψ_0 is related to c_{Ag} through a Nernst relationship, the capacity can be found if an adsorption isotherm is measured. A particular reason why such isotherms are experimentally accessible is that the surface area can here be made very large, the adsorbent consisting of AgI particles immersed in a solution containing indifferent electrolyte of a given concentration. The adsorbate is here contained in a solution of $AgNO_3$ or KI in water. This solution must be slowly added to the adsorbent. The p_{Ag} in the medium containing the adsorbent is measured by means of an AgI/Ag electrode, its initial value brought to its pzc value $p_{Ag} = 5\cdot6$. Plots of σ against ψ_0 are given in Fig. 4.2 and the differential capacity as a function of ψ_0 is given in Fig. 4.3. Comparison

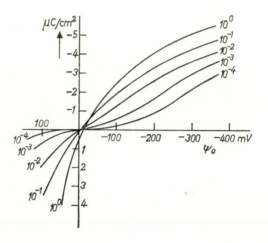

FIG. 4.2. Surface charge density σ against surface potential ψ_0 for five different concentrations (M) of KNO_3. After J. Lijklema[6].

with the Hg/aqueous solution interface is interesting. In both cases there is a steep rise of C when the surface is made more positive and also in both cases the capacity curves show a minimum which is especially pronounced at low ionic concentrations. This minimum is readily explained in terms of the Gouy theory. To this purpose we assume the same model as for the Hg/aqueous solution interface: there is a Gouy layer with capacity C_G and an inner layer with capacity C_1. Then, if no specific adsorption has taken place:

$$\frac{1}{C} = \frac{1}{C_1} + \frac{1}{C_G}. \tag{4.2.2}$$

The differences between the capacities at the two interfaces are:

(a) if the surface is made more negative, the capacity of the AgI/soln. interface decreases much faster than the capacity in the Hg case;

(b) there is no "hump" in the inner layer capacity curve in the AgI case;

(c) the pzc is little varied when indifferent electrolytes are added.

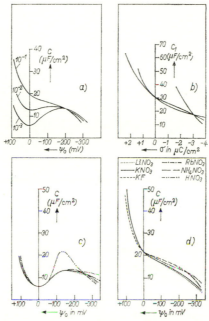

FIG. 4.3. Capacity curves after J. Lijklema thesis, Utrecht, 1957. (a) Differential capacity of the double layer for aqueous KF solutions (10^{-3}, 10^{-2} and 10^{-1} M) versus potential ψ_0. (b) Inner region differential capacity for the same solutions versus surface charge density σ. (c) Differential capacity in various 10^{-3} M aqueous monovalent nitrate solutions versus surface potential ψ_0. Key to curves appears on (d), which is the same as (c), but in 10^{-1} M solutions.

These differences may be largely due to the specific nature of the AgI surface. The cell-dimensions of the AgI lattice and of ice resemble each other quite closely. The first layer of water molecules adhering to the AgI surface may be an "ice-like" layer. The high capacity at the positive side and the absence of a hump may be explained by assuming that anions may come close to the surface or rather to excess Ag$^+$ ions present there and that the water molecules have turned their negative side to the AgI surface. A kind of hydrated salt molecule may be formed by an excess Ag^+ ion, an adhering water molecule and an anion. It is difficult to explain the small capacity value at the negative side. There seems to be a limiting value for the negative charge at the surface of about -5 $\mu C/cm^2$, which corresponds to an excess of 2×10^{13} I^- ions per cm^2. A final explanation of this limiting value may be sought in a detailed consideration of the solid state properties of the AgI crystal and its behaviour near the surface. In other words, the chemical part of the electrochemical potential of the I^- ions in and at the surface must be considered more closely.

Equations for the contributions of the cations and the anions to the total double-layer capacity C were derived by Lijklema[21] the method being analogous to that followed by Grahame for the double layer in the Hg case. Assuming the presence of an indifferent salt CA in the solution which is in contact with the AgI crystal, Lijklema found for the cation contribution C_+ and for the anion contribution C_-:

$$C_+ = -F\left(\frac{\partial \sigma}{\partial \mu_{CA}}\right)_{\psi_0} + \tfrac{1}{2}C, \qquad (4.2.3)$$

$$C_- = F\left(\frac{\partial \sigma}{\partial \mu_{CA}}\right)_{\psi_0} + \tfrac{1}{2}C. \qquad (4.2.4)$$

where F is the Faraday.

From the analysis of the experimental data it appeared that at high negative potentials there was an anomalous tendency toward an increased adsorption of the co-ions (the anions). Levine et al. (see also Subsection 4.3.6) invoked the influence of the discreteness of the charges to explain this anomaly. However, the anomaly is small and may be due to experimental error.

Cations appeared to be specifically adsorbed, although to a much lesser extent than are the anions in the Hg case. For different cations Lijklema[6] found that the tendency to be adsorbed in the inner layer (for the adsorption equations Stern's theory was used) increased with increasing radius of the cations. This was explained by assuming that the dielectric constant in the inner layer, ε_1, follows the lyotropic order. The small Li^+ ion has a relatively high field around it and this lowers ε_1 and consequently also C_1. This in turn means, that the adsorption of ions in the inner layer is lowered. The mean thickness of the inner layer was assumed the same for different ions, because this will be chiefly determined by the ice-like structure of the water molecules adhering to the surface.

The capacities discussed in this subsection were found in systems obtained in the "classical" way. It was tacitly assumed that there was no space charge region at the AgI side of the interface. On the other hand, the experimental results by Ottewill and Woodbridge suggest that, at least for their monodisperse particles, space charges in the solid phase may be important. To investigate this point we recall briefly the theory of Verwey and Niessen[22] and of Grimley.[23] These authors considered space charges present at the two sides of an interface.

Assume the interface placed at $x = 0$. The space charges extend from $x = 0$ to large negative values of x and from $x = 0$ to large positive x-values. We ignore here the χ-potential, i.e. we have put it zero for all Ag^+ ion concentrations. Furthermore, we assume the absence of Stern layers. The whole system is electrically neutral. One has at $x = 0$

$$\varepsilon_s\left(\frac{\partial \psi_s}{\partial x}\right)_{x=0} = \varepsilon_a\left(\frac{\partial \psi_a}{\partial x}\right)_{x=0} \tag{4.2.5}$$

where the subscript "s" denotes quantities related to the space charge in the region where x is negative. This space charge is placed in the solid. The subscript "a" is reserved for quantities in the region where x is positive; this is the aqueous solution side. We assume only monovalent charge carriers. Equation (4.2.5) can be rewritten as (see Section 2.3)

$$-\sqrt{(n_s\varepsilon_s)} \sinh \tfrac{1}{2}z_s = \sqrt{(n_a\varepsilon_a)} \sinh \tfrac{1}{2}z_a. \tag{4.2.6}$$

The total potential drop ψ_0 (see Fig. 4.4) across the interface is $(z_a - z_s)(kT/e)$ where the sign convention for z_s is chosen such as to be in agreement with the extension of the space charge s in the direction of negative x-values. Although, now specifying the case of the AgI/aq. soln. system, the potential drop in the solution is given by $z_a(kT/e)$ $(= \psi_{a0})$, the Nernst law requires that

$$z = z_a - z_s = \ln c_{Ag}/c_{0Ag}. \tag{4.2.7}$$

Combination of eqs. (4.2.6) and (4.2.7) gives

$$e^{z_a} = \frac{\alpha + (1/\gamma)}{\alpha + \gamma} \tag{4.2.8}$$

where

$$\alpha = \sqrt{\left(\frac{n_a\varepsilon_a}{n_s\varepsilon_s}\right)} \quad \text{and} \quad \gamma = e^{-\frac{1}{2}z}.$$

If α is large there is only a weak relationship between z_a and γ. identifying the potential ψ_{a0} with the zeta potential the result obtains that large α-values weaken the relationship between the relationship between the electrophoretic mobility and c_{Ag}. This was the argument used by Davies and Holliday[20] and by Ottewill and Woodbridge.[3] The capacity of the whole system, C, is given as

$$\frac{1}{C} = \frac{\partial \psi_0}{\partial \sigma} = \frac{\partial \psi_{s0}}{\partial \sigma} + \frac{\partial \psi_{a0}}{\partial \sigma} = \frac{1}{C_s} + \frac{1}{C_a} \tag{4.2.9}$$

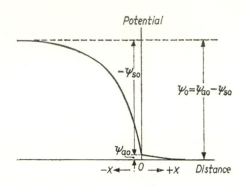

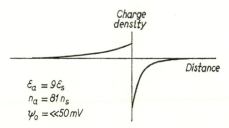

$$\varepsilon_a = 9\varepsilon_s$$
$$n_a = 81 n_s$$
$$\psi_0 = \ll 50\,mV$$

Fig. 4.4. Double diffuse double layer, potential distribution and charge distribution (E. J. W. Verwey[22]) $n_a = 81 n_s$; $\varepsilon_a = 9\varepsilon_s$. Therefore $\alpha = 27$. Equation (4.2.6), together with the Debye–Hückel approximation $\sinh x = x$, gives $-\psi_{s0} = 27\psi_{a0}$.

where $C_s = \varepsilon_s \kappa_s / 4\pi \cosh \frac{1}{2} z_s$ and $C_a = \varepsilon_a \kappa_a / 4\pi \cosh \frac{1}{2} z_a$. Here κ_s is the inverse Debye–Hückel length in the solid and κ_a is that quantity in the solution. The interpretation of the capacity measurements has generally been given with the (tacit) assumption that

$$C_s \gg C_a. \tag{4.2.10}$$

According to the data by Ottewill and Woodbridge this may not be justified for their monodisperse sols.

Introducing for the solid phase $\varepsilon_s \cong 15$, $n_s \cong 10^{18}$ cm^{-3} (there is much uncertainty concerning this value[18]) and for the solution phase $\varepsilon_a \cong 80$, $n_a = 6 \times 10^{17}$ cm^{-3} (a 10^{-3} M monovalent electrolyte) one has $\alpha \cong 1\cdot7$, $\kappa_s^{-1} = 3 \times 10^{-7}$ cm and $\kappa_a^{-1} = 10^{-6}$ cm. With these values the capacities C_s and C_a would both become of the order of 10 μF/cm^2. This is justifiable for the "monodisperse" sols, but not for the "classical" sols. Certainly for the classical sols an important modification of the model is needed. Such a modification is implicit in a later publication by Verwey,[24] where he considered two space charge regions as Verwey and Niessen did, but where in addition a layer of charges was placed at the plane $x = 0$, the whole system being made electrically neutral. We shall meet such a situation in Chapter 6 (see also Chapter 1) where the charges at the plane $x = 0$ are ascribed to surface states. The introduction of a charged layer at $x = 0$ means a partial restoration which was, and still is, at the basis of the interpretation of the capacity data: one space charge compensated by a thin charged layer. In the revised model there will be fewer states in the "monodisperse" case than in the "classical" case.

Another set of experiments is relevant here: Frens[25] found that the adjustment of the double layer at the AgI/aq. soln. interface (classical case) to variations of the ionic concentration was slow, the rate-determining step being the slow exchange of Ag^+ and (or) I^- ions between the AgI phase and the solution phase. Thus, in the cell AgI/electrolyte/calomel (0·1 N) the e.m.f. was recorded as a function of the time elapsed after the addition of a solution of 1 M KNO_3 and some KI or $AgNO_3$. The data indicated that during the first seconds the surface charge had the tendency to remain constant. Thus, even when KI was added simultaneously with KNO_3, the first effect was a shift of the e.m.f. in a positive direction. The adjustment to the equilibrium of the surface charge required almost a minute when working near the equivalence concentration ($pAg = 8$) and was quite rapid at extreme values of the pAg ($pAg = 5$ and $pAg = 11$). Relaxation effects in the same range of pAg values have been found in measurements of the capacity of the AgI/aq. soln. interface at frequencies above 100 Hz.[25] These time effects are probably too large to be accounted for by RC times which can be calculated on the basis of the original double space-charge model. Now in Chapter 6 we shall meet the concept of "slow surface states". For the interface under consideration here the existence of these states would mean the existence of trapping centres at the interface where the exchange of potential determining ions can take place at a slow rate. This concept may be invoked here for the description of the time effects found by Frens and by Oomen[25]. The concept may also be helpful for the understanding of the difference of the pzc values of the classical sols and the monodisperse sols. The (more numerous) states in the classical case may show a preference to trap positive charges. Some evidence for what is called here surface states was also given by Honig.[20] According to him there is a surface charge at the

AgBr/aq. soln. interface, which, at the pzc, has a value of about -10^{-8} C/cm² (or, about -6×10^{10} electrons per cm²). This surface charge is compensated by a positive space charge inside the crystal. Honig argued that its value does not depend on the impurity content, because it was only in this case that Frenkel's model appeared to be consistent with the analysis.

4.3.   Stability theory of the colloidal solution

4.3.1.   *Introduction*

The criteria of the stability of colloidal solutions (or colloids or sols) are, at least for hydrophobic colloids, largely governed by the interplay of repulsive forces and attractive forces between the particles. The idea of such an interplay was suggested by London,[27] by Kallmann and Willstätter,[26] by De Boer,[28] and more fully by Hamaker[29] who constructed already many curves of the total potential energy of two particles versus their distance. Detailed consideration of this interplay is at the basis of the theory of the stability of hydrophobic colloids as given by Derjaguin and Landau,[30] and by Verwey and Overbeek[31] ("DLVO"-theory). This theory is outlined in this section.

The main ideas are as follows:

The repulsive force is due to the electrical double layer at, and around, each particle. As we have seen, the charge on AgI particles may be positive or negative, depending on the Ag^+ ion concentration in the solution. In either case (we exclude the pzc) the particles are surrounded by counter ions residing partly in the Stern layer and partly in the Gouy layer. The total charge of the counter ions is equal but opposite in sign to that on the particle and it is assumed that this remains the case whatever the interparticle distance. There is no interaction when the particles are far apart but as soon as they come into each others' proximity such that there is interpenetration of the two Gouy layers there is a tendency to restore the original situation and this leads to a repulsion. In the case of AgI particles having a reversible interface, counter ions are "squeezed out" between the particles. This reduces the (absolute value) of the total charge of Stern and Gouy layers and consequently it reduces also the charge *on* the particles. This amounts to a forced desorption of potential determining ions. If desorption is impossible (ideally polarizable interface) the Gouy layers are compressed and the total entropy is reduced. In both cases the result is an inter-particle repulsion. The range of the extension of the double-layer repulsion is roughly equal to the "thickness" of the Gouy layer and a good estimate of this "thickness" is the Debye–Hückel length. Therefore salt addition leads to a decrease of the range of extension of the repulsion.

The attraction is ascribed to long-range (London–Van der Waals) forces. Although there is a rapid decay with increasing distance of the attractive force between two atoms, this is no longer the case for two particles, each consisting of, for instance, 10^6 atoms. Integration procedures lead to the result that the attractive force in that case is still appreciable at a distance which is of the order of the Debye–Hückel length. London–Van der Waals forces are independent of the salt concentration. Since the total interaction is the difference of the (absolute values) of the double-layer repulsion and the London–Van der Waals attraction, and since these are of the same magnitude, it must be expected that the ionic concentration is an important parameter in the stability theory, an increased concentration leading to a decreased stability, and eventually to coagulation.

The DLVO model uses concepts which are well founded in thermo-dynamics and electrochemistry. These concepts are combined with the theory of long-range attractive forces which is based on first principles. Nevertheless, the model was rejected by some authors. Langmuir[32] considered London–Van der Waals forces an unnecessary hypothesis. (His objection that the London forces between particles in a medium, and not in vacuum, are too small can easily be refuted. The medium has the effect of reducing the magnitude of the London-forces, but the reduction factor is of the order of n^4, where n is the refractive index of the medium. With $n = 1.4$, the reduction factor $n^4 = 4$.)

Tezak[33] criticized the model because it would provide only an apparent coherence of a number of selected and relatively few experi-mental facts. This criticism is unjustified. The DLVO model and the theory based on it do not pretend to give a clue to all phenomena in colloid chemistry (see, for example, Section 4.4) and it is also true that in the original presentations a number of over-simplifications are contained. However, the DLVO model has met two criteria which are essential for any model:

(a) a rapid progress has been gained in the understanding of numerous phenomena in colloid chemistry and related topics;

(b) much new work has been done which would not have been under-taken if the DLVO model would not have been available. This new work has led to modifications of the original model and it has provided new insights.

Tezak considered any model which inquired to some detail into the nature of stability and flocculation as premature and used only very general concepts. One of these concepts is the "methorical layer". This is the transition region between two bulk phases, e.g. the particle and the bulk of the solution. Not much specification was given to the methorical layer. The methorical layer must be involved in the coagulation mechanism. It was assumed that coagulation was taking place after direct contact between two particles, the only specification being that a discrete ion with a charge opposite to that of the particles could act as a "bridge".† Mirnik[34] has recently proposed a more detailed theory based on these concepts. However, the presentation of this theory is rather poor[35] and it will not be considered here.

One of the main objectives of any stability theory must be the explanation of the rule of Schulze[36] and Hardy.[37] This rule has been established long ago as a result of the determination of critical flocculation concentrations (cfc) of ions of different valencies. According to the rule of Schulze and Hardy the ratio of the cfc values of salts containing mono-, di-, and trivalent cations for negative sols, and for mono-, di-, and trivalent anions for positive sols is approximately $1:10^{-2}:10^{-3}$. The valency of the ions of the same sign as the particle charge is only of secondary importance. The cfc of 1–1 electrolytes is of the order of $0 \cdot 1$ M. The strong dependence of the cfc upon the valency suggests that the situation is more complicated than reflected in the Debye–Hückel theory, because this theory contains only functions which are linearly dependent upon the valency.

Tezak assumed an exponential relationship between cfc and valency; the DLVO theory leads to a z^{-6} (z is valency) relationship.

4.3.2. The double-layer interaction

The system is reduced to two particles, a and b, immersed in a medium. Each particle is provided with a double layer. The potential energy of these particles can be written as

$$V_R = F_{2d}{}^a + F_{2d}{}^b - F_\infty{}^a - F_\infty{}^b \qquad (4.3.1)$$

where $F_{2d}{}^a$ and $F_{2d}{}^b$ are the Helmholtz free energies of the particles a and b at a distance $2d$. $F_\infty{}^a$ and $F_\infty{}^b$ are the Helmholtz free energies when the particles are a large distance apart (although they remain immersed in the same medium). Most calculations have been carried out for identical particles. In that case there is often symmetry with respect to the plane at the half-distance d. For this reason the distance

---

† The "discrete-ion" concept and the "bridge" concept are very useful. Their use has been suggested also elsewhere in colloid chemistry. (Subsection 4.3.6.1 and Section 4.4.)

is taken as $2d$. For identical particles where this symmetry exists, eq. (4.3.1) reduces to

$$V_R = 2(F_{2d} - F_\infty). \qquad (4.3.2)$$

This case will often be considered. Introduction of eq. (2.2.92) for the Helmholtz free energies leads to

$$V_R = -2A[(\int_0^{\psi_0} \sigma \, d\psi_0)_d - (\int_0^{\psi_0} \sigma \, d\psi_0)_\infty]. \qquad (4.3.3)$$

For both distances, $2d$ and ∞, the concentration of the ions is the same. Therefore there is no contribution of the Gibbs free energy to V_R. For particles with unlike wall potentials an equation similar to eq. (4.3.3) can be written down. We shall see that in that case it is possible to obtain a double-layer *attraction* instead of a repulsion, depending on the values of the wall potentials and the distance.

For the flat plate approximation and for identical wall potentials Verwey and Overbeek have shown that in eq. (4.3.3) instead of ψ_0, the potential ψ_δ at the Gouy–Stern interface can be taken. This facilitates the calculations. The error so committed is usually small (see, however, Levine for a critical review[38]). Even in this case, i.e. even when considering ψ_δ as a constant with respect to d-variations, only the second integral of eq. (4.3.3) can be handled mathematically. It is easy to show that this integral becomes, if ψ_δ is used instead of ψ_0,

$$(\int_\sigma^{\psi_\delta} \sigma \, d\psi_\delta)_\infty = 8n\kappa^{-1}kT\left(\cosh \frac{e\psi_\delta}{2kT} - 1 \right). \qquad (4.3.4)$$

Here σ is the charge opposing the Gouy layer, n is the number of cations or anions per cm^3 in the bulk of the solution and only the presence of monovalent ions is assumed.

In view of the mathematical difficulties Verwey and Overbeek proposed a second method for the calculation of the free energy. This method is essentially the same method as that used by Debye[39] for the calculation of activity coefficients in dilute electrolytes. We now again consider the more general case of particles of any shape with or without a Gouy–Stern interface.

All the charge carriers participating in the double layers are assumed to be discharged. This discharged state is the initial state of the system. The charge carriers are given their charge by small steps $z_i \, e \, d\lambda_i$ where λ_i is a dimensionless parameter, $0 \leq \lambda_i \leq 1$. After each small step thermodynamic equilibrium is assumed to be re-established. As pointed out by Overbeek,[40] the charging process can be carried out in a number of different but equivalent ways. We prefer here to carry out the charging process in the following way:

1. The interfaces are charged up such that their wall potentials have the value which they must have in the state of the system in which all the ions have their full charge. An amount of charge, equal but opposite in sign, is spread in the solution. As a general rule this step does not require the charging parameter of each ion to be unity. An extreme example is provided by the AgI sol at $p_{Ag} = p_I = 8\cdot1$. Here $\psi_0 = 150$ mV. The surface charge is positive and the negative charge in the solution is solely given by the I^- ions, which all have their full charge. An indifferent salt, say KNO_3, may be present in the solution. These ions are still in the discharged state. In more general cases ($p_{Ag} \neq p_I$) the charge in the solution is distributed over more than one kind of ion.

2. At constant wall potential all the charge carriers are given their full charge. This part of the process involves also desorption or adsorption of potential determining ions.

Concerning the first step, this is closely related to the process described in Subsection 2.2.7 and thereore the Helmholtz free energy change of the system which is connected with the first step may be written at once as

$$F_1 = -\sum \{A_j[\int \sigma \, d\psi_{j0}]_{\lambda\min}\} \tag{4.3.5}$$

where we have taken the sum over the particles j in the system. The subscript "$\lambda\min$" indicates that as soon as the final value of the wall potential is reached, the first step is completed.

Concerning the second step, this resembles Debye's original treatment. Consider a volume element dV at a position where the potential is ψ' at a certain stage of the process. The work, necessary to bring a charge $z_i \, e \, d\lambda_i$, belonging to charge carrier i from a part of the solution where the potential is zero, to the considered volume element, is $z_i \, e \, d\lambda_i \psi'$. This process should be carried out for all the ions involved keeping in mind that electroneutrality should be preserved. After integration over all dV and over the λ_i's from an initial value given by the completed first step, to unity, the Helmholtz free energy contribution F_2 of the second step, which has a purely electrostatic character can be written as

$$F_2 = \int \int_{\lambda_i}^{1} \int_V z_i n_i' \, e\psi' \, d\lambda_i \, dV \tag{4.3.6}$$

where n_i' is the number per volume unit of ions i at the position of dV. Equation (4.3.6) can conveniently be written as

$$F_2 = \int_0^1 \int_V \frac{\rho'}{\lambda} \, \psi' \, d\lambda \, dV - \sum \int_0^{\lambda_i} \int_V z_i n_i' \, e\psi' \, d\lambda_i \, dV \tag{4.3.7}$$

where the first integral describes the charging of all the ions from $z_i e\lambda = 0$ to $z_i e\lambda = z_i e$ and where $\rho' = z_i n_i' e\lambda$ is the charge density in dV. The total result can be written as

$$F_1 + F_2 = F_{\lambda\min} + \int_0^1 \int_V \frac{\rho'}{\lambda} \psi' \, d\lambda \, dV \qquad (4.3.8)$$

where

$$F_{\lambda\min} = -\sum \left(A_j\left[\int \sigma \, d\psi_{j0}\right]_{\lambda\min}\right) - \sum \int_0^{\lambda_i} \int_V z_i n_i' e\psi' \, d\lambda_i \, dV. \qquad (4.3.9)$$

If we reduce the system to two particles a distance $2d$ apart, we get for V_R

$$V_R = (F_{\lambda\min})_{2d} - (F_{\lambda\min})_\infty + \left(\int_0^1 \int_V \frac{\rho'\psi'}{\lambda} \, d\lambda \, dV\right)_{2d}$$

$$- \left(\int_0^1 \int_V \frac{\rho'\psi'}{\lambda} \, d\lambda \, dV\right)_\infty. \qquad (4.3.10)$$

The difference $(F_{\lambda\min})_{2d} - (F_{\lambda\min})_\infty$ can be interpreted as the electric interaction energy between two particles with non-zero wall potentials in a chargeless medium. We consider constant wall potentials and therefore the surface charge varies with variations of d. In this respect it differs from a purely Coulombic interaction. For two parallel plates of infinite extension and the same wall potentials this interaction is evidently zero, and the force of interaction is zero at any distance. Levine[38,41] pointed out that intricate difficulties arise if flat plates are considered. In that case, namely, there is no place in the system where the potential is zero. Therefore, if the flat plate approximation is discussed it should be assumed that the surfaces are slightly curved, the radii of curvature being so large that the requirement of zero potential in the solution can be fulfilled, but also so small that only a negligible surface charge is needed to give the particle surfaces their final potential value. In that case

$$(F_{\lambda\min})_{2d} - (F_{\lambda\min})_\infty = 0 \text{ (flat plates, same wall potential).} \qquad (4.3.11)$$

Casimir[42] showed by vector analysis that

$$-\int_A \int \int \sigma \, d\psi_0 \, dA = +\int_0^1 \int_V \frac{\rho'\psi'}{\lambda} \, d\lambda \, dV \qquad (4.3.12)$$

Levine[41] and Ikeda[43] showed that the proof was only valid for the flat plate approximation and discussed what is denoted here as $F_{\lambda\min}$.

There are two cases in which $F_{\lambda\min}$ plays a dominant role:

1. Particles in a medium where the charge density is low. A non-aqueous medium with a low solubility for ionized salts, may serve the purpose. Van der Minne and Hermanie[44] and later Koelmans[45] and numerous other workers (Subsection 4.3.6.2) considered such systems of, for instance, suspensions of metal oxides in xylene, stabilized with oleates or other ionic stabilizers, or emulsions of water droplets in benzene, stabilized with oleates. In these cases the Debye–Hückel length is 5–10 microns. The integral

$$\int_0^1 \int_V \rho'\psi' \frac{d\lambda}{\lambda}\, dV$$

could be neglected and the interaction could be described by a Coulombic potential energy $(R)^{-1}\psi_0^2\varepsilon a^2$, where a is the radius of a particle or a droplet ($a \cong 1$ micron) and R is the centre–centre distance of the particles.

2. Particles with different wall potentials.[46,47] Consider again the case of two parallel plates in a chargeless medium, the wall potentials being ψ_0^a and ψ_0^b. The surface charge on plate a is

$$\sigma^a = \frac{\varepsilon}{8\pi d}(\psi_0^a - \psi_0^b) \tag{4.3.13}$$

and that on plate b is $-\sigma^a$. The flat plate approximation gives

$$F_1 = (F_{\lambda\min})_{2d} = -\frac{\varepsilon A}{16\pi\, d}(\psi_0^a - \psi_0^b)^2. \tag{4.3.14}$$

This represents an attraction and it has the same origin as the condenser plate attraction. For small d this attraction can dominate contributions arising from F_2.

Verwey and Overbeek worked out the second integral of eq. (4.3.8) for the flat plate approximation and identical wall potentials. We follow their reasoning here but include the case of different wall potentials.

For $0 < \lambda < 1$ the charge density ρ' is

$$\rho' = -\frac{\varepsilon}{4\pi}\frac{d^2\psi'}{dx^2}. \tag{4.3.15}$$

For monovalent electrolytes the charge density is also given as

$$\rho' = -2n\,e\lambda\,\sinh\frac{\lambda\,e\psi'}{kT}.\qquad(4.3.16)$$

The Poisson–Boltzmann equation is (see eq. (2.2.7))

$$\frac{d^2\psi'}{dx^2} = \frac{8\pi n\,e\lambda}{\varepsilon}\,\sinh\frac{\lambda\,e\psi'}{kT}.\qquad(4.3.17)$$

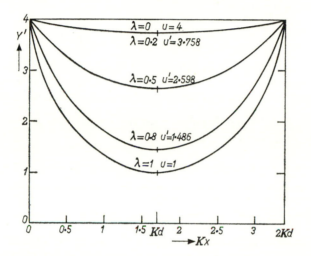

FIG. 4.5. Potential ψ' versus distance x at various values of the charging parameter λ. Half-distance $x = d$ chosen at $1\cdot702\ \kappa^{-1}$. From M. J. Sparnaay, *Rec. Trav. Chim. Pays-Bas*, 1962, **81**, 395.

Integration gives

$$\left(\frac{\partial\psi'}{\partial x}\right)^2 = 32\pi n\varepsilon^{-1}kT\left(2\cosh\frac{\lambda\,e\psi'}{kT}+C'\right)\qquad(4.3.18)$$

where C' is a constant of the integration, the value of which depends on the boundary conditions. As a general rule C' depends on the inter-particle distance, on λ and on the value of ψ_0 or of $\psi_0{}^a$ and $\psi_0{}^b$. Figure 4.5 illustrates the behaviour of ψ' with respect to x for identical wall potentials and different values of λ. Here $C' = -2\cosh u'$ where $u' = (\lambda\,e/kT)\psi_{x=d}'$.

The integration of eq. (2.2.7) would have given an equation analogous to eq. (4.3.18) but with $\lambda = 1$:

$$\left(\frac{d\psi}{dx}\right)^2 = 32\pi n\varepsilon^{-1}kT\left(2\cosh\frac{e\psi}{kT}+C\right)\qquad(4.3.19)$$

or, in dimensionless parameters,

$$\left(\frac{dy}{d\xi}\right)^2 = 2\cosh y + C; \quad \left(y = \frac{e\psi}{kT}\right): \quad (\xi = \kappa x). \qquad (4.3.20)$$

Again C is a constant of the integration. Previously only the value $C = -2$, which obtains if $d\psi/dx = 0$ at $x = \infty$, was considered. For two parallel flat plates with the same wall potential the potential gradient is zero at $x = d$, i.e. midway the plates. Denoting the potential at $x = d$ as ψ_d we get for C

$$C = -2\cosh u; \quad \left(u = \frac{e\psi_d}{kT}\right). \qquad (4.3.21)$$

For positive ψ_0 the potential gradient is negative when $0 \le x < d$ and positive when $d < x \le 2d$. For negative wall potentials these signs are reversed. For unequal wall potentials three cases[46] can be distinguished:

(a) $\psi_0{}^a$ and $\psi_0{}^b$ have like signs and there is a value $x = d_a$ where the potential gradient is zero. Then

$$C = -2\cosh u_a; \quad \left(u_a = \frac{e\psi_{da}}{kT}\right) \qquad (4.3.22)$$

where ψ_{da} is the potential at $x = d_a$.

(b) $\psi_0{}^a$ and $\psi_0{}^b$ have like signs but either there is no x-value where the potential gradient is zero, or the potential gradient is zero at $x = 0$ or $x = 2d$. We assume that $\psi_0{}^b$ is closer to zero than $\psi_0{}^a$ and write for C

$$C = \frac{\varepsilon}{32\pi nkT}\left(\frac{d\psi}{dx}\right)^2_{x=2d} - 2\cosh z_b; \quad \left(z_b = \frac{e\psi_0{}^b}{kT}\right). \qquad (4.3.23)$$

(c) $\psi_0{}^a$ and $\psi_0{}^b$ have unlike signs. There is a value, $x = d_c$, where the potential is zero. There is no x-value where the potential gradient is zero. Now we get

$$C = \frac{\varepsilon}{32\pi nkT}\left(\frac{d\psi}{dx}\right)^2_{x=d_c}. \qquad (4.3.24)$$

To obtain the final expression of the free energy the root of eq. (4.3.19) or eq. (4.3.20) is needed. Remembering what we have just said concerning the sign of $d\psi/dx$, the root of eq. (4.3.20) is

$$d\xi = \frac{-dy}{\sqrt{(2\cosh y + C)}}; \quad 0 \le \xi < \kappa d_a, \qquad (4.3.25)$$

$$d\xi = \frac{+dy}{\sqrt{(2\cosh y + C)}}; \quad \kappa d_a < \xi \le 2d. \qquad (4.3.26)$$

If there exists no distance d_a then either sign may obtain depending on the sign of $(\psi_0{}^a - \psi_0{}^b)$.

For the flat plate approximation we must replace dV by $A\,dx$ and the integrations run from $x = 2d$ to $x = 0$.

Verwey and Overbeek wrote

$$\iint \frac{\rho'\psi'}{\lambda}\, d\lambda\, dV = I_1 + I_2 \qquad (4.3.27)$$

where, using eq. (4.3.16),

$$I_1 = \iint \rho' \frac{\partial(\lambda\psi')}{\lambda\,\partial\lambda}\, d\lambda\, dV = -2n\,e \iint \sinh \frac{\lambda\,e\psi'}{kT} \frac{\partial(\lambda\psi')}{\partial\lambda}\, d\lambda\, dV \quad (4.3.28)$$

and, using eq. (4.3.15),

$$I_2 = -\iint \rho' \frac{\partial\psi'}{\partial\lambda}\, d\lambda\, dV = +\frac{\varepsilon}{4\pi} \iint \frac{\partial^2\psi'}{\partial x^2} \frac{\partial\psi'}{\partial\lambda}\, d\lambda\, dV. \quad (4.3.29)$$

Integration of eq. (4.3.28) over all λ-values gives

$$I_1 = -2nkT \int (\cosh y - 1)\, dV. \qquad (4.3.30)$$

For the integration of eq. (4.3.29) the following mathematical relation is needed:

$$\frac{\partial^2\psi'}{\partial x^2} \frac{\partial\psi'}{\partial\lambda} = \frac{\partial}{\partial x}\left[\left(\frac{\partial\psi'}{\partial x}\right)\left(\frac{\partial\psi'}{\partial\lambda}\right)\right] - \tfrac{1}{2}\frac{\partial}{\partial\lambda}\left(\frac{\partial\psi'}{\partial x}\right)^2. \qquad (4.3.31)$$

We now use the flat plate approximation. The part in square brackets, introduced into eq. (4.3.29), leads to

$$\frac{\varepsilon A}{4\pi} \int \left(\frac{\partial\psi'}{\partial x}\right)\left(\frac{\partial\psi'}{\partial\lambda}\right)_{x=2d} d\lambda - \frac{\varepsilon A}{4\pi} \int \left(\frac{\partial\psi'}{\partial x}\right)\left(\frac{\partial\psi'}{\partial\lambda}\right)_{x=0} d\lambda. \qquad (4.3.32)$$

Since at $x = 0$ and at $x = 2d$ the potentials are independent of λ, the integrals are zero. Therefore I_2 becomes

$$I_2 = -\frac{\varepsilon A}{8\pi} \int_0^{2d} \left(\frac{\partial\psi}{\partial x}\right)^2_{\lambda=1} dx + \frac{\varepsilon A}{8\pi} \int_0^{2d} \left(\frac{\partial\psi}{\partial x}\right)^2_{\lambda=0} dx. \qquad (4.3.33)$$

For $\lambda = 0$ the potential gradient is given by $(\psi_0{}^a - \psi_0{}^b)/2d$. Inspection of eq. (4.3.14) which gives F_1 leads to the result that the second integral of eq. (4.3.33) and F_1 just cancel each other in the sum $F_1 + F_2$. Therefore in the flat plate approximation the free energy is, irrespective of the fact whether the wall potentials are equal or non-equal, given by

$$F_1 + F_2 = -2nkTA \int_0^{2d} (\cosh y - 1) \, dx - \frac{\varepsilon A}{8\pi} \int_0^{2d} \left(\frac{d\psi}{dx}\right)^2 dx. \quad (4.3.34)$$

To integrate eq. (4.3.34) we insert the eqs. (4.3.25) and (4.3.26). This procedure provides expressions for $F_{2d}{}^a$ and $F_{2d}{}^b$:

$$F_{2d}{}^a = \frac{2nkTA}{\kappa} \left[\int_{z_a}^{u_a} \left(\frac{\cosh y - 1}{\sqrt{(2 \cosh y + C)}} + \tfrac{1}{2}\sqrt{(2 \cosh y + C)} \right) dy \right], \quad (4.3.35)$$

$$F_{2d}{}^b = \frac{2nkTA}{\kappa} \left[\int_{z_b}^{u_a} \left(\frac{\cosh y - 1}{\sqrt{(2 \cosh y + C)}} + \tfrac{1}{2}\sqrt{(2 \cosh y + C)} \right) dy \right]. \quad (4.3.36)$$

If the potential gradient is nowhere zero between the plates, we do not distinguish separate contributions $F_{2d}{}^a$ and $F_{2d}{}^b$ and the ingration runs from z_b to z_a ($z_a > z_b$). For $d = \infty$ we have in any instance

$$F_\infty{}^a = -\frac{8nkTA}{\kappa} (\cosh \tfrac{1}{2}z_a - 1), \quad (4.3.37)$$

$$F_\infty{}^b = -\frac{8nkTA}{\kappa} (\cosh \tfrac{1}{2}z_b - 1). \quad (4.3.38)$$

To obtain V_R, eq. (4.3.1) must be used. If a and b are identical, we use eq. (4.3.2). The integration of eqs. (4.3.35) and (4.3.36) requires knowledge of elliptic integrals. Tables and graphs of V_R for identical parallel plates were given by Verwey and Overbeek.† We have reproduced such a plot in Fig. 4.6 together with a plot which is the result of an approximation which is useful at not too small κx.

This approximation is based on the consideration that the value of $d\psi/dx$ at and near the Stern–Gouy interface (and therefore also σ) is hardly affected by the presence of an opposing plate. In eq. (4.3.34) we insert therefore

$$\left(\frac{d\psi}{dx}\right)^2 = \frac{8\pi nkT}{\varepsilon} (2 \cosh y - 2). \quad (4.3.39)$$

---

† Honig and Mul[31a] have given tables of V_R for two parallel flat plates and two spheres, both at constant surface potential and at constant surface charge. Computation of ratios of coagulation concentrations for mono-, di-, and trivalent ions were also presented. Devereux and de Bruyn[31b] have tabulated data concerning also *un*equal surface potentials.

The potential energy for identical plates becomes

$$V_R = 2nkT \int_{\infty}^{d} 4 \sinh^2 \tfrac{1}{2}y \, dx. \qquad (4.3.40)$$

We now use the approximation $(\cosh y - 1) = 2 \sinh^2 \tfrac{1}{2}y \to \tfrac{1}{2}y^2$. In passing we note that this approximation is not equivalent to a Debye–Hückel approximation. In the latter case we only have $\sinh y \to y$ and $\cosh y \to 1$.

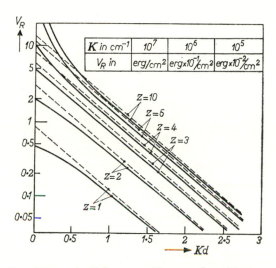

FIG. 4.6. Repulsive potential energy V_R on a logarithmic scale versus half-distance d between the plates. Dotted lines: eq. (4.3); full lines: theory (Verwey and Overbeek[31]). (1 erg $= 10^{-7}$ J.)

Equation (4.3.41) becomes

$$\left(\frac{d\psi}{dx}\right)^2 = \kappa^2\psi^2. \qquad (4.3.41)$$

The potential is an exponential function of x (see also Section 2.3). At $x = d$ the potential is built up of equal contributions of both double layers. Therefore

$$u = 8\gamma \, e^{-\kappa d} \qquad (4.3.42)$$

where $\gamma = \tanh z/4$ as derived in Section 2.3. Then

$$V_R = \frac{64nkT}{\kappa} \gamma^2 A \, e^{-2\kappa d}. \qquad (4.3.43)$$

Derjaguin preferred to consider the force $f_R = -\partial V_R/\partial 2d$ rather than the potential energy because the force equation is particularly simple. It reads

$$f_R = 2nkT(-\tfrac{1}{2}C - 1) \qquad (4.3.44)$$

where C is given by eq. (4.3.21) for identical plates and by eqs. (4.3.22), (4.3:23) or (4.3.24) for non-identical plates. Verwey and Overbeek have carried out the differentiation for the particular case of identical plates. The result can then be written as

$$f_R = 2nkT(\cosh y - 1) = kT(n_+ + n_- - 2n) \qquad (4.3.45)$$

where n_+ and n_- are the concentrations in numbers per cm^3 of cations and anions at $x = d$.

Derjaguin discussed, in a paper in 1954,[46] the case of non-identical plates on the basis of eq. (4.3.44). His general conclusion was that for different wall potentials a decreased repulsion obtains compared with the case of equal wall potentials. There is attraction when $C > -2$. Bierman also considered non-equal wall potentials and arrived at similar conclusions.[47]

Derjaguin[48] and Verwey and Overbeek derived the force equation directly by considering all the forces in a system of two charged walls placed in an electrolyte. Langmuir[32] also derived the force equation by comparing f_R with an osmotic pressure. Defay and Sanfeld[49] gave a detailed thermodynamic analysis of the problem and reviewed critically all the derivations. They arrived at the conclusion that the force equation as applied in the stability theory is correct.

Derjaguin[50] measured the force necessary for establishing a conducting contact between two metal wires, placed in an electrolyte, as a function of their potential, their surface potential and the concentration. There was qualitative agreement between the experimental data and eq. (4.3.45).

4.3.3. The interparticle attraction

In the original version of the DLVO theory the concept of the interparticle attraction was derived from the concept of the (London–Van der Waals) attraction between individual atoms. Later developments have shown that this approach conceals a number of uncertainties and that a complete electrodynamic treatment of the problem, embracing right from the beginning the particles as a whole, is more appropriate. However, the original, atomistic, approach leads in many cases to results which are probably of the correct order of magnitude. Therefore we illustrate our arguments with examples borrowed from the "atomistic" approach.

The potential energy v_A of two atoms or molecules, placed at a distance r, where r is large compared with molecular dimensions, can be given as

$$v_A = -\frac{\lambda}{r^6} \qquad (4.3.46)$$

where λ is a constant, pertaining to the atoms or molecules under consideration. Theory predicts λ-values of 10^{-77}–10^{-79} J m^6 ($=10^{-58}$–10^{-60} erg cm^6). Keesom[51] was the first to give an explicit expression for the intermolecular interaction, but he considered only the case of two rotating polar molecules. When these were placed in vacuum, λ turned out to be

$$\lambda = \frac{p_1{}^2 p_2{}^2}{3kT} \qquad (4.3.47)$$

where p_1 and p_2 are the dipole moments. For two water molecules $p_1 = p_2 = 1\cdot8 \times 10^{-18}$ e.s.u. cm or $0\cdot6 \times 10^{-29}$ Cm and λ is, at room temperature, $0\cdot9 \times 10^{-77}$ J m^6 ($= 0\cdot9 \times 10^{-58}$ erg cm^6).

Debye[52] considered the more general case of the interaction between two atoms characterized by a polarizability and a quadrupole moment and found an r^{-8} law.† Falkenhagen[53], following the lines indicated by Debye, considered the case of two molecules, each characterized by a polarizability and a dipole moment. Simplifying this case to the interaction (in vacuum) between a rotating dipole with moment p (and no polarizability) and a molecule or atom with polarizability α (and no dipole) λ becomes

$$\lambda = \alpha p^2. \qquad (4.3.48)$$

With $\alpha = 10^{-30}$ m^3 (10^{-24} cm^3) and $p = 0\cdot6 \times 10^{-29}$ Cm ($=1\cdot8 \times 10^{-18}$ e.s.u. cm) we have $\lambda = 3 \times 10^{-79}$ J m^6. The Debye–Falkenhagen or induction effect is usually smaller than the Keesom effect. (The word "induction" is used because the effect can be explained by the polarization of the first atom induced by the electric field originating from the quadrupole or dipole of the second atom or molecule.)

London[56] developed a theory for the interaction between two neutral non-polar atoms,[57–59] which he elucidated by means of the oscillator model. This model is simple and it reveals characteristics of the theory of (second-order) intermoleculer interaction. Therefore we reproduce here London's argument and extend it somewhat later on.

According to the oscillator model there is an outer electron which is the oscillating mass (m). Its position with respect to the remainder of the

---

† Contrary to a statement made by Margenau,[54] Keesom's work preceded that of Debye. See also Chu's book on molecular forces.[55,59]

atom, the nucleus and the other electrons is given by three coordinates x, y and z in cartesian space and its movement is described by p_x, p_y and p_z which are the three components in space of the momentum. London found for two oscillators in the ground state in vacuum

$$\lambda = \tfrac{3}{4}\sqrt{(h\nu_1)}\alpha_1\sqrt{(h\nu_2)}\alpha_2 \qquad (4.3.49)$$

where ν_1 and ν_2 are the frequencies and

$$\alpha_1 = \frac{e^2}{m(2\pi\nu_1)^2} \quad \text{and} \quad \alpha_2 = \frac{e^2}{m(2\pi\nu_1)^2}$$

are the polarizabilities of the oscillators. Usually λ is of the order of 10^{-78} J m^6 ($= 10^{-59}$ erg cm^6) because $h\nu_1$ and $h\nu_2$, being about equal to the ionization energies of the atoms, are roughly 10^{-18} J ($= 10^{-11}$ erg), whereas the polarizabilities are of the order of 10^{-30} m^3 (10^{-24} cm^3) London obtained his result as follows:

It is well known that the Hamiltonian function (H^0), the sum of the kinetic energy containing the momenta and the potential energy containing the coordinates, of a single three-dimensional oscillator can be written as an expression, which is purely quadratic in the momenta and the coordinates

$$H^0 = \frac{1}{2m}(p_x{}^2 + p_y{}^2 + p_z{}^2) + \tfrac{1}{2}m(2\pi\nu)^2(x^2 + y^2 + z^2). \qquad (4.3.50)$$

The classical limit of the total energy U (for U, see Section 2.2) is equal to $3kT$, i.e. the frequency ν has vanished. The quantum limit provides the well-known zero-point energy: $U = \tfrac{3}{2}h\nu$.

If there are two oscillators, 1 and 2, present which have interaction the Hamiltonian function of this set is no longer purely quadratic. We write the Hamiltonian function, H, of the system as

$$H = H_1{}^0 + H_2{}^0 + V_{1,2} \qquad (4.3.51)$$

where $V_{1,2}$ represents the interaction term.

$$V_{1,2} = \frac{e^2}{r^3}(x_1 x_2 + y_1 y_2 - 2z_1 z_2) \qquad (4.3.52)$$

where the subscripts refer to the oscillators 1 and 2 respectively. The Hamiltonian function H can also be written as

$$H = H_x + H_y + H_z \qquad (4.3.53)$$

where

$$H_x = \frac{1}{2m}\,(p_{x1}{}^2 + p_{x2}{}^2) + \tfrac{1}{2}(2\pi)^2 m(v_1{}^2 x_1{}^2 + v_2{}^2 x_2{}^2) + \frac{e^2}{r^3}\,x_1 x_2, \quad (4.3.54)$$

$$H_y = \frac{1}{2m}\,(p_{y1}{}^2 + p_{y2}{}^2) + \tfrac{1}{2}(2\pi)^2 m(v_1{}^2 y_1{}^2 + v_2{}^2 y_2{}^2) + \frac{e^2}{r^3}\,y_1 y_2, \quad (4.3.55)$$

$$H_z = \frac{1}{2m}\,(p_{z1}{}^2 + p_{z2}{}^2) + \tfrac{1}{2}(2\pi)^2 m(v_1{}^2 z_1{}^2 + v_2{}^2 z_2{}^2) - \frac{2e^2}{r^3}\,z_1 z_2. \quad (4.3.56)$$

London was able to express H_x, H_y and H_z in purely quadratic form by introducing new momenta and coordinates by means of the transformation

$$s_+ = \tfrac{1}{2}\sqrt{2}(x_1 + x_2'); \quad s_-' = \frac{v_2}{v_1}\,s_- = \tfrac{1}{2}\sqrt{2}(x_1 - x_2') \quad (4.3.57)$$

where

$$x_2' = \frac{v_2}{v_1}\,x_2$$

and similarly for the other coordinates and momenta. The whole set of transformations is in agreement with the theory of linear transformations.[60] Combination of eqs. (4.3.54) and (4.3.57) gives

$$H_x = \frac{1}{2m}\,(p_{x+}{}^2 + p_{x-}{}^2) + \tfrac{1}{2}(2\pi)^2 (v_{x+}{}^2 s_+{}^2 + v_{x-}{}^2 s_-{}^2)m \quad (4.3.58)$$

where

$$v_{x+} = v_1\sqrt{(1+b)} \quad \text{and} \quad v_{x-} = v_2\sqrt{(1-b)} \quad (4.3.59)$$

with

$$b = \frac{\sqrt{(\alpha_1 \alpha_2)}}{r^3}. \quad (4.3.60)$$

It is seen that H_x is of purely quadratic form and that it contains two new frequencies v_{x+} and v_{x-}. Similar transformations, carried out for H_y and H_z, provide frequencies

$$v_{y+} = v_1\sqrt{(1+b)}; \quad v_{y-} = v_2\sqrt{(1-b)} \quad (4.3.61)$$

$$v_{z+} = v_1\sqrt{(1-2b)}; \quad v_{z-} = v_2\sqrt{(1+2b)}. \quad (4.3.62)$$

The zero point energy is $\tfrac{1}{2}h(v_{x+} + v_{x-} + v_{y+} + v_{y-} + v_{z+} + v_{z-})$. Expansion of the frequencies in powers of b and comparison with the situation at infinite r (i.e. at $b = 0$) leads to eq. (4.3.49). It is seen that the contribution to λ arising from H_z is four times that arising from H_x or H_y.

In the classical limit the energy is equal to $6kT$ for any value of r. However, the *free* energy F is dependent on r. For one linear oscillator one has

$$F = -kT \ln \frac{kT}{h\nu} .$$ (4.3.63)

In the case considered here the frequencies of eqs. (4.3.59), (4.3.61) and (4.3.62) appear in the free energy expression. Expansion in powers of b finally leads to a value of λ given by

$$\lambda = 3\alpha_1 \alpha_2 kT.$$ (4.3.64)

Comparing the expressions for λ, given in eq. (47) (Keesom, dipole-dipole), eq. (4.3.48) (Debye–Falkenhagen, dipole-polarizable atom) and eq. (4.3.64) (classical oscillator–classical oscillator) we see that the Debye–Falkenhagen case is just intermediate between the other two. In fact, by writing down the Hamiltonian function of a dipole, which is in interaction with a three-dimensional oscillator, one would, in the classical limit, have obtained eq. (4.3.48).

Casimir and Polder[61] have shown that at $r \gg \nu/c$, where c is the velocity of light, a complete electrodynamic treatment leads to an r^{-7} law instead of an r^{-6} law. For two atoms in the ground state and in vacuum they found

$$v_A = -\frac{23hc\alpha_1\alpha_2}{8\pi^2 r^7} \quad \left(r \gg \frac{\nu}{c} ; \text{retardation case} \right).$$ (4.3.65)

For $r < \nu/c$ London's expression was found. Casimir and Polder argued, that in the pertinent Hamiltonian function contributions due to radiation oscillators should appear. Their influence can be neglected at $r < \nu/c$. Overbeek[62] was the first to discuss the possible effect of a finite velocity of the transmission of electromagnetic waves, now known as the retardation effect. Thus, the field created by an unequal charge distribution in one atom is not instantaneously "felt" by the other atom but only after a time $\tau = r/c$. On the basis of this (over-simplified) view Overbeek envisaged a faster decay of the attractive energy with increasing distance than was provided for by London's theory. Experimental data concerning the stability of suspensions led Overbeek to assume such a faster decay.

Roughly speaking the r^{-6} law is adequate for $r \lesssim 1000$ Å. The most simple way to obtain the attractive potential energy between two colloidal particles where we use the r^{-6} law is to add all the contributions of pair interactions, a pair consisting of one atom in one particle and one atom in the other. For flat plates the result is found by simple integration to be

$$V_A = -\frac{A}{48\pi\, d^2} \qquad (4.3.66)$$

where d is the half-distance between the parallel flat plates, one of which having unit area and the other having a surface area which is large compared with d^2. Furthermore, it is assumed that the plate thickness is large compared with d. Finally A is a constant. It is called the Hamaker–De Boer constant, because Hamaker[63] and De Boer[28] were among the first to carry out this type of integration. It is relevant to note here that their work was preceded by that of Newton[64] who started from an r^{-n} force law between two mass units. He not only considered the gravitational case ($n = 2$) but paid also attention to cases in which $n \geq 3$, because in those cases the force between a mass unit and a macroscopic body approaches (minus) infinity when the distance approaches zero. It is possible that Newton was acquainted with the work of Hawksbee on the rise of liquids in capillaries and that he considered the case $n \geq 3$ relevant here. If so, Newton must be considered as a precursor of modern students of intermolecular forces.

Returning to the Hamaker–De Boer constant, this is given by

$$A = \pi^2 q_1 q_2 \lambda \qquad (4.3.67)$$

where q_1 and q_2 are the numbers of atoms per unit of volume of the two particles. Usually q_1 and q_2 are of the order of 2×10^{22} cm^{-3}. Therefore A becomes usually

$$A \cong 10^{-19}\text{--}10^{-20} \text{ J } (10^{-12}\text{--}10^{-13} \text{ erg}) \qquad (4.3.68)$$

Values of this order are often used in the DLVO theory.

Starting from eq. (4.3.65) an integration procedure analogous to that carried out by Hamaker and De Boer provides for two flat plates

$$V_A = -\frac{3B}{(2d)^3} \quad \text{(retarded case)} \qquad (4.3.69)$$

or for the force (per unit area)

$$f = -\frac{B}{(2d)^4}\,. \qquad (4.3.70)$$

In eqs. (4.3.69) and (4.3.70) the constant B is (we follow the usual notation)

$$B = \frac{\pi q_1 q_2}{10} \cdot \frac{23hc\alpha_1\alpha_2}{8\pi^2}\,. \qquad (4.3.71)$$

It appears that B is usually of the order of

$$B \approx 10^{-23}\text{--}10^{-24} \text{ J m } (10^{-18}\text{--}10^{-19} \text{ erg cm}). \qquad (4.3.72)$$

Equations (4.3.66) and (4.3.69) and (4.3.70) have been obtained along the lines of the "atomistic approach". The "macroscopic approach" was introduced by Casimir[65] who calculated the zero-point pressure of electromagnetic waves between two perfectly conducting parallel flat plates. Total reflection of the electromagnetic waves was assumed here and for a penetration depth which is small compared with the interplate distance Casimir obtained for the force f per unit area between the plates

$$f = -\frac{\pi hc}{480(2d)^4}, \qquad (4.3.73)$$

i.e. for a distance $2d = 10^{-4}$ cm the force per cm^2 is $0 \cdot 013 \times 10^{-5}$ N $(= 0 \cdot 013$ dyne) or also, the force constant is equal to $1 \cdot 3 \times 10^{-23}$ N $(= 1 \cdot 3 \times 10^{-18}$ dyne) per cm^2.

Corrections for imperfect reflection and for temperature effects have been introduced respectively by Hargreaves[66] and by Sauer.[67] Casimir's expression gives essentially the attractive force between two metal plates $(\varepsilon = \infty)$. Lifshitz[68] has given the expression between two dielectrics and has shown that the force constant $\pi hc/480$ must be considered as an upper limit for high values of the dielectric constants. Hargreaves[66] has examined the Lifshitz expression and has corrected it for some numerical errors.

It is seen that both the atomistic and the macroscopic approach lead to a force–distance expression, which is for two flat plates given by eq. (4.3.70) with a force constant B which is for the two ways of approach given by eq. (4.3.72) although the Lifshitz expression usually leads to somewhat lower B-values.

Experiments[69–75] have been carried out to verify directly eq. (4.3.70) or eq. (4.3.66). These experiments of course required a sensitive mechanical equipment. Apart from rather trivial but troublesome difficulties, such as flaws in the mechanical set-up (uncertainty whether mechanical equilibrium can be attained at all) and dust particles between the macroscopic bodies, a more fundamental source of error is given by the fact that the opposing surfaces usually have different potentials. When electrons are able to pass from one surface to another in a reversible manner, i.e. when there is a uniform electrostatic potential of the electrons through the system, then the same considerations apply as those given in the preceding subsection and the plates attract each other according to eq. (4.3.14). For the case considered here this equation is rewritten as

$$f = -4\cdot43 \times 10^{-6} \frac{(\Delta V)^2}{(2d)^2} \text{ N m}^{-2} \begin{cases} \Delta V \text{ in mV} \\ d \text{ in microns} \end{cases} \quad (4.3.74)$$

where f is the force and ΔV the potential difference. For $\Delta V = 17$ mV and $2d = 1$ μm the force is equal to that given by Casimir's equation (4.4.73) at the same d-value. Even between apparently identical surfaces ΔV may amount to 1 volt. This points to a non-uniformity of the surfaces. The experimental difficulties, inherent in these phenomena, are of special importance in the study of clean surfaces as is pointed out in Section 6.3. The source of error, due to a non-zero ΔV, may be called the potential effect.

On the other hand, in the case of insulating materials, the presence of spurious electric charges fixed on the surfaces (the electrochemical potential of the electrons is not uniform in that case) provides another source of error which we call the electrostatic effect. Derjaguin and Abrikossova[69] were the first to observe a force–distance law which was in fair agreement with the Lifshitz theory. They measured, by means of an ingeneous balance system, the attraction between a quartz lens and a quartz plate at distances between 0·1 and 0·7 μm. Prior to these measurements, Overbeek and Sparnaay[70] found, by means of a simple spring system, rough agreement with eq. (4.3.66) at a distance of about 0·02 μm whereas their results at distances larger than 0·6 μm between glass plates and silvered glass plates can be explained largely by the potential effect, the mean value of the exponent in the force–distance law being slightly larger than 2 and ΔV being about 150 mV. Kitchener and Prosser[71] confirmed by means of the spring system the results of Derjaguin and Abrikossova for two borosilicate glass plates. Measurements by Sparnaay[72] of the attraction between two steel plates ($0\cdot5 \mu < 2d < 2\mu$) were in agreement with Casimir's expression eq. (4.3.73). Agreement with the Lifshitz theory was also found by the Russian workers[73] for the attraction between two Tl-halide samples (a lens and a plate) and between a quartz lens and a chromium plate and so were the results concerning the attraction between a fused quartz lens and a fused quartz plate, obtained by Black et al.[74] To date the most accurate measurements have been carried out by Van Silfhout.[75] He used a spring system and observed the attraction between a lens and a plate. Results were given for the following combinations: quartz–quartz (fused or cryst.), cryst. quartz–quartz lens coated with a chromium layer of 1200 Å thickness, cryst. quartz + 380 Å chromium–quartz lens + 1200 Å chromium, cryst. quartz + 80 Å chromium–quartz lens + 1200 Å chromium. The B-values for the quartz–quartz case appeared to be considerably smaller ($B = 0.84 \times 10^{-29}$ J m $(0.84 \times 10^{-19}$ erg cm), experiments by Van Silfhout and Rouweler) than in the presence of chromium. ($B = 4.9 \times 10^{-28}$ J m ($= 4.9 \times 10^{-19}$ erg cm.) This is in agreement with the Casimir–Lifshitz theory.

We see that there is a fairly large number of investigations presently available. This number gives confidence to the belief that the measured effects were really those predicted by the theory of intermolecular forces and are not due to a partial compensation of other effects. Complete certainty is difficult to obtain. Thus, for one thing the elimination of the electrostatic effect by means of the action of a radio-active agent may bring about the appearance of the potential effect (Sparnaay[72]).

For the application of London–Van der Waals forces in colloid-chemistry the following points should be mentioned:

1. The particles are placed in a medium with a static dielectric constant ε and a refractive index n.

To apply London–Van der Waals forces in colloid chemistry the fact must be considered that the particles are placed in a medium and not in vacuum. The medium has a relative dielectric constant, whose value as a general rule depends on the frequency of the electromagnetic field used for the measurement (dispersion). At low frequencies the relative dielectric constant is $\varepsilon(0)$, the static relative dielectric constant and at light frequencies it becomes equal to n^2 (n refractive index). The dispersion is particularly important for dipolar liquids such as water. The dispersion properties of the medium interfere with the dispersion properties of the particles. By using quantum field theory Dzyalozinski et al.[76] worked out the problem for two flat plates of different material, placed in a medium of a third kind of material. Pitaevski[77] had already obtained a result for the case of two molecules placed in a foreign medium. The same problem was also solved by McLachlan[78] (see also Linder[79]) in a more elementary way (quantum field theory was avoided). By way of illustration results are given for two molecules 1 and 2 placed in a medium. London's approximation (eq. 4.3.49) combined with eq. (4.3.46) must be replaced by

$$v_A = -\frac{3h}{2\pi^2 r^6} \int_0^\infty \frac{\alpha_1^*(i\xi)\alpha_2^*(i\xi)}{\varepsilon^2(i\xi)}\,d\xi \qquad (4.3.75)$$

where $\varepsilon(i\xi)$ is related to the relative dielectric constant measured at real frequencies ε by means of the Kramers–Kronig relation

$$\varepsilon(i\xi) - 1 = \frac{2}{\pi}\int_0^\infty \frac{\omega\varepsilon''(\omega)}{\omega^2 + \xi^2}\,d\omega, \qquad (4.3.76)$$

$\varepsilon''(\omega)$ being the imaginary part of the relative dielectric constant. The quantities $\alpha_1^*(i\xi)$ and $\alpha_2^*(i\xi)$ are "excess polarizabilities" whose values can be found by comparing the relative dielectric constant of the medium before and after the addition of a molecule or a molecule 2:

$$\alpha_1^* = \frac{1}{4\pi}\frac{\partial\varepsilon}{\partial n_1} \quad \text{and} \quad \alpha_2^* = \frac{1}{4\pi}\frac{\partial\varepsilon}{\partial n_2}. \qquad (4.3.77)$$

The square term ε^2 in eq. (4.3.75) arises because the electric field and therefore also $V_{1,2}$ in eq. (4.3.52) is reduced by a factor ε compared with vacuum. The London expression contains essentially the square of the field strength times a constant (b^2 in eq. (4.3.61) and (eq. 4.3.62) after expanding).

For the medium effect eq. (4.3.65) must be replaced by

$$v_A = -\frac{23\lambda c}{8\pi^2 r^7}\frac{\alpha_1^*(0)\alpha_2^*(0)}{\varepsilon(0)^{3/2}}. \qquad (4.3.78)$$

The term $\varepsilon^{3/2}$ in this equation arises because the electromagnetic field is reduced by a factor ε compared with vacuum and because the velocity of light is reduced by $\varepsilon^{1/2}$.

The concept of "excess polarizabilities" was anticipated by Hamaker who introduced a constant to be denoted here by A_m for the inter-molecular interaction between two particles 1 and 2 in a medium 3. This constant is

$$A_m = A_{21} + A_{33} - (A_{13} + A_{23}) \qquad (4.3.79)$$

where

$$A_{12} = \pi^2 q_1 q_2 \lambda_{12m}; \quad A_{13} = \pi^2 q_1 q_3 \lambda_{13m},$$
$$A_{23} = \pi^2 q_2 q_3 \lambda_{23m}; \quad A_{33} = \pi^2 q_3^2 \lambda_{33m}. \qquad (4.3.80)$$

The only difference between eq. (4.3.80) and eq. (4.3.67) is the use of the constants λ_{jkm} $(j, k = 1, 2, 3)$ instead of λ_{12}. When the electromagnetic field, created by the molecules in particles 1 and 2, fluctuates more rapidly than can be followed by any dipolar movement in the medium it will be a good approximation to put $\lambda_{12m} = \lambda_{12}n^{-4}$. When it fluctuates very slow one may put $\lambda_{12m} = \lambda_{12}\varepsilon^{-2}$. It is difficult to predict values of λ_{13m}, λ_{23m} and λ_{33m}, especially when material 3 is a dipolar liquid such as water.

Let us consider the constant A_{33m}. It is related to the attraction which would be obtained when a "particle" of material 3 is in inter-action with another "particle" of material 3 under the condition that the electromagnetic field is transmitted across a medium which has exactly the same dispersion properties as those of the "particles". We introduce a reduction factor f and assume that $\varepsilon > f > n^2$. For a dipolar liquid we distinguish in λ_{33} a dispersion contribution and a dipole–dipole contribution. (We leave aside contributions due to the induction effect. These contributions are usually small.) The reduction factor f_{disp} for the dispersion contribution will be closer to n^2 than the reduction factor f_{dip} for the dipole–dipole contribution. However, it would be dangerous to ignore the dipole–dipole contribution because Keesom's expression is derived for isolated dipoles whereas in a medium like water there is a strong interaction between neighbouring dipolar molecules.

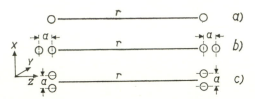

FIG. 4.7. Atomic arrangements. Each circle represents an atom.

For particles which are both of material 1 it is usually concluded, with Hamaker, that $A_m > 0$ because $\lambda_{13}^2 = \lambda_{11}\lambda_{33}$. This remains true if the influence of the medium can be accounted for by putting $\lambda_{13m}^2 = \lambda_{11m}\lambda_{33m}$. In that case we may write

$$A_m = \pi^2(q_1\sqrt{\lambda_{11}}\,n^{-2} - q_3\sqrt{\lambda_{33}}\cdot f^{-1})^2 \cong 10^{-20}\ \mathrm{J}\ (=10^{-13}\ \mathrm{erg}). \quad (4.3.81)$$

Only a qualitative meaning must be attached to this expression and to the given A_m-value.

Finally we discuss briefly some anisotropy aspects of the dispersion (London) forces. These are connected with the non-additivity of these forces.

Consider two pairs of three-dimensional harmonic oscillators as drawn in Fig. 4.7b and 4.7c.[70,80] It is assumed that there is only electric interaction. The pair distance is r, the distance between the oscillators in one pair is a. We assume that $r \gg a$, for instance $r = 30$ Å and $a = 3$ Å. The oscillators of the first pair are labelled 1 and 2 and those of the second pair are labelled 3 and 4. The Hamiltonian function H can be written as

$$H = \sum_{i=1}^{i=4} H_i^0 + (V_{13} + V_{23} + V_{14} + V_{24}) + (V_{12} + V_{34}) \quad (4.3.82)$$

where, for V_{13}, V_{23}, V_{14} and V_{24},

$$V_{ij} = \frac{e^2}{r^3}(x_i x_j + y_i y_j - 2z_i z_j); \quad ij = 13,\ 23,\ 14 \text{ and } 24. \quad (4.3.83)$$

When the oscillators are placed as in Fig. 4.7b then, for V_{12} and V_{34},

$$V_{kl} = \frac{e^2}{a^3} (x_k x_l + y_k y_l - 2z_k z_l) ; \quad \text{Fig. 4.7b} ; \quad kl = 12, 34 \quad (4.3.84)$$

whereas the situation sketched in Fig. 4.7c provides

$$V_{kl} = \frac{e^2}{a^3} (-2x_k x_l + y_k y_l - z_k z_l) ; \quad \text{Fig. 4.7c} ; \quad kl = 12, 34. \quad (4.3.85)$$

The contribution V_{kl} accounts for the non-additivity. It also accounts for an anisotropy effect since different orientations in space provide different forms for V_{kl}.

For simplicity we assume that the four oscillators are identical. The following linear transformation leads again to a Hamiltonian function which is purely quadratic:

$$s_1 = \tfrac{1}{2}(x_1 + x_2 + x_3 + x_4) ; \quad s_3 = \tfrac{1}{2}(x_1 - x_2 + x_3 - x_4)$$

$$s_2 = \tfrac{1}{2}(x_1 + x_2 - x_3 - x_4) ; \quad s_1 = \tfrac{1}{2}(x_1 - x_2 - x_3 + x_4) \quad (4.3.86)$$

and similarly for the other coordinates and momenta.

The Hamiltonian function H can be written as the sum $H_x + H_y + H_z$. Denoting the transformed momenta in the x-direction as p_{si} ($i = 1, 2, 3$) H_x becomes

$$H_x = \sum_{i=1}^{i=4} \left(\frac{p_{si}^2}{2m} + 2\pi m v_{xi}^2 s_i^2 \right) \quad (4.3.87)$$

where the frequencies v_{xi} are, for the situation given in Fig. 4.7b,

$$v_{x1} = v\sqrt{(1 + c + 2b)} \qquad v_{x3} = v\sqrt{(1 - c)}$$

$$v_{x2} = v\sqrt{(1 + c - 2b)} \qquad v_{x4} = v\sqrt{(1 - c)} \quad (4.3.88)$$

where, as before, $b = \alpha/r^3$ and where $c = \alpha/a^3$ represents the non-additivity effect. Further studies concerning non-additivity problems were carried out by Nijboer and Renne.[80a] It is reasonable to assume that $c \sim 0.1$ ($\alpha \approx 10^{-24}$ cm^3, $a \approx 10^{-23}$ cm^3). It is important to note that the polarizability of a pair of oscillators is only slightly modified compared with that of two isolate doscillators, the modification being of the order of c^2.

Only the frequencies v_{x1} and v_{x2} depend on r. At infinite r they are not given by v but by $v\sqrt{(1 + c)}$. The dispersion energy connected with H_x is

$$\Delta E_x = \tfrac{1}{2}h(v_{x1} + v_{x2}) - \tfrac{1}{2}h(v_{xc} + v_{xc}) \quad (4.3.89)$$

where $v_{xc} = v\sqrt{(1 + c)}$.

In an analogous way expressions for ΔE_y and ΔE_z can be derived. The frequencies ν_{yi} are, for the situation of Fig. 4.7b, the same as the ν_{xi}. For the ν_{zi} one has

$$\nu_{z1} = \nu\sqrt{(1-2c-4b)} \qquad \nu_{z3} = \nu\sqrt{(1+2c)},$$

$$\nu_{z2} = \nu\sqrt{(1-2c+4b)} \qquad \nu_{z4} = \nu\sqrt{(1+2c)}. \qquad (4.3.90)$$

The situation sketched in Fig. 4.7c leads to

$$\nu_{x1} = \nu\sqrt{(1-2c+2b)} \qquad \nu_{x3} = \nu\sqrt{(1+2c)},$$

$$\nu_{x3} = \nu\sqrt{(1-2c-2b)} \qquad \nu_{x4} = \nu\sqrt{(1+2c)}. \qquad (4.3.91)$$

The frequencies ν_{yi} and ν_{zi} can be derived in analogous way. Expansion of the expressions for the frequencies in powers of r^{-3}, and writing $(1+c)^{\frac{1}{2}} = 1 + \frac{1}{2}c$, leads finally to

$$\nu_A = -4\,\frac{3h\nu\alpha^2}{4r^6}\,(1+\tfrac{3}{2}c) \quad \text{(Fig. 4.7b, increased attraction),} \qquad (4.3.92)$$

$$\nu_A = -4\,\frac{3h\nu\alpha^2}{4r^6}\,(1-\tfrac{3}{4}c) \quad \text{(Fig. 4.7c, decreased attraction).} \qquad (4.3.93)$$

Depending on the spatial orientation λ becomes $\frac{3}{4}h\nu\alpha^2(1+\frac{3}{2}c)$ or $\frac{3}{4}h\nu\alpha^2(1-\frac{3}{4}c)$. The non-additivity effect becomes of the order of 10–20% It is noteworthy that the A- and the B-values derived from the Hamaker–De Boer procedure (assumed additivity) and those derived from the Lifshitz theory show up also a difference of 10–20%. This gives confidence to the belief that the simple model considered here reflects the physical reality.

4.3.4.   *The total potential energy*

According to the DLVO theory the total potential energy V_T,

$$V_T = V_R + V_A \qquad (4.3.94)$$

dictates the behaviour of a colloidal solution. It is usually sound to write

$$V_T = C\,e^{-2\kappa d} - \frac{A'}{d^g} \qquad (4.3.95)$$

where for the flat plate approximation ignoring retardation

$$C = 64n\kappa^{-1}kT\gamma^2\,; \quad A' = \frac{A_m}{48\pi}: \quad g = 2. \qquad (4.3.96)$$

In the next subsection it is shown that an equation of the type of eq. (4.3.95) remains valid for particles of other shape, although the interparticle distance must be redefined.

Curves of V_T versus d are given in Fig. 4.8. At sufficiently low electrolyte concentration a potential barrier is present which prevents two particles from coming close together, i.e. which prevents coagulation.

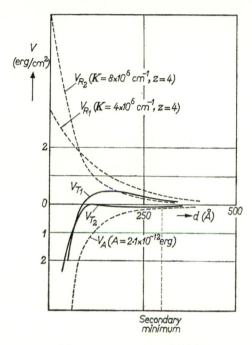

FIG. 4.8. Total potential energy V_T against half-distance d between two plates for two concentrations of the indifferent electrolyte. Wall potential $\psi_\delta = 100$ mV ($z = 4$). The value of A is chosen such that the secondary minimum in V_{T2} is at $\kappa d = 2\cdot1$ and the maximum at $\kappa d = 1$.

The barrier separates two minima. The first minimum (the primary minimum) lies at small d-values. According to the DLVO theory so far given it would lie at $d = 0$ and its depth would be infinite. However, we have omitted the Born-repulsion which becomes important when the particles, or rather the hydration shells of the adsorbed ions on them, have a distance d of the order of atomic dimensions. Furthermore, the flat plate model will be unrealistic at these small distances under discussion, irregularities of the (solid) surfaces preventing the particles from coming close together over the whole area. Finally, in a number of cases molecules will be adsorbed, which indeed lift up the primary minimum at a level where coagulation becomes impossible (stabilization). Such cases will be discussed in Section 4.4.

The second minimum is very shallow. It is found at $d \sim \kappa^{-1}$. Its shallowness prevents a straightforward experimental test of its existence in colloidal solutions, thermal movements of the particles being too intense. It is only in emulsions and suspensions, where large particles are present, that adhesion in the secondary minimum is found.[81,82] The most direct evidence for a secondary minimum is found in the equilibrium thickness of soap lamellae as is pointed out in Chapter 5.

The potential extrema are determined by the condition

$$\frac{\partial V_T}{\partial d} = 0 \qquad (4.3.97)$$

or

$$\frac{gA'\kappa^g}{2C} = (\kappa d)^{g+1} e^{-2\kappa d}. \qquad (4.3.98)$$

Leaving aside the case $d = 0$, two values of d can be found where eq. (4.3.98) is obeyed, a relatively small value, d_m, where the potential is at its maximum, and a larger d-value, at the secondary minimum. A convenient way to see this is to plot $(\kappa d)^{g+1} e^{-2\kappa d}$ against κd for a given value of g. This is done in Fig. 4.9 for $g = 2$. This function has a maximum M say, for $\kappa d = \frac{1}{2}(g+1)$ (for $g = 2$, one has $M = 0 \cdot 16g$, $\kappa d = 1 \cdot 5$). Here not only $\partial V_T/\partial d$, but also $\partial^2 V_T/\partial d^2$ is zero. At values of $gA'\kappa^g/2C$, which are larger than this maximum, no potential barrier is present and rapid coagulation takes place. For the flat plate approximation, see eq. (4.3.96), one finds a limiting value

$$\frac{A_m}{48\pi} \cdot \frac{\kappa^3}{64nkT\gamma^2} \cong 0 \cdot 159. \qquad (4.3.99)$$

For a z–z electrolyte κ becomes

$$\kappa = \sqrt{\left(\frac{8\pi n(z\,e)^2}{\varepsilon kT}\right)} \qquad (4.3.100)$$

Combination of eqs. (4.3.99) and (4.3.100) shows that the critical value of n, i.e. the value of the ionic concentration which is necessary for coagulation, is proportional to z^{-6}. Expressing the coagulation concentration c_z of a z–z electrolyte in mmol/l., one finds

$$c_z = 5 \times 10^{-22} \frac{\gamma 4}{A_m{}^2 z^6} \qquad (4.3.101)$$

or

$$c_1:c_2:c_3:\ \ldots\ =\ 1:\tfrac{1}{64}:\tfrac{1}{729}:\ \ldots \tag{4.3.102}$$

which is in good agreement with the rule of Schultze and Hardy. It may be stressed that the pertinent equation for V_R is only found when Debye–Hückel approximations are avoided. (This indeed is the crux of the theory.) Therefore it was always assumed that co-ions (ions bearing charges with the same sign as the particle charge) were absent. Consequently their valency was of little importance. Therefore for simplicity z–z electrolytes were chosen in eq. (4.3.100), whereas coagulation experiments are usually carried out with unsymmetrical electrolytes.

So far we have considered rapid coagulation. The kinetics of rapid coagulation has been worked out by Smoluchowski,[83] who assumed that every collision was "successful", in other words, he assumed that colliding particles remained together. This is no longer the case for slow coagulation. Here only a fraction of the collisions is successful and this is ascribed to the presence of a potential barrier. Its height V_m is determining for the fraction of successful collisions or rather for the stability. Reerink and Overbeek[84] showed, on the basis of a theory by Fuchs,[85] that the factor with which the rate of coagulation is decreased is practically proportional to exp V_m/kT. This factor is the stability factor and is often denoted as W. The DLVO theory predicts that V_m decreases almost linearly with increasing ln n. This will now be shown by finding an expression for $\partial V_m/\partial \ln \kappa$. This is justified because for the addition to the solution of a given electrolyte at a given temperature we have $\partial \ln n = 2\ \partial \ln \kappa$.

For the potential maximum we have

$$V_m = C\left(1 - \frac{2}{g}\,\kappa\,d_m\right)e^{-2\kappa d_m}. \tag{4.3.103}$$

We consider the differentiation of κd at $d = d_m$. By virtue of eq. (4.3.98) we have

$$\frac{\partial \kappa\,d_m}{\partial \kappa} = d_m + \kappa\,\frac{\partial d_m}{\partial \kappa} = \frac{g-f}{g+1-2\kappa\,d_m}\,d_m \tag{4.3.104}$$

where $f = \partial \ln C/\partial \ln \kappa$.

Reerink and Overbeek considered spherical particles. In their case (see also next subsection) $f = 0$ and $g = 1$. For flat plates (eq. (4.3.96)) $f = 1$ and $g = 2$. We see that $V_m = 0$ for $\kappa d_m = \tfrac{1}{2}g$. However, in that case we still have a secondary minimum. Only when $\kappa d_m = \tfrac{1}{2}(g+1)$ is there a monotonous decrease of the potential energy with increasing distance. For small κd_m the potential barrier is high and accordingly the coagulation must be slow. Equation (4.3.104) shows that in that case κd_m is fairly insensitive to variations of κ.

For $\partial V_m / \partial \ln \kappa$ we have

$$\frac{\partial V_m}{\partial \ln \kappa} = C\, e^{-2\kappa d_m}(f - 2\kappa\, d_m). \qquad (4.3.105)$$

This differential quotient, which is proportional to $\partial \ln W / \partial \ln n$ must be fairly constant for small d_m. Indeed, coagulation experiments indicate that, to a good approximation, $\ln W$ decreases linearly with increasing $\ln n$ until the region of rapid coagulation is reached. However, problems remain because variations of particle size and of ionic valency in the experiments were not sufficiently reflected in the theory. The reason is not clear[84a] but it is not probable that it must be sought in the hydrodynamic basis of the theory of slow coagulation.[84a] This leaves as an important alternative the time effects connected with the process of adsorption and of desorption of potential determining ions as studied by Frens *et al*.

4.3.5.   *Other shapes, constant charge, constant potential*

Curves for both the repulsive and the attractive potential energy versus distance are available for the interaction between identical spherical[86] and identical cylinder-shaped[87] particles. For particles of various other shapes, such as cubes,[88] ellipsoids,[89] parallelepipeds,[88] only attractive potential energies have been calculated. On the whole no new insights concerning the stability theory are gained. For completeness we give here expressions for spheres and for cylinders which involve some approximations. For V_R, consider the curved particle surfaces as composed of flat parallel portions. These are rings for the sphere–sphere interaction and rectangles for the cylinder–cylinder interaction. Opposing rings and opposing rectangles are considered to obey the same laws as those found for the flat plate approximation dealt with above. Integration yields expressions of V_R. This method is due to Derjaguin.[90] Here essentially the free energy expression

$$\iint \rho' \psi' \frac{d\lambda}{\lambda}\, dV$$

is used. Levine and Dube[91] proposed another method. They used the free energy expression $-\int \sigma\, d\psi_0$ and, owing to mathematical difficulties, applied it to small ψ_0-values. Verwey and Overbeek,[86] Hoskin and Levine[92] and Honig and Mul[31a] worked out the sphere–sphere case and compared the two methods. Sparnaay[87] did the same, although in a less elaborate way, for cylinders. The result was that Derjaguin's method often provided a good approximation. By using eq. (4.3.43) and carrying out the integration according to Derjaguin one has

$$V_R = 64nkT\gamma^2 l \frac{\sqrt{(\pi\kappa a)}}{\kappa^2} e^{-\kappa H_0} \quad \text{(parallel cylinders)}, \quad (4.3.106)$$

$$V_R = 64nkT\gamma^2 \frac{a}{\kappa^2} e^{-\kappa H_0} \qquad \text{(spheres)}. \qquad (4.3.107)$$

In eq. (4.3.106) l is the length of one cylinder (the length of the other cylinder being infinite). In eqs. (4.3.106) and (4.3.107) a is the radius of a cylinder and of a sphere and H_0 is the closest distance of the opposing surfaces. The case of crossed cylinders is similar to that of two spheres, the only difference being that the numerical constant 64 in eq. (4.3.107) must be replaced by 128. This feature, a factor 2 between the sphere–sphere case and the crossed-cylinder case, has appeared quite general. Concerning the case of parallel cylinders, this is as a general rule intermediate between the flat plate case and the sphere–sphere case.

For identical spheres two other important approximations are often used. They both use the Debye–Hückel approximation ($\nabla^2\psi = \kappa^2\psi$). First, for large κa, Derjaguin derived in 1939[90]

$$V_R = \tfrac{1}{2}\varepsilon a\psi_0{}^2 \ln (1 + e^{-\kappa H_0}). \qquad (4.3.107a)$$

This approximation leads to values which are too high, but the error is only appreciable (i.e. > 10%) for $|\psi_0| \gtrsim 50$ mV, compared with the more exact values.[86,92]

Second, for small κa, where the method of Levine and Dube is practical, one gets

$$V_R = \frac{\varepsilon a\psi_0{}^2}{R} e^{-\kappa H_0}\beta \qquad (4.3.107b)$$

where β is a numerical factor which accounts for the deviation from the spherical symmetry of the double layers which arises when the particles approach each other. The exponential accounts for the screening of the (spherical) Gouy layers. The value of β is smaller than but often close to unity. When the particle charge, and not the surface potential, is held constant, at varying $R = H_0 + 2a$ the factor β must be replaced by another numerical factor denoted[86] by γ, which again is smaller than, but close to, unity. The calculations have shown that $\gamma > \beta$, i.e. the repulsion at constant charge is larger than at constant potential. This is understandable: when the surface potential is constant the surface charge diminishes upon a decrease of the distance. When the opposite particles have *different* surface potentials (although of the same sign) there even is *attraction* at a sufficiently small distance. This, of course, is not the case when the particle charge is constant. Here the (absolute value) of the surface potential increases when the distance decreases. Sets of values of σ and H_0 correspond to certain values of ψ_0. These values of ψ_0, inserted in an equation for V_R, give the repulsion at the considered values of H_0. This consideration holds good for particles of any shape.

Attraction laws can be found by direct integration following the Hamaker–De Boer procedure. For cylinders this integration is more difficult to carry out than for spheres but expressions are available[87] on the basis of an r^{-7} force law between the atoms. For small separations ($H_0 \ll 2a$) one obtains:

$$V_A = -\frac{A}{24}\frac{1}{a}\left(\frac{a}{H_0}\right)^{\frac{3}{2}} \quad \text{(parallel cylinders)}, \qquad (4.3.108)$$

$$V_A = -\frac{A}{12}\frac{a}{H_0} \qquad \text{(spheres)}. \qquad (4.3.109)$$

Again the attraction between two crossed cylinders follows, at short separations, the same law as that for spheres, but is twice as large. At large separations one has an H_0^{-4} potential energy law for crossed cylinders, provided $1 \gg H_0$, an H_0^{-5} law for parallel or crossed cylinders when $1 \ll H_0$ (or rather, for an atom and a cylinder of infinite length) and an H_0^{-6} law for spheres ($H_0 \gg a$).

4.3.6.   *Further aspects. Limits of the theory*

It is impossible to pay due attention to all the electrochemical aspects related to colloids. Only a few points can be mentioned here.

4.3.6.1. DISCRETE-ION EFFECT

Just as in Chapter 3 macro- and micropotentials should be distinguished. The macropotentials, ψ_0 and ψ_δ, are potentials pertinent to macroscopic measurements. The micropotential, which we shall consider in the region $0 < x < \delta$, is the potential which is "felt" by a unit charge which is transported along the x-axis at a given location (y, z). If at the plane $x = \beta$ (the inner Helmholtz plane or IHP) this location is just that of a site where specific adsorption of an ion A can take place, we consider the micropotential ψ_A which is determining[93] for the electrostatic part of the adsorption energy of the ion A. The micropotential ψ_A must be distinguished from the macropotential ψ_β at the plane $x = \beta$.

The discrete-ion effect was used by Levine *et al.*[95] to explain such phenomena[94] as the maximum of the electrophoretic velocity and that of the colloidal stability which are often found when ψ_0 is increased. Such a maximum points to a maximum of ψ_δ when ψ_0 is increased. In the following we assume ψ_0 positive. Intuitively one might expect ψ_δ to rise monotonously with rising ψ_0. However, in this intuitive view the role of a specific adsorption of counterions in the inner region ($0 < x < \delta$) is overlooked. This adsorption strongly increases when the particle charge is increased thus causing the potential drop $\psi_0 - \psi_\delta$ to increase

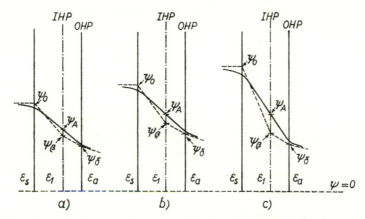

FIG. 4.9. Schematic representation of macropotential (dotted line) and micropotential (continuous line) in inner region between two dielectrics. ε_s, ε_1, ε_a relative dielectric constants. ($\varepsilon_1 < \varepsilon_s$ and $< \varepsilon_a$) IHP inner Helmholtz plane; OHP outer Helmholtz plane. Three different values of ψ_0 showing a maximum of ψ_β and of ψ_δ.

strongly. This behaviour is due to the peculiar behaviour of ψ_A. The same line of thought can be followed here as in Chapter 3 for the explanation of the Esin and Markov effect, the difference φ_β between ψ_A and ψ_β increasing when σ is increased. A qualitative illustration is given in Fig. 4.9 which assumes the same model for the inner region as used for the Hg/soln. interface. According to elementary electrostatics

$$\psi_0 - \psi_\beta = \frac{\beta}{K\delta}\,\sigma, \tag{4.3.110}$$

$$\psi_\beta - \psi_\delta = \frac{\gamma}{K\delta}\,(\sigma + \sigma_A), \tag{4.3.111}$$

where K is the integral capacity of the inner region and where σ_A is the charge per unit area of the ions A adsorbed at the plane $x = \beta$. In Chapter 3 we have represented results of calculations concerning ψ_A which we rewrite as

$$\psi_A - \psi_\delta = \frac{\gamma}{\delta}\,(\psi_0 - \psi_\delta) + (1-g)\,\frac{\beta\gamma}{\delta^2}\,\sigma_A \tag{4.3.112}$$

where g (Levine's notation) is a factor which depends on various parameters (Fig. 4.10). Combination of eqs. (4.3.110), (4.3.111) and (4.3.112) gives

$$\varphi_A = \psi_A - \psi_\beta = -\frac{\beta\gamma}{K\,\delta^2}\,g\sigma_A.$$
$$(4.3.113)$$

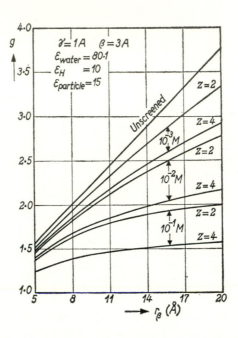

FIG. 4.10. The factor g of eq. (4.3.43) against r_β. (From S. Levine, G. M. Bell and D. Calvert, *Canad. J. Chem.*, 1965, **40**, 518, and ref. 95.) The parameter r_β is an average value of the radius of the area of an adsorbed ion, but it would be an over-simplification to identify it with $(z_A e/\pi\sigma_A)^{\frac{1}{2}}$.

In Chapter 3, g was essentially taken unity. In Fig. 4.10, in which the parameters apply to the AgI/aq. soln. interface, g may be considerably greater than unity. Consequently the discrete-ion effect is more pronounced here than it is in the Hg/soln. case. Without presenting details of the calculation this is feasible. The effectiveness of imaging in the electrolyte depends on the concentration. The situation can best be described by means of a (screening) Debye–Hückel ionic atmosphere (a hemisphere in this case) which forms at the solution side of each individual adsorbed ion A. This explains why a dilute solution (less effective screening) provides higher g-values than a concentrated solution. For high σ_A-values (low r_β†) the proportionality law between φ_A and σ_A is not obeyed, as pointed out by the Russian workers and by Grahame (Chap. 3). This is manifested by the non-zero slope of the curves. For higher r_β-values the curves flatten off, leading to constant g-values. In this respect we recall that the first interpretation of the Esin and Markov effect led to an "over-explanation", no imaging being taken into account. Imaging in the AgI surface ($\varepsilon = 15$) is less complete than in mercury. This contributes to the importance of the discrete-ion effect in colloidal solutions, compared to that at the Hg/soln. interface. The "self-atmosphere" model of Levine *et al.* provides an insight into the problems of the discrete-ion effect which is not qualitatively different from that offered by models chosen by other workers.

Assuming g constant, φ_A is proportional to σ_A. The potential drop $\psi_0 - \psi_\beta$ increases more rapidly than envisaged on the basis of $\varphi_A = 0$. Therefore a curve of ψ_β against ψ_0 can show up a maximum. According to Levine *et al.* a maximum of ψ_δ then occurs at the same value of ψ_0. As stated by these authors, other peculiarities than given here concerning the stability of colloidal systems also find their explanation in the discrete-ion effect although some care is needed.[96]

4.3.6.2. NON-AQUEOUS SYSTEMS

It has been shown, by the work of Fuoss and Kraus,[97] that in liquids with a low dielectric constant ($\varepsilon < 10$) ions can exist. The ionic concentration in apolar media such as benzene or xylene can be made of the order of 10^{-10} M. Under these circumstances the Debye–Hückel length is 5 μm or thereabouts.

Van der Minne and Hermanie[44] were the first to carry out reliable electrophoresis measurements of suspended particles (carbon black, containing 10–11% oxygen) in benzene with small additions of a Ca-soap or of tetra iso amyl ammonium picrate ("tiap"). It was essential to keep the electric field strength E in their cell below 100 V/cm in order to avoid electrostatic effects which are proportional to E^2. The additions stabilized the suspensions. With stabilizer the zeta potential, calculated from the Hückel equation,

$$\xi = \frac{6\pi\eta}{\varepsilon} v \qquad (4.3.114)$$

(v electrophoretic velocity, η viscosity, ε^{rel} dielectric constant) was

---

† r_β is smaller than $(z_A e/\pi\sigma_A)^{1/2}$ certainly at low adsorption.[95] The reason is that, especially at low adsorption, other ions A than those adsorbed can penetrate temporarily into the inner region around an ion A that is (in the model) held fixed. (Chap. 3, pp. 118-19).

positive and of the order of 25 mV. Only with "tiap" was a negative zeta potential obtained.

Without stabilizer the zeta potentials were small and irreproducible. Positive zeta potentials ranging from 10 mV to 70 mV were reported by Koelmans[45] for metaloxide (Fe_2O_3, Al_2O_3) and $BaSO_4$ particles suspended in xylene and stabilized in a way analogous to that of Van der Minne and Hermanie. Again, small and irreproducible zeta potentials were found when a stabilizing agent was absent (or "tiap" was present). Thus, the addition of a stabilizer resulted in a (usually positive) particle charge Q of roughly 10–40 proton charges. The DLVO theory was successfully applied here. Ignoring the role of the ions in the medium (see Subsection 4.3.3) the repulsive potential energy, V_R, assumes a simple Coulombic form

$$V_R = \frac{Q^2}{\varepsilon R} = \frac{\psi_0^2 \varepsilon a^2}{R} \qquad (4.3.115)$$

where R is the distance between the centres of the (supposedly) spherical particles and a their radius ($a \approx 1$ μm). The wall potential ψ_0 will not be very much different from the zeta potential. Equation (4.3.115), in combination with the Hamaker equation for the attraction between spheres, leads to energy barriers of 20–30 kT. Evidently for small particles the barrier is much lower. In that case another stabilizing factor may be found in the presence of large adsorbed molecules preventing the particles from coming close together (steric hindrance, see next section). Another consequence of small particles is their small particle charge, which may be formed by only few proton charges. Charge fluctuations may then be important, i.e. one proton charge "too much" or "too 'little" may effect the behaviour of the particles considerably.

Koelmans explained the positive particle charge by an adsorption of the small metal ions of the stabilizer, these ions being "squeezed-out" by the low-ε medium. In later work, by Parfitt et al.,[98] this explanation was modified and it was suggested that the adsorbability of the metal ions was primarily determined by small amounts of water adsorbed on the particle surface. Thus, dried suspensions of rutile (TiO_2) in xylene, stabilized by "Aerosol OT" (Na di-2-ethylmethyl sulfosuccinate) contained only negatively charged particles, whereas the addition of water caused the particles to become positive. Further evidence for the specific role of water was given by Micale et al.[99] and by Romo. Romo[100] found that Al_2O_3 particles in (C_3–C_8) alcohols were negative when dry and positive when covered with at least a monolayer of water, no stabilizer being present. The physical origin of the particle charge here probably differs from that in the previous cases. Ionization of surface hydroxyl groups may here be determining for the particle charge.[101] The ionization may take place according to M—OH $\rightarrow$ M$^+$ + OH$^-$ or to M—OH $\rightarrow$ MO$^-$ + H$^+$ depending on the physicochemical (acidic or basic) nature of material of the particle and of the medium.

Micale *et al.*[99] obtained, for rutile in heptanol, a zeta potential of 90 mV, the water adsorption amounting to 3% of a monolayer whereas the zeta potential was only 5 mV at 30% coverage, suggesting a decrease of the ionization of surface hydroxyl groups when water is added. The role of traces of water was stressed by Young and Chessick[102] who observed an increased aggregation of quartz particles in oil upon the addition of water, this addition leading to an adsorption which was only a small fraction of a monolayer. Qualitatively the same observation was already made by Kruyt and Van Selms[103] but these authors considered a monolayer coverage of water essential for the aggregation: the interfacial oil/water area is diminished by aggregation and this is thermodynamically favourable. The work of Young and Chessick suggests that "bridges" between the particles are formed (see Section 3.4) and that water molecules are essential for their formation.

When the average interparticle distance is of the order of 10 μm, then V_R becomes nowhere less than about kT (take eq. 4.3.116), assume that $Q = 20e$, $R = 10^{-3}$ cm and $\varepsilon = 2$). That means: for sufficiently high particle concentrations the potential barrier begins to diminish and it can become so low that the suspension or emulsion is no longer stable. This appeared to be the case in water in oil emulsions stabilized by metal oleates, as studied by Albers.[104] The concentration of the water droplets was here of the order of 10^9 per cm^3. The possibility of a decreased stability is inherent in the general treatment of colloidal solutions by Bell and Levine.[105] The zeta potentials, measured by Albers, ranged from 10 mV to 70 mV. A remarkable, and unexplained, feature of his experiments was a relatively easy redispersion (which could be brought about by moderate stirring), leading to a Hamaker constant of only 10^{-14} erg.[104a]

4.3.6.3. LIMITED COAGULATION

Gillespie[106] found a limited coagulation of latex emulsions in water, the latex droplets all having the same diameter of 0·58 μm (some salt and a commercial emulsifier were also present). Coagulation could be observed during the first hour after preparing the emulsion. Then the concentration of single droplets remained stable at about half the initial concentration ($\sim 10^7$ per cm^3). Gillespie assumed a coagulation–decoagulation equilibrium. Such an equilibrium fits the DLVO theory when it is assumed that the first minimum is shallow and that the potential barrier is low. The adsorption of large molecules of the emulsifier might lead to a shallow first minimum. Gillespie mentioned old experiments by Westgren[107] who studied stability properties of gold sols. In these experiments too, a large number of single particles would remain. Large adsorbed complexes, involving Au and Cl (gold sols are prepared by reducing $HAuCl_4$), may provide here for a shallow first minimum.

4.3.6.4. ANISOMETRIC PARTICLES

If the colloidal particles are highly anisometric (rods such as V_2O_5 particles or tobacco mosaic virus particles (T.M.V.); sheets such as bentonite particles or tungstic acid particles) they may hinder each other's rotational movements. Onsager[108] showed that in such a case entropy may be gained, when only part of the anisometric particles is present in a dilute phase where they can move freely, whereas the remaining particles are present in a concentrated phase, where they are orderly arranged. Experimental evidence for such a phase separation in available.[1] The concentrated phase is present in the form of spindles, containing large numbers of parallel particles. These spindles are called tactoids. No Van der Waals forces need be invoked for the explanation of this phase separation, although they may be helpful to give a better quantitative agreement concerning the particle concentrations than that obtained from Onsager's theory. The anisotropic aspects of Van der Waals forces may be invoked here.

The tungstic acid suspensions studied by Hachisu and Furusawa[82] (see Subsection 4.3.4) do not exhibit this phase separation. These workers have studied the secondary minimum.

4.3.6.5. MAGNETIZED PARTICLES

Colloidal particles may be magnetized (ironoxides). Then there is, in addition to the forces considered so far, a magnetic force, which, being attractive, decreases the stability of the colloid. Experimental evidence for such a decrease was found by Jonker et al.[109] As to the magnitude of the magnetic interaction, consider two particles at a distance $2d$ and each having a magnetic moment M of 10^{-15} μJ/gauss. The interaction energy E_M, which is written as

$$E_M = -\frac{M^2}{(2d)^3},\qquad (4.3.116)$$

then becomes, with $2d = 10^{-5}$ cm, of the order of 10^{-14} μJ, which is about 2·5 kT. This estimate seems reasonable (one Bohr magneton is about 0.9×10^{-21} μJ/gauss; one particle has a diameter of a few times 100 Å). Magnetic interaction can thus become equally important as double-layer interaction or Van der Waals interaction.[110]

4.4. The adsorption of organic molecules

In recent years much attention has (again[111]) been paid to the effect of adsorbed organic molecules upon the behaviour of colloidal particles. Here the DVLO theory can only partly account for the observed phenomena. The reason is that this theory is developed for the effect of only small-size ions. To explain the role of large (i.e. mol. wt. 10^2–10^7) organic molecules and ions new concepts should be invoked. Thus, their adsorption may promote the stability and this is ascribed to steric hindrance. The principle of steric hindrance is applied to the behaviour of adsorbed molecules both in aqueous and in non-aqueous systems. On the other hand, the addition of large-size polyelectrolytes (mol. weight $n \times 10^6$) to colloidal systems may lead to a decrease of the stability. LaMer and co-workers[112] explained this decrease by a "bridging"-concept: a single polyelectrolyte molecule is adsorbed simultaneously on the surfaces of *two* particles, thus forming a "bridge". LaMer distinguished between flocculation and coagulation, flocculation being the consequence of the addition of polyelectrolytes whereas coagulation resulted from the addition of small-size ions. A "flocculate" forms a loose network and this is important for filtration purposes; a coagulate has a larger final density. However, the distinction has met with objections[113] because the terms have been used indiscriminately for so many years.

4.4.1. *Adsorption of polar molecules*

We mentioned already Mackor's result concerning the shift of the zero point of charge after the addition of acetone to a AgI sol. This shift could be interpreted as a shift of the χ-potential, the maximum value of $\Delta\chi$ being about 185 mV. Qualitatively the same result was obtained by Bijsterbosch,[114] when instead of acetone, alcohols were added. It was found in these cases, that the adsorption caused the χ-potential to change 130–175 mV in the same direction. The fact that for a number of varying adsorbed substances qualitatively the same $\Delta\chi$ is found indicates that it is the removal of a layer of oriented water molecules rather than the introduction of the (polar) organic molecules adjacent to the AgI surface which determines the value of $\Delta\chi$.

Just as in the case of the Hg/aq. soln. system the adsorbed organic molecules will have turned their polar heads toward the aqueous phase. Bijsterbosch pointed out that on the whole the adsorption properties of these organic molecules at the AgI/aq. soln. interface have much in common with the adsorption properties of the same molecules at the Hg/aq. soln. interface. He determined in a way similar to that followed by Lijklema (and Van Laar) the surface charge σ as a function of the e.m.f. of a cell containing a Ag/AgI electrode and a calomel electrode both in contact with an aqueous 0·1 M KNO$_3$ solution containing a

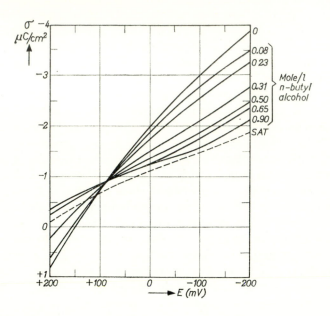

FIG. 4.11. Surface charge density σ as a function of the cell-e.m.f., E, for various concentrations of n-butyl alcohol ($cKNO_3 = 10^{-1}$ N — 10^{-1} M). (From Bijsterbosch and Lyklema.[114])

known amount of AgI precipitate and of a known alcohol concentration. The cell-e.m.f. was varied through the addition of aqueous 0·01 M AgNO$_3$ or KI solutions. The surface potential ψ_0 is equal to the measured e.m.f. minus the e.m.f. at the zero point of charge. Figure 4.11 gives results obtained by Bijsterbosch for n-butyl alcohol. In order to analyse the data, essentially Fig. 3.18 can be applied. However, whereas in the Hg/aq. soln. system complete desorption was obtained for high surface potentials, the potential range in the AgI/aq. soln. system is too limited and we are restricted to the middle regions of Fig. 3.18, the experimental values of σ lying in between $+1·8$ μC/cm^2 and $-1·2$ μC/cm^2.

The common point of intersection of all the curves in Fig. 4.11 is the point of maximum adsorption according to the relation (see Subsection 2.2.7)

$$\left(\frac{\partial \sigma}{\partial \mu_{\text{alc}}}\right)_{\psi_0} = \left(\frac{\partial \Gamma_{\text{alc}}}{\partial \psi_0}\right)_{\mu_{\text{alc}}} \tag{4.4.1}$$

where we assume that $\mu_{\text{alc}} = \mu_{\text{alc}}{}^0 + kT \ln c_{\text{alc}}$. For each curve (except the curve $c_{\text{alc}} = 0$ of course) the amounts of desorption could be given. The absolute value of the maximum adsorption, which is different for each curve, remained unknown, because complete desorption could not at all be accomplished. Therefore an adsorption isotherm could not be obtained. However, there appeared to be a close analogy between the adsorption characteristics of alcohols at the Hg/aq. soln. interface and at the AgI/aq. soln. interface. Thus, at the same (small and constant) values of σ the adsorption properties were alike provided the alcohol-concentrations in the AgI/aq. soln. case were taken about 7 times larger than in the Hg/aq. soln. case. In particular the n-butyl alcohol adsorption curves in the two cases could well be fitted. In both cases a Langmuir law was obeyed. Since for the Hg/aq. soln. system an adsorption free energy has been found[115] of about -16 kJ mol^{-1} ($= -3.9$ Kcal. mol^{-1}) it may be inferred that the adsorption free energy in the AgI/aq. soln. system is about -13 kJ mol^{-1} ($= -3.1$ kcal mol^{-1}). Furthermore it could be estimated, that at $c_{\text{alc}} = 1$ mole/l. (saturation) the maximum degree of adsorption was about 0.8. This was in agreement with estimates made on the basis of the change of the χ-potential. Finally the investigations made by Bijsterbosch show that the choice of the electrical parameter which must be held constant (the surface charge or the surface potential) will be fairly irrelevant, in agreement with a statement made by Parsons (Subsection 2.2.7) for the adsorption of organic molecules at the Hg/aq. soln. interface.

4.4.2. *The adsorption of surface active cations on negatively charged particles and of surface active anions on positively charged particles*

The role of these ions can only incompletely be described by the standard DLVO theory. Ottewill and co-workers[116,118] assumed that large ions are almost entirely adsorbed in the Stern layer. It was furthermore assumed that at constant pI the charge σ was also a constant irrespective of the concentration of adsorbed organic ions. The effect of this adsorption is therefore merely a shift of the charge from the Gouy layer to the Stern layer, the organic ions being adsorbed in what might still be called the inner Helmholtz plane. In the experiments described by Ottewill and co-workers organic salts such as octyl-, dodecyl- and cetylpyridinium bromides were added to colloidal solutions of negative AgI particles ($pI = 4$). These systems were studied by means of turbidity measurements and of electrophoretic measurements. Concerning the optical measurements an expression was used, derived by Oster,[117]

$$(dD/dt)_{t\to 0} = 0\cdot 434 A k_0 (V_0 n_0)^2 \frac{1}{W} \qquad (4.4.2)$$

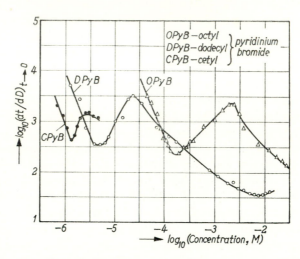

FIG. 4.12. Stability of AgI sol ($pI = 4$) in the presence of organic salts. Log $(dt/dD)_{t\to 0}$ measures the stability ratio (eq. (4.4.2)), D is the optical density, t the time. Horizontal: salt concentration. (Ottewill and Rastogi, ref. 118, p. 86.)

where D is the optical density ($= 0\cdot 434\tau$, where τ is the turbidity) t is the time, A is an optical constant which can be calculated when the refractive indices of particles and solvent are known and when the wavelength of the light used (*in vacuo*) is known, n_0 is the number per unit of volume of the sol particles at time $t = 0$, V_0 is the volume of a particle and W is the stability ratio. The product $V_0 n_0$ is the sol concentration, c_s. Finally, k_0 is a constant, $k_0 = 8kT/3\eta$ (η is the viscosity of the solvent). This constant is used in Smoluchowski's theory of rapid flocculation. At 20°C $k_0 = 1\cdot 08 \times 10^{-11}$ cm³/sec. For rapid flocculation ($W = 1$) and given c_s, the measurements permit the calculation of $A \times k_0$. Since A is known, the constant k_0 can be found. On the whole k_0 was somewhat lower than the theoretical value. Therefore, in the case of rapid flocculation of this system W may be still somewhat larger than unity.

Upon increasing the concentration of the organic salt, there was an initial decrease of the stability and of the electrophoretic mobility. Upon the addition of more organic cations the stability passed a minimum and increased at higher concentrations (Fig. 4.12). Measurements of the electrophoretic mobility (Fig. 4.13) indicated a change of sign. Depending on the nature of the organic ion, the stability minimum occurred at a concentration of 10^{-6}–10^{-4} M. The point of minimum stability can be expected to coincide with zero electrophoretic mobility. However, it appeared that in the stability minimum the charge attached

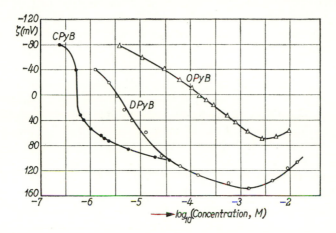

Fig. 4.13. Zeta potentials AgI sols (pI = 4). Same systems as in Fig. 4.12. (Ottewill and Rastogi, ref. 118, p. 880.)

to the particles is slightly positive. (Therefore W may be slightly larger than unity in the stability minimum.) Prolonged addition of the organic cations makes the particles more positive and more stable but the stability remains lower than the stability prior to the addition of any organic salt, a maximum occurring at roughly 10^{-2} M. At higher concentrations the stability steadily decreases. Finally the critical micelle concentration (cmc) is reached, i.e. cluster formation of the organic ions amongst themselves is taking place (see Chapter 5). The concentration of individual organic ions is then no longer dependent upon the amount of organic salt added and accordingly the stability is independent of further addition. Ottewill et al. estimated that the free energy of adsorption of these and similar cations is of the order of $41 \cdot 9$ kJ mol^{-1} ($= 10$ kcal mol^{-1}).

The adsorption of the organic cations will preferably take place on excess negative charges in the top layer of the AgI lattice, where they will be present in lattice imperfections. As the adsorption increases, the surface is becoming more oleophilic. The addition of dodecyl pyridinium bromide produces a stability minimum when about $3 \cdot 7 \times 10^{13}$ ions per cm^2 are adsorbed. The excess I$^-$ ions is, at pI $= 4$, about 2×10^{13} per cm^2. Assuming that the adsorbed organic ions lie flat on the surface, Ottewill et al. estimated that in the stability minimum about 30% of surface is covered whereas at the point of neutralization this figure is less than 20%. A 30% coverage may have rendered the surface sufficiently oleophilic (hydrophobic) to explain the lack of stability by means of the "hydrophobic bonding" concept. If hydrophobic bonding occurs between hydrocarbon chains of different particles, a force of attraction exists which should be added to the Van der Waals inter-particle attraction. However, more experiments are needed to put this hypothesis on a firm basis. This is also pointed out by Ottewill et al.

The addition of organic ions at concentrations beyond that where the stability minimum is found leads to multilayer formation. Individual organic ions added to the solution are now adsorbed with their hydro-carbon tails placed on the already covered surface, the origin of this adsorption being (partly at least) due to hydrophobic bonding between the hydrocarbon tails lying flat on the surface and those of the newly added ions. The surface becomes now more hydrophilic, the ionized polar groups tending to the aqueous phase. The colloidal dispersion loses its hydrophobic character. The decrease of the stability upon the addition of more organic ions can therefore not be explained by the DVLO theory. The reason of this decrease may be found in a premicellization phenomenon.[119–20]

Experiments analogous to those just described have been carried out by Somasundaran et al.[119] They used α-Al$_2$O$_3$ particles dispersed in an aqueous solution of 2×10^{-3} M NaCl at pH $= 6 \cdot 9$ and pH $= 7 \cdot 2$. Here the surface active agent was sodium dodecyl sulphonate. Since the α-Al$_2$O$_3$ particles are negative below pH $= 9 \cdot 1$ we have here the case of adsorbed positive ions on a negative surface. The results were only partly in keeping with those obtained by Ottewill and co-workers (see Fig. 4.14). Thus, for the dodecyl sulphonate adsorption three stages were found. At low concentrations the ions are adsorbed as individual entities. As soon as their surface concentration reaches a certain value ($\sim 5 \times 10^{13}$ ions per cm^2) the adsorption changes abruptly its character, the isotherm becoming much steeper. This behaviour is ascribed to a "hemi-micelle" formation and the concentration where this sets in is the hemi-micelle concentration (hmc). The hmc decreases quite marked-ly with decreasing pH. This shows that the adsorption is of an electro-static character because at lower pH values the surface charge is rendered more negative. At 25°C the hmc decreases from $8 \cdot 5 \times 10^{-5}$ M to $3 \cdot 2 \times 10^{-5}$ mole/l. if the pH is decreased from $7 \cdot 2$ to $6 \cdot 9$. The temperature effect is less pronounced. In all cases the "hemi-micelle" *surface* concentration is about 5×10^{13} ions per cm². Stage 3 sets in, in all cases, at a surface concentration of $1\cdot2 \times 10^{14}$ ions per cm². As shown by electrophoretic mobility data, the particles are here already positive. This indicates that the co-operation between the organic ions is here

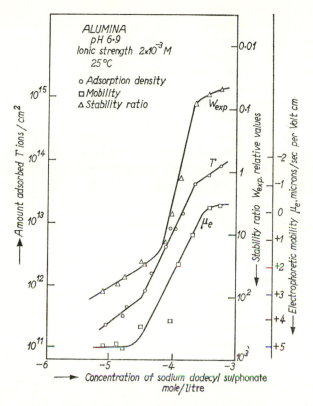

FIG. 4.14. Adsorption (Γ), electrophoretic velocity (μ_e) and stability ratio W_{exp} for α-alumina, against the equilibrium concentration of sodium dodecyl sulphonate. (P. Somasundaran *et al.*[119])

determining for the adsorption mechanism. In stage 3 the adsorption isotherm shows saturation behaviour.

The stability decreases in all three stages upon increasing the dodecyl sulphonate concentration. An increase such as reported by Ottewill *et al.* was not observed. This indicates that the attractive forces between the particles with the dodecyl sulphonate ions on them are now more important. This is not so surprising. In the AgI case the point where the charge of the adsorped (dodecyl pyridinium) ions just balances the particle charge is at an adsorption of about 2×10^{13} per cm^2 whereas this point in the Al$_2$O$_3$ case was reached at a dodecyl sulphonate adsorption of about 6.7×10^{13} per cm^2. It may be expected that at higher pH values (where the hmc is shifted to a higher value or may be altogether absent) a stability rise of the α—Al$_2$O$_3$ suspensions may be observed. Alternatively, AgI sols at lower pI values (more negative particles) may show up a hmc and (or) disappearance of the region of increased stability. The "pre-micellization" discussed by Ottewill and Rastogi may be related to hemi-micelle formation both phenomena occurring at concentrations lower than cmc. The difference is that hemi-micelles are assumed to be restricted to a single particle whereas pre-micellization embraces more particles. However, this distinction is probably too sharp because in both cases the stability is lower than in the absence of organic material.

Finally we note that the adsorption of surface activity ions as discussed here probably leaves the first water layer attached to the solid surface largely intact because changes of the χ-potential[118] do not seem to be very important.

4.4.3.   *The adsorption of (non-ionic) polyoxyethylene glycol monoalkyl ethers*

These ethers have the general formula $C_nH_{2n+1}(CH_2CH_2O)_x$ or briefly C_nE_x. These compounds are presently available to a high degree of purity. Mathai and Ottewill[121] used ethers with $n = 8, 10, 12$ and 16 and $x = 6$. Some experiments were done with the $C_{16}E_9$ ether. The experiments consisted of the measurement of adsorption isotherms on negative (pI = 4) colloidal AgI particles. Some experiments were carried out at pAg = 3 (positive particles). Also the sol stability and the electrophoretic mobility were measured.

Glazman[122] carried out similar measurements but on the whole his n- and x-values were higher. Examples were given by him for $C_{18}E_x$ ethers with $x = 12, 18, 27, 48, 98$, and $C_{16}E_x$ ethers with $x = 6, 9, 48, 98$. He measured zeta potentials of negative AgI sols whereas also other sols (As$_2$S$_3$, positive AgI) were tested.

In all these experiments the addition of C_nE_x ethers led to a decrease of the electrophoretic mobility whereas the sol stability increased. Evidence was reported for the assumption that the surface charge and the χ-potential remained unchanged. Glazman[122] found, in contrast to Ottewill and Mathai, that upon addition of the ether there was an initial decrease of the stability especially for positive sols. This may be due to bridge-formation or, as pointed out by Ottewill,[123] to the presence of impurities, notably fatty acids in the added ether.

Apart from this experimental disagreement the explanation of the observations was sought in an increased hydrophilicity when non-ionic detergent was added. The slipping plane will be displaced in the direction of the solution and this brings the zeta potential closer to zero. Just before the detergent concentration reaches the cmc the adsorption rises sharply and, consistent with this observation, the mobility shows a sharp drop. At higher detergent concentrations both the adsorption isotherm and the mobility vs. concentration curve show a plateau. The amount

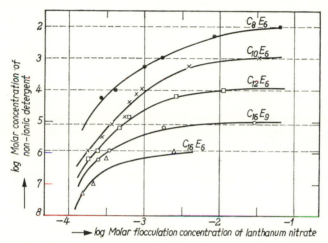

FIG. 4.15. Stability of AgI sols ($pI = 4$) in the presence of non-ionic polyoxyethylene glycol monoalkyl ethers C_nE_x, a non-ionic detergent. Concentrations C_nE_x on vertical axis; flocculation concentrations of $La(NO_3)_3$ on horizontal axis. Dotted horizontal lines indicate the cmc of the non-ionic detergents. (K. G. Mathai and R. H. Ottewill, ref. 121, p. 759.)

adsorbed is considerable. Mathai and Ottewill found an upper limit of 5×10^{14} molecules adsorbed per cm². This indicates multilayer adsorption, perhaps micelle formation on the surface. The value of the maximum adsorption increases with increasing n. This is understandable if it is assumed that it is the hydrocarbon chain which is primarily adsorbed. An increase of the value of x led to a lower adsorption plateau. This is probably due to steric hindrance of the oxyethyl chains extending in the solution. Figure 4.15 gives flocculation data of AgI sols at $pI = 4$ with $La(NO_3)_3$ solutions in the presence of non-ionic detergents, taken from the work of Mathai and Ottewill. There is here the problem of the competition for adsorption sites between the trivalent lanthanum ion and the detergent molecules, but for a broad view upon the effect of the ethers upon the stability this problem can be ignored. We see that the transition, hydrophobic–hydrophilic, is clearly demonstrated, the sol becoming quite stable near the cmc of each detergent.

4.4.4. *"Bridge" formation*

LaMer and Smellie[112] found in 1951–2 that the addition of potato starch to stable suspensions or sols which may consist of negative or positive particles produced flocculation. The addition of corn starch instead of potato starch had no effect. The reason for this difference is the presence in potato starch of (negatively charged) phosphate groups, such groups being absent in corn starch. Corn starch becomes effective when phosphorylated. The phosphate groups are here responsible for the "bridge"-formation. Bridge-formation is the basic concept of the flocculation mechanism proposed by LaMer *et al.*[112] Upon increasing the concentration of the flocculant (i.e. the potato starch) the flocculation action increases, reaches a maximum and then decreases. At this maximum the number of bridges will be maximal and this consideration led LaMer *et al.* to a $\theta(1-\theta)$ term in the expression for the number of bridges in a given system, θ being the fraction of a surface occupied by (in this case) phosphate groups and $(1-\theta)$ being the fraction which is free.

Bridge formation is not only due to phosphate groups. Thus, LaMer found already that corn starch, autoclaved at 115°C in the presence of caustic alkali, had turned into a good flocculant.

Nowadays a number of synthetic water-soluble polymeric flocculants is known. They are often of the polyacrylamide type. Their molecular weight is usually more than a million. Linear chain polymers are on the whole better flocculants than branched-chain polymers of the same molecular weight.[124] Little is known of the nature of the specific interaction between the active (anionic, cationic or neutral but polar) groups of the macromolecule and the surface. It is not at all necessary that the interaction is purely electrostatic, because the negative phosphate groups are able to link negatively charged particles as was inferred from experiments by LaMer *et al.* However, the particle surface may contain localized positive charges where the adsorption can take place. Further possibilities are hydrogen bonding, dipole–dipole interaction and specific chemisorption.

LaMer and his school used polymer flocculants for filtration purposes. A theory was worked out for the mechanism of the filtration of flocculated systems. This theory was challenged by Slater and Kitchener,[124] but the basic concepts, the formation of "bridges" and the $\theta(1-\theta)$ term, were confirmed.

4.4.5. *The stability*

Steric hindrance as a means for stabilization was introduced by Mackor,[125] and Mackor and Van der Waals[126] for the explanation of experimental data by Van der Waarden.[127] Van der Waarden found that carbon black particles, dispersed in liquid paraffins, were protected against flocculation if aromatic molecules with a long aliphatic chain were added, the protection being more effective with increasing chain length. Evidently in this case the aromatic part of the molecule is attached to the particle surface whereas the side chain is tending away from the surface and is relatively free to move with the point of attachment as its pivot. This model was used by Mackor in his first approach to the stability problem. When two surfaces approach each other at a distance smaller than twice the chain length, steric hindrance occurs and the entropy of the system increases. This must be interpreted as a repulsion. In a second approach Mackor and Van der Waals modified this model. The solvent molecules were considered as spheres and the solute molecules were simplified to dumb-bells consisting of two joint spheres of the same radius as that of the solvent molecules. It was assumed that one end of the dumb-bell tended to adsorption. The stability problem was completely solved for this model. At interparticle distances smaller than 4 times the sphere diameter the repulsive potential energy could very well compete with the Van der Waals attraction for values of a parameter X between $0 \cdot 1$ and 1 where $X = a_B/a_A{}^2 \exp\left(-\Delta\mu_0/kT\right)$. Here a_B and a_A are the activities of the solute and the solvent respectively and $\Delta\mu_0$ is the free energy of adsorption.

Two aspects can be distinguished in the mechanism of the repulsion. First we have the steric hindrance aspect, laid down by Mackor in his first approach. Second, if the interparticle distance is decreased the less favourable position of the adsorbed large molecules promotes desorption. Both aspects are interconnected as was well recognized by Mackor and Van der Waals. Recently, Verwey[128] (and Haydon[128]) stressed the resemblance of the second aspect with the origin of the repulsion in the DVLO theory. In that theory the surface potential is kept constant upon variation of the interparticle distance. A decreasing distance leads to a decreasing surface charge on the opposite faces, i.e. ions present on the surfaces are desorbed. In both cases (adsorbed ions and adsorbed neutral organic molecules) the desorption leads to an entropy gain but in both cases there is a loss of chemical adsorption (free) energy and, as explained by Verwey, in both cases the repulsion is determined by the work that must be performed against the adsorption forces.

The first approach (entropy loss purely through steric hindrance) corresponds to the constant charge case.

The free energy of adsorption of an organic molecule (ion) such as the dodecyl pyridinium ion (Ottewill) is of the order of 10–20 kT units per ion. Comparing this with a monovalent ion situated at an interface where the potential is 100 mV, the adsorption free energy is only 4 kT per ion. This suggests that the desorption aspect in the protection mechanism by organic molecules will not be preponderant. Indeed, the theoretical considerations so far given deal with the steric hindrance aspect.

Thus, Fischer[129] applied the Flory–Huggins equation[130-3] to polymers adsorbed on two particles a distance of 0·3 μm and less, apart. The Flory–Huggins equation for the free energy of mixing of a polymer (component 2) and a solvent (component 1) is

$$\Delta F = kT(N_1 \ln \varphi_1 + N_2 \ln \varphi_2 + \chi \varphi_1 \varphi_2 (N_1 + m N_2)). \qquad (4.4.3)$$

Here N_1 and N_2 are the numbers of solvent and polymer molecules, m is the ratio of the molar volumes, $\bar{v}_2/\bar{v}_1$ (for polymers $m \gg 1$), φ_1 and φ_2 are volume fractions, $\varphi_1 = N_1/(N_1 + m N_2) = 1 - \varphi_2$ and χ is a parameter accounting for the intermolecular interaction in the system. For non-polar solvents containing non-polar solutes, χ will be determined by London dispersion forces. A reasoning, analogous to that given for the sign of the Hamaker constant for particles in a medium, then leads to the conclusion that χ must be positive. According to Tompa[132] this is usually, though not invariably, the case. Equation (4.4.3) expresses a "zeroth approximation" to the problem of mixing (Tompa,[132] Morawetz,[133] Guggenheim[134]). For non-polar components this will be nearer to the real situation than for aqueous solutions, where a re-arrangement of the water molecules in the neighbourhood of a solute molecule will often have taken place, but we assume here the validity of eq. (4.4.3) also for this case. From eq. (4.4.3) we derive

$$\Delta \mu_1 = kT \ln \varphi_1 + kT \left(1 - \frac{1}{m}\right) \varphi_2 + \chi \varphi_2^2 \qquad (4.4.4)$$

which, for $\varphi_2 \ll 1$, becomes

$$\Delta \mu_1 = -kT \frac{\varphi_2}{m} - kT(\tfrac{1}{2} - \chi) \varphi_2^2. \qquad (4.4.5)$$

Fischer proceeded as follows: eq. (4.4.5) is the expression of the osmotic pressure of a non ideal solution. That part of the osmotic pressure, which depends on the non-ideality of the solution, is

$$\pi_E = \{(\Delta \mu_1)_{id} - \Delta \mu_1\}/\bar{v}_1^{-1} = kT(\tfrac{1}{2} - \chi) \varphi_2^2/\bar{v}_1 \qquad (4.4.6)$$

and the repulsive potential energy is

$$F_R = 2 \int_{\delta} \pi_E \, \delta V \qquad (4.4.7)$$

where δV is the volume of interpenetration of two adsorbed layers on the surfaces of two particles in close proximity. This procedure is analogous to that of Langmuir for the case of two interpenetrating Gouy layers.

Fischer gives a thickness of the adsorbed layer of 1500 Å. This means that there is a repulsion at an interparticle distance smaller than 3000 Å. To give a numerical example, replace the integral in eq. (4.4.7) by a product of π_E and δV and assume that the particles are so close together, that the overlap volume δV is $10^{-10} \times 2 \times 10^{-5}$ cm^3 (i.e. the interpenetration is 2×10^{-5} cm over an area of 10^{-10} cm^2). Assume $\bar{v}_1 \cong 10^{-22}$ cm^3, $\chi = 0$ and $\varphi_2 \cong 10^{-2}$. Then $F_R \cong 10^3\ kT$. Values of this order were given by Fischer for particles of 0·1–1 μm diameter. We note that positive χ-values tend to diminish F_R. This is understandable. In solutions values of χ greater than, roughly, $\frac{1}{2}$, lead to demixing. If, in eq. (4.4.5), we take $\chi > \frac{1}{2}$, we have an attraction instead of a repulsion. This attraction, the solvent is then "squeezed out" between the particles, can also be considered as demixing. In this way it is possible to connect the Flory–Huggins parameter χ with the change in stability of hydrophobic colloids after organic compounds are added to the system.

Clayfield and Lumb[135] approached the same problem in a different way. They assumed a thermal mixing ($\chi = 0$), and considered the configuration entropy of a polymer molecule, which at one side was attached to a surface and was free otherwise. This configuration entropy was compared with that which was found when the restriction of a limited extension in a direction normal to the plane of the surface was imposed upon the system. In the latter case the configurational entropy was lower. The limitation of the extension being due to the presence of a second particle nearby this decrease must be interpreted as a repulsion. Clayfield and Lumb found approximately an exponential relationship between F_R and the interparticle distance. This is not surprising because it is a well-known fact[136-7] that the density of the polymer segments of a random-coiled polymer molecule decays exponentially with increasing distance away from the centre of the molecule. Clayfield and Lumb defined a root mean square thickness of an adsorbed polymer molecule and found for an interparticle distance, which was about equal to this thickness, that F_R per adsorbed molecule was about 0·3 kT this value varying as different models were taken.

A systematic investigation concerning the problems raised in this section was undertaken by Fleer and Lijklema.[138] They studied the flocculation and coagulation properties of AgI sols in the presence of small amounts of polyvinylalcohol (PVA) with a molecular weight of about 50,000 (and occasionally 30,000 or higher, up to 160,000). The adsorption of PVA on the AgI sol particles was saturated at less than

10 ppm PVA in the (aqueous) solutions. Moreover, it was completely irreversible. The adsorption was given in mgPVA per mmol of the AgI sol, the saturation amounting to nearly 2 mgPVA/mol sol, or, since the surface area of the particles was 1·8 m² in a 1 mmol sol, to about 1·5 mg PVA per m². This gives an average thickness of about 16 Å of the adsorbed PVA layer. However, loops, protruding from the surface, are probably present and can act as "bridges" between covered and uncovered sol particles. This is inferred from Fig. 4.16 which shows the flocculation of x cm³ of a PVA-covered sol when $(3-x)$ cm³ $(x \leqslant 3)$

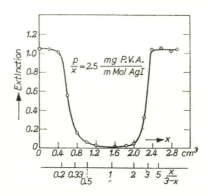

FIG. 4.16. Interaction between covered and uncovered sol particles. p mg PVA covers mol AgI sol until saturation $(p/x \approx 2\cdot5)$. Then $(1-x)$ mol AgI sol was added and the extinction measured. (Fleer and Lijklema.[138]

of an uncovered sol is added. Maximum flocculation (minimum extinction) is obtained at $x/(3-x) \approx 1$. This is in agreement with the $\theta/(1-\theta)$ rule of LaMer and Smellie although the interpretation is now slightly different: "bridges" exist in the case of Fleer and Lijklema only between covered and uncovered particles and not between particles which are each half-covered. Coagulation (the loss of stability after salt addition) should be distinguished from flocculation as pointed out by LaMer *et al.* Coagulation values increase sharply only when the PVA adsorption reaches saturation. In that case the loops are sufficiently numerous to provide for an entropic repulsion.

REFERENCES

1. J. TH. G. OVERBEEK in Kruyt's *Colloid Science*, I, Elsevier, 1952.
2. J. A. W. VAN LAAR, thesis Utrecht, 1952, ref. 1, p. 161.
3. R. H. OTTEWILL and R. F. WOODBRIDGE, *J. Coll. Sci.*, 1964, **19**, 606.
4. E. L. MACKOR, *Rec. Trav. Chim. Pays-Bas,* 1951, **70**, 747 (acetone capacity, p. 763).

5. J. TH. G. OVERBEEK, A. WATILLON and J. M. SERRATOSA, *Rec. Trav. Chim. Pays-Bas.*, 1957, **76**, 549.
6. J. LIJKLEMA, thesis Utrecht, 1957, *Kolloid-Z.*, 1961, **175**, 129.
7. E. J. W. VERWEY, *Chem. Revs.*, 1935, **16**, 363; E. J. W. VERWEY and J. H. DE BOER, *Rec. Trav. Chim. Pays-Bas,* 1936, **55**, 681.
8. S. A. TROELSTRA, thesis Utrecht, 1941; ref. 1, p. 175.
9. A. N. FRUMKIN, *Phys. Z. d. Sovjet-Union*, 1933, **4**, 239; also ref. 7; see ref. 1, p. 168 for more references.
10. R. H. OTTEWILL and R. W. HOME, *Koll-Z.*, 1956, **149**, 122, for photographs.
11. E. J. W. VERWEY and H. R. KRUYT, *Z. physik. Chemie* A, 1933, **167**, 137.
12. G. H. JONKER, thesis Utrecht, 1943, ref. 1, p. 67.
13. R. H. OTTEWILL and R. F. WOODBRIDGE, *J. Coll. Sci.*, 1961, **16**, 581.
14. V. K. LaMER and R. H. DINEGAR, *J. Am. Chem. Soc.*, 1950, **72**, 4847.
15. V. B. VOUK, J. KRATOHVIL and B. TEZAK, *Arkiv. Kem.*, **25**, 219, 1953; J. KRATOHVIL, B. TEZAK and V. B. VOUK, *ibid.*, 1954, **26**, 191.
16. C. TUBANDT, *Z. Anorg. u. Allgem. Chemie*, 1921, **115**, 105.
17. K. WEISS, W. JOST and H. J. OEL, *Z. Physik. Chemie*, N.F., 1958, **15**, 429.
18. See F. A. KRÖGER, *Chemistry of Imperfect Crystals*, North Holland Publ. Co., 1964, p. 433.
19. J. N. BRADLEY and P. D. GREENE, *Trans. Faraday Soc.*, 1966, **62**, 2069.
20. E. P. HONIG and J. H. T. HENGST, *J. Coll. Intert. Sci.*, 1969, **31**, 545; E. P. HONIG, *Trans. Faraday Soc.*, 1969, **65**, 2248. See also K. N. DAVIES and A. K. HALLIDAY, *Trans. Faraday Soc.*, 1952, **48**, 1061, 1066.
21. J. LIJKLEMA, *Trans. Faraday Soc.*, 1963, **59**, 418.
22. E. J. W. VERWEY and K. F. NIESSEN, *Phil. Mag.*, 1939, (7), **28**, 435.
23. T. B. GRIMLEY and N. F. MOTT, *Discussions Faraday Soc.*, 1947, no. 1, 3; T. B. GRIMLEY, *Proc. Roy. Soc. London* A, 1950, **201**, 40.
24. E. J. W. VERWEY, *Proc. Koninkl. Nederland Akad. Wetenschap.*, 1950, **53**, 376.
25. G. FRENS, D. J. C. ENGEL and J. TH. G. OVERBEEK, *Trans. Faraday Soc.*, 1967, **63**, 418; J. J. C. OOMEN, thesis Utrecht, 1966.
26. H. KALLMANN and M. WILSTÄTTER, *Naturwissenschaften*, 1932, **20**, 952.
27. F. LONDON; KALLMANN and WILLSTÄTTER attribute the idea of attractive forces in colloid chemistry to London (ref. 26).
28. J. H. DE BOER, *Trans. Faraday Soc.*, 1936, **32**, 21.
29. H. C. HAMAKER, *Rec. Trav. Chim. Pays-Bas*, 1936, **55**, 1015; 1957, 56, 3, 727.
30. B. V. DERJAGUIN and L. LANDAU, *Acta Physica Chim. U.R.S.S.*, 1941, **14**, 633; *J. Expl. Theor. Phys. (Russ.)*, 1941, **11**, 802.
31. E. J. W. VERWEY and J. TH. G. OVERBEEK, *Theory of the Stability of Lyophobic Colloids*, Elsevier, 1948; E. J. W. VERWEY, *Chem. Weekblad,* 1942, **39**, 563; see also ref. 1.
31a. E. P. HONIG and P. M. MUL, *J. Coll. Interf. Sci.*, 1971, **36**, 258
31b. O. F. DEVEREUX and P. L. DE BRUYN, *Interactions of Plane-Parallel Double Layers*, M.I.T. Press, Cambridge, Mass., 1963.
32. J. LANGMUIR, *J. Chem. Phys.*, 1938, **6**, 837.
33. B. TEZAK, see, for instance, *Arkiv. Kem.*, 1950, **22**, 26; *Discussions Faraday Soc.*, 1966, **42**, 175 for literature from 1932 onwards.
34. M. MIRNIK, *Nature*, 1961, **190**, 689; *Croat. Chem. Acta,* 1962, **34**, 97. M. MIRNIK, F. FLASJMAN and B. TEZAK, *Kolloid-Z. u. Z. Polymere*, 1962, **185**, 138; N. MIRNIK, *XX Congress of Pure and Applied Chemistry*, Moscow, U.S.S.R., July 1965; M. MIRNIK, discussion remarks in *Discussions Faraday Soc.*, 1966, **42**, 14, 209.

35. S. LEVINE and E. MATIJEVIĆ, *J. Coll. and Interface Sci.*, 1967, **23**, 188; see also discussion remark by Overbeek, *Discussions Faraday Soc.*, **42**, 1966, 212 and Derjaguin's remarks, same page.

36. H. SCHULZE, *J. prakt. Chem.*, 1882, (2), **25**, 431; 1883, **27**, 320; 1885, **32**, 390.

37. W. B. HARDY, *Proc. Roy. Soc.*, 1889, **66**, 110; *J. Phys. Chem.*, 1900, **4**, 255.

38. S. LEVINE, *J. Coll. Sci.*, 1951, **6**, 1.

39. P. DEBYE, *Phys. Z.*, 1924, **25**, 97.

40. OVERBEEK, ref. 1, p. 141.

41. S. LEVINE, *Phil. Mag.*, 1950, **41**, 53.

42. H. B. G. CASIMIR, in ref. 31, p. 63.

43. Y. IKEDA, *J. Phys. Soc. Japan*, 1953, **8**, 49.

44. J. L. VAN DER MINNE and P. H. J. HERMANIE, *J. Coll. Sci.*, 1953, **8**, 38; 1952, **7**, 600.

45. H. KOELMANS and J. TH. G. OVERBEEK, *Discussions Faraday Soc.*, 1954, **18**, 52; H. KOELMANS, *Philips Research Reports*, 1955, **10**, 161.

46. B. V. DERJAGUIN, *Discussions Faraday Soc.*, 1954, **18**, 85.

47. A. BIERMAN, *Proc. Nat. Acad. Sci. U.S.*, 1955, **41**, 245; *J. Coll. Sci.*, 1955, **10**, 231.

48. B. V. DERJAGUIN, *Trans. Faraday Soc.*, 1940, **36**, 203.

49. R. DEFAY and A. SANFELD, *J. Chim. Phys.*, 1963, p. 634.

50. T. VOROPAJEVA, B. V. DERJAGUIN and B. KABANOV, *C.R. Acad. Sci. U.R.S.S.*, 1959, **128**, 981.

51. W. H. KEESOM, *Versl. Kon. Ned. Akad. Wetenschap.*, 1912, 20^2, 1414; 1912, 21^1, 492; 1915, 24^1, 614; 1916, 24^2, 1699; 1920, 29^1, 722; *Phys. Z.*, 1921, **22**, 129.

52. P. DEBYE, *Phys. Z.*, 1920, **21**, 178.

53. H. FALKENHAGEN, *Phys. Z.*, 1922, **23**, 87.

54. H. MARGENAU, *Rev. Mod. Phys.*, 1939, **11**, 1; H. MARGENAU and N. R. KESTNER, *The Theory of Intermolecular Forces*, 2nd ed., Pergamon Press, 1971.

55. B. CHU, *Intermolecular Forces,* Interscience, 1966.

56. F. LONDON, *Z. physik. Chemie* B, 1930, **11**, 222.

57. Slightly different theories are given by: J. C. SLATER and J. G. KIRKWOOD, *Phys. Rev.*, 1932, **37**, 682; TH. NEUGEBAUER, *Z. Physik*, 1937, **107**, 785; R. A. BUCKINGHAM, *Proc. Roy. Soc. (London)*, A, 1937, **160**, 112; J. K. KNIPP, *Phys. Rev.*, 1939, **55**, 1244.

58. K. S. PITZER, *Adv. Chem. Phys.*, 1959, **II**, 59, many-electron problems. Interactions between large molecules.

59. P. DEBYE in *Adhesion and Cohesion*, Elsevier, 1963, p. 1.

60. See textbooks such as: M. BORN and P. JORDAN, *Elementare Quantenmechanik*, Springer, 1930; E. S. KEMBLE, *The Fundamental Principles of Quantum Mechanics,* McGraw-Hill, 1937.

61. H. B. G. CASIMIR and D. POLDER, *Phys. Rev.*, 1948, **73**, 360; *Nature*, 1946, **158**, 787.

62. J. TH. G. OVERBEEK, see ref. 61.

63. H. C. HAMAKER, *Physica*, 1937, **4**, 1058.

64. I. NEWTON, *Principia etc.*, London, 1686, Prop. LXXV, Theorema XXXV, Prop. LXXVI; also *Opticks*, London, 1717, Query 31.

65. H. B. G. CASIMIR, *Proc. Kon. Akad. Wetenschap., Amsterdam*, 1948, **51**, 793.

66. C. M. HARGREAVES, *Proc. Kon. Akad. Wetenschap., Amsterdam*, 1965; **68B**, 231; and discussion remark in *Discussions Faraday Soc.*, 1966, **42**, 279.

67. F. SAUER, thesis Göttingen, 1962.
68. E. M. LIFSHITZ, *Doklady Nauk Akad. S.S.S.R.*, 1954, **97**, 643; 1955, **100**, 879; *Zhur. Eks. Teor. Fiz.*, 1955, **29**, 94; *Soviet Physics JETP*, 1956, **2**, 73.
69. B. V. DERJAGUIN and L. I. ABRIKOSSOVA, *Discussions Faraday Soc.*, 1954, **18**, 24; *Proc. IInd international Conference on Surface Activity*, 1957, **III**, 398 (A and D).
70. J. TH. G. OVERBEEK and M. J. SPARNAAY, *Proc. Kon. Akad. Wetenschap. Amsterdam*, 1951, **54**, 387; *J. Coll. Sci.*, 1952; **7**, 343; *Discussions Faraday Soc.*, 1954, **18**, 12.
71. J. A. KITCHENER and A. P. PROSSER, *Proc. Roy. Soc. A*, 1957, **242**, 403.
72. M. J. SPARNAAY, *Physica*, 1958, **24**, 751; *Nature*, 1957, **180**, 334.
73. B. V. DERJAGUIN and I. I. ABRIKOSSOVA, *J. Phys. Chem. Solids*, 1958, **5**, 1.
74. W. BLACK, J. G. V. DE JONGH, J. TH. G. OVERBEEK and M. J. SPARNAAY, *Trans. Faraday Soc.*, 1960, **56**, 1597.
75. A. VAN SILFHOUT, *Proc. Kon. Nederland. Akad. Wetenschap. B*, 1966, **69**, 501.
76. I. E. DZYALOSHINSKII, I. E. E. M. LIFSHITZ and L. P. PITAEVSKII, *Adv. in Physics*, 1961, **10**, 165.
77. L. P. PITAEVSKII, *Zhur. Eks. Teor. Fiz.*, 1959, **37**, 577; *Soviet Physics JETP*, 1959, **10**, 295.
78. A. D. McLACHLAN, *Proc. Roy. Soc. A*, 1963, **274**, 80; *Discussions Faraday Soc.*, 1965, **40**, 239.
79. B. LINDER, *J. Chem. Phys.*, 1961, **35**, 37; 1960, **33**, 668.
80. M. J. SPARNAAY, *Physica*, 1959, **25**, 217; for large *a*-values the case becomes identical to that considered by AXILROD and TELLER, *J. Chem. Phys.*, 1943, **11**, 299.
80a. M. J. RENNE and B. R. A. NIJBOER, *Chem. Phys. Lett.*, 1967, **1**, 317; B. R. A. NIJBOER and M. J. RENNE, *ibid.*, 1968, **2**, 35; M. J. RENNE and B. R. A. NIJBOER, *ibid.*, 1970, **6**, 601; M. J. RENNE, Thesis, Utrecht, 1971.
81. J. H. SCHENKEL and J. A. KITCHENER, *Trans. Faraday Soc.*, 1960, **56**, 161.
82. S. HACHISU and K. FURUSAWA, *Science of Light*, 1963, **12**, 1, 157.
83. M. VON SMOLUCHOWSKI, *Physik. Z.*, 1916, **17**, 557, 585; *Z. physik. Chemie*, 1918, **92**, 129.
84. H. REERINK and J. TH. G. OVERBEEK, *Discussions Faraday Soc.*, 1954, **18**, 74.
84a. E. P. HONIG, G. J. ROEBERSEN and P. H. WIERSEMA, *J. Coll. Interf. Sci.*, 1971, **36**, 97.
85. N. FUCHS, *Z. Physik*, 1934, **89**, 736.
86. E. J. W. VERWEY and J. TH. G. OVERBEEK, *Theory of the Stability of Lyophobic Colloids*, Elsevier, 1948, Part III.
87. M. J. SPARNAAY, *Rec. Trav. Chim. Pays-Bas*, 1959, **78**, 680.
88. A. G. DE ROCCO and W. G. HOOVER, *Proc. Natl. Acad. Sci.*, 1960, **46**, 1057.
89. M. J. VOLD, *J. Coll. Sci.*, 1954, **9**, 451.
90. B. V. DERJAGUIN, *Koll. Z.*, 1934, **69**, 155; *Acta Physicochim. U.R.S.S.*, 1939, **10**, 333.
91. S. LEVINE and G. P. DUBE, *Phil. Mag.*, 1940, (7), **29**, 105; *J. phys. Chem.*, 1942, **46**, 239.
92. N. E. HOSKIN, *Phil. Trans. A*, 1956, **248**, 433; N. E. HOSKIN and S. LEVINE, *Phil. Trans. A*, 1956, **248**, 449.
93. C. A. BARLOW Jr., *Am. J. Physics*, 1963, **31**, 247.
94. B. TEZAK, E. MATIJEVIĆ and K. F. SCHULZ, *J. Phys. Chem.*, 1951, **55**, 1567 and 1955, **59**, 769; H. R. KRUYT and M. A. M. KLOMPÉ, *Kolloidbeihefte*, 1943, **54**, 484.

95. S. Levine, J. Mingins and G. M. Bell, *J. Electroanal. Chem.*, 1967, **13**, 280 (a review paper); first paper: S. Levine, G. M. Bell and D. Calvert, *Nature*, 1961, **191**, 395.

96. J. Th. G. Overbeek, *Pure and Applied Chem.*, 1965, **10**, 359.

97. R. M. Fuoss, *Chem. Rev.*, 1935, **17**, 27.

98. D. N. L. McGown, G. D. Parfitt and E. Willis, *J. Coll. Sci.*, 1965, **20**, 650.

99. F. J. Micale, Y. K. Lui and A. C. Zettlemoyer, *Discussions Faraday Soc.*, 1966, **42**, 238.

100. L. A. Romo, *Discussions Faraday Soc.*, 1966, **42**, 232.

101. E. J. W. Verwey, *Rec. Trav. Chim. Pays-Bas*, 1941, **60**, 625, and ref. 7.

102. G. J. Young and J. J. Chessick, *J. Coll. Sci.*, 1958, **13**, 358.

103. H. R. Kruyt and F. G. Van Selms, *Rec. Trav. Chim. Pays-Bas*, 1942, **62**, 407.

104. W. Albers, *J. Coll. Sci.*, 1959, **14**, 501, 510; 1960, **15**, 489.

104a. G. M. Fair and R. S. Gemmell, *J. Coll. Sci.*, 1964, **19**, 360, reconsider Smoluchowski's theory.

105. G. M. Bell and S. Levine, *Trans. Faraday Soc.*, 1957, **53**, 143; 1958, **54**, 785, 975.

106. T. Gillespie, *J. Coll. Sci.*, 1960, **15**, 313.

107. A. Westgren, *Arkiv Kemi Mineral. Geol. Mo.*, 1918, **6**, 7 and ref. 1.

108. L. Onsager, *Phys. Rev.*, 1942, **62**, 558; *Ann. N.Y. Acad. Sci.*, 1949, **51**, 627.

109. H. G. Jonker, private communication.

110. R. E. Rosenzweig, *Int. Sci. Techn.*, July 1966, p. 48 ("Magnetic fluids").

111. H. Freundlich, *Kapillarchemie* II, Leipzig, 1932, pp. 378, 447.

112. A review of his work in this field by: V. K. LaMer, *Discussions Faraday Soc.*, 1966, **42**, 248. See also V. K. LaMer and R. H. Smellie, *J. Coll. Sci.*, 1956, **11**, 704, 710; R. H. Smellie and V. K. LaMer, *ibid.*, 1958, **13**, 589.

113. J. Th. G. Overbeek, *Discussions Faraday Soc.*, 1966, **42**, 277.

114. B. H. Bijsterbosch and J. Lijklema, *J. Coll. Sci.*, 1965, **20**, 665, and thesis Bijsterbosch, Utrecht, 1965.

115. M. A. V. Devanathan, *Proc. Roy. Soc.* A, 1962, **195**, 150.

116. R. H. Ottewill, M. C. Rastogi and A. Watanabe, *Trans. Faraday Soc.*, 1960, **56**, 854.

117. G. Oster, *J. Coll. Sci.*, 1947, **2**, 291.

118. R. H. Ottewill and M. C. Rastogi, *Trans. Faraday Soc.*, 1960, **56**, 866, 880.

119. P. Somasundaran, T. W. Healy and D. W. Fuerstenau, *J. Coll. and Interf. Sci.*, 1966, **22**, 599.

120. P. Somasundaran and D. W. Fuerstenau, *J. Phys. Chem.*, 1966, **70**, 90.

121. K. G. Mathai and R. H. Ottewill, *Trans. Faraday Soc.*, 1966, **62**, 750, 759.

122. Yu. Glazman, *Discussions Faraday Soc.*, 1966, **42**, 255, and in *Research in Surface Forces* II (ed. B. V. Derjaguin), Plenum Press, New York, 1966, p. 232.

123. R. H. Ottewill, *Discussions Faraday Soc.*, 1966, **42**, 284.

124. R. W. Slater and J. A. Kitchener, *Discussions Faraday Soc.*, 1966, **42**, 267.

125. E. L. Mackor, *J. Coll. Sci.*, 1951, **6**, 492.

126. E. L. Mackor and J. H. Van der Waals, *J. Coll. Sci.*, 1952, **7**, 535.

127. M. Van der Waarden, *J. Coll. Sci.*, 1950, **5**, 317; 1952, **6**, 443.

128. E. J. W. Verwey, *Discussions Faraday Soc.*, 1966, **42**, 314.

129. E. W. FISCHER, *Koll. Z.*, 1958, **160**, 120, especially p. 139.
130. P. J. FLORY, *J. Chem. Phys.*, 1942, **10**, 51.
131. M. L. HUGGINS, *J. Chem. Phys.*, 1941, **9**, 440; *Ann. New York Acad. Sci.*, 1942, **43**, 1.
132. H. TOMPA, *Polymer Solutions*, Butterworth, 1956, chaps. 3 and 4.
133. H. MORAWETZ, *Macromolecules in Solution*, Interscience, 1965, chap. II (Vol. **21** of the series: *High Polymers*).
134. E. A. GUGGENHEIM, *Mixtures*, Clarendon Press, Oxford, 1952.
135. E. J. CLAYFIELD and E. C. LUMB, *J. Coll. and Interf. Sci.*, 1966, **22**, 269.
136. J. J. HERMANS in *Colloid Science*, II, Elsevier, 1949, chapts. III and IV.
137. W. KUHN and F. GRÜN, *Kolloid Z.*, 1942, **101**, 248.
138. G. J. FLEER and J. LIJKLEMA, *Vth International Congress on Surface Activity*, Barcelona, 1969, p. 247.

CHAPTER 5

SURFACTANTS

5.1. Introduction

This chapter deals with double-layer systems which are formed in solutions of surfactants in water or in aqueous electrolytes. In the case of insoluble surfactants, they are formed by spreading them on a liquid surface. Three different types of system can be distinguished: monolayers, micelles and thin films. Because of their similarity to thin films, membranes will also be included. Figure 5.1 gives an illustration of the structure of these systems. The general views upon which these are based is due to the classical work of Gibbs[1] (see Mysels[2]), Pockels,[3] Adam,[4] Rideal,[5] Langmuir,[6] Harkins[7] and Hartley[8] (micelles). The electrochemical aspects of these systems have, however, largely been considered at a later date.

Dissolution of a surfactant leads to a sharp decrease of the surface tension, or if a non-polar liquid is in contrast the aqueous phase, of the interfacial tension. According to the Gibbs adsorption isotherm this decrease is to be interpreted as the consequence of a strong positive adsorption, i.e. of an accumulation of the surfactant molecules at the surface or the interface. Gibbs was already fully aware of this consequence of the adsorption equation. Figure 5.2 gives a curve[9] of the surface tension (γ) versus the logarithm of the concentration (c) of sodium dodecyl sulphate. This curve is characteristic of the whole class of soluble surfactants. It can be divided into three portions. First, at low c-values the slope increases with increasing concentration. This points to an increasing adsorption when c is increased. Second, at a certain c-value, to be denoted as c_s, the slope becomes constant. Assuming ideal behaviour of the solute in solution, this means that the adsorbed amount is constant with respect to variations of c. In this saturation region as it is called, the number of adsorbed sodium dodecyl sulphate molecules per cm^2 is $26 \cdot 4 \times 10^{13}$, which means that, assuming the adsorbed particles to form a monolayer, each particle occupies an area of 29 $Å^2$. The third portion of the γ-log c curve has an almost zero slope: addition of sodium dodecyl sulphate leaves γ almost unaffected. The added molecules now no longer act as individuals. They form clusters in the bulk of the solution. These clusters are called micelles. They are

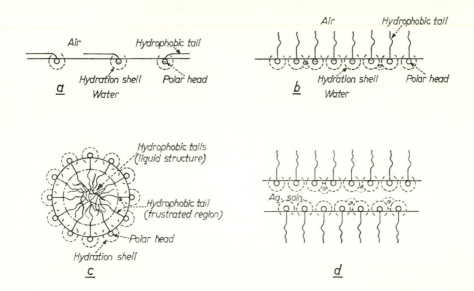

FIG. 5.1. (a) Non-saturated monolayer. Gegen-ions not shown. (b) Saturated monolayer. Some specifically adsorbed cations are shown. Gegen-ions
not shown. (c) Micelle, largely according to Hartley.[8] For "liquid structure"
and "frustrated region" see text. (d) Thin film.

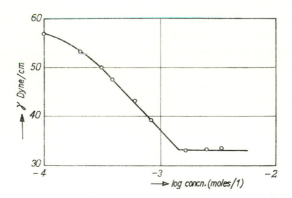

FIG. 5.2. Surface tension against log concentration of Na-dodecylsulphate
solutions, pH = 6·5, c_{Na^+} = 0·1 M. (1 dyne/cm = 10^{-3} N m^{-1}) (from Roe
and Brass[9]).

often considered as a separate phase, but this concept has met with objections.[10] Micelle formation can also be detected by other methods,[11] for example by conductance measurements (in the case of ionized surfactants), or by light scattering. The number of monomers per cluster is of the order of 30–100 and the clusters are roughly spherical. The concentration at which micelle formation sets in is called the critical micelle concentration (cmc). For many different surfactants the ratio c_s/cmc has roughly the same value as is shown in Table 5.1 (this table is taken from the work of Van Voorst Vader[12]). This constancy suggests that the formation mechanisms of saturated monolayers and of micelles are essentially the same.

TABLE 5.1

Surfactant	c_s/cmc	Interface	Refs.
K-laurate	0·15	air	9
K-palmitate	0·15	air	9
Na-dodecyl sulphate	0·20	air	9
Dodecyl amine hydrochloride	0·25	air	9
Na-decyl sulphate	0·20	n-heptane	13
Na-dodecyl sulphate	0·15	n-heptane	13
Na-myristyl sulphate	0·30	n-heptane	13
Na-laurate	0·20	paraffin oil	14

Another rule, found long ago by Traube,[14] is that the concentration needed for a certain reduction of γ of the solution/air interface decreases by a factor of about 2 for each —CH_2— group added to the paraffin chain. Traube's rule applies best to non-ionic systems. The rule points to a non-specific role of the —CH_2— groups, each added —CH_2— group increasing the free energy of adsorption by about $kT \ln 2$ (i.e. 2600 J mol^{-1} ($= 625$ cal mol^{-1})). In this connection Table 5.2, from Mukerjee,[11] is useful. The transfer from an aqueous solution to the gas phase is accompanied by a relatively small free energy loss which is ascribed (see also Mukerjee) to a rearrangement of water molecules. As pointed out in the Introduction, Chapter 1, the water structure in the neighbourhood of dissolved monomer molecules is usually assumed to be somewhat more "ice-like" than the water structure of bulk water. The free energy loss of the transition aqueous solution → gas phase is therefore explained in terms of an entropy increase: the "ice" "melts" when the —CH_2— groups are removed from the aqueous phase. Table 5.2 shows that this entropy effect does not play a preponderant role in the other transitions. In these cases the action of Van der Waals forces is more important. Another entropy factor of importance in micellization may be mentioned: isolated monomer molecules in water will tend to be curled up to minimize the area of contact of the hydrocarbon chains with water. The number of configurations which a monomer molecule can assume increases when brought into contact with other monomers. The interior of a micelle will then be liquid-like. This is shown by the phenomenon of solubilization: compounds which are soluble in non-polar liquids can also be dissolved in the interior of a micelle.

TABLE 5.2

Free energy change per —CH₂— group on transfer from an aqueous solution to various final states

Final state	Free energy change per $-CH_2-$ group (J mol⁻¹)
Gas phase	-710
Air/water	-2600
Hydrocarbon/water	-3390 to -3430
Micelles	-2720 to -3000
Hydrocarbon solution	-3450

The table is based on experiments with $C_nH_{2n+1}(OCH_2CH_2)_6OH$ compounds. (From Mukerjee, ref. 10).

The hydrophilic parts of the surfactant molecules, i.e. those parts which tend to surround themselves with water molecules, work against the formation of monolayers and micelles. If two hydrophilic groups are forced together to within, say, 10 Å, there is a repulsion because there will be hardly any penetration possible of their hydration shells and, when ionized, we have in addition the Coulombic repulsion. If in a saturated monolayer or a micelle, we pass from the water phase to the paraffin chains, we cross a transition region which we call the frustrated region for the following reason. The water molecules of the hydration shells are pressed together and some of them are in contact with the hydrophobic parts. Close to the hydrophilic heads the liquid state of the hydrophobic chains will not be realized, and this will be due to steric reasons provided through the presence of the water molecules. They give the hydrophilic heads an area larger than required for the hydrophobic tails directed to the liquid interior. When the hydrophilic heads are ionized, the total repulsion can be affected by varying the ionic concentration. If ions are added, screening of the charges of the hydrophilic heads is more effective. Therefore, in concentrated electrolytes, the formation of micelles and monolayers from ionized surfactants is easier than in dilute electrolytes, i.e. the cmc and the c_s are lowered when electrolyte is added.

Thin films can be considered as a systems of two parallel saturated monolayers, separated by an aqueous layer of 100–1000 Å thickness with the hydrophilic heads turned inward. The easiest accessible thin films are soap bubbles. Study of thin films is important because their equilibrium thickness as a function of the electrolyte concentration has turned out to be given by essentially the same theory as that applied to explain the stability of hydrophobic colloids.

Another class of thin films is that of the "inverted soap films". These films separate two aqueous phases and have their hydrophilic parts turned outward, i.e. toward these phases. These films are supposed to resemble real biological membranes.[15] In fact, model membranes have been constructed whose behaviour shows much similarity with those of real membranes. On the other hand, recent investigations of real membranes have shown that current ideas concerning their structure need revision. Much work is done in this field nowadays. Some of it will be presented in Section 5.5.

5.2. Monolayers

5.2.1. *Experimental methods*

The formation of a monolayer leads to a change of the potential drop across the interface. This can be shown by means of experiments such as shown in Fig. 5.13. One of the methods (Fig. 5.3a) is essentially the Kelvin vibrating plate method. This method can not only be used for potential measurements at gas/liquid and gas/solid interfaces (see, for instance, Zisman[16]) but also at liquid/liquid interfaces. For this purpose the liquid in which the vibrating plate is immersed must be insulating. This is the case for most non-polar liquids. Numerous experiments have been carried out by Davies.[17,17a]† To avoid disturbances, Davies kept the vibrating plate about 0·5 mm away from the interface and held the vibrational amplitude and the frequency to within 10 μm and a few hundreds Hz respectively. The Kelvin method has an accuracy of about 1–3 mV.

The second method, illustrated in Fig. 5.3b, is the liquid jet method originated by Kenrick[19] in 1896 and used by Frumkin,[20] by Kamiènski[21] and by Randles.[22] The best measurements probably stem from Randles.

---

† A critical examination of the experimental difficulties encountered in measurements of interfacial potentials at polar and non-polar oil/water interfaces was presented by Mingins *et al.*[18]

In Randles' version the two surfaces, A and B, say, were (A) is that of a liquid wetting the inside of a glass tube of about 5 mm diameter and (B) that of a liquid emerging from a capillary which had an orifice of internal diameter of about 70 μm, the capillary being placed in concentric position with respect to the outer tube. If the surface potentials are different the surfaces become oppositely charged. When the jet breaks into droplets the charge on the surface (B) of the jet is carried away and the reservoir supplying the jet then develops a charge of opposite sign. The process goes on until the surface potentials are equal. The potential

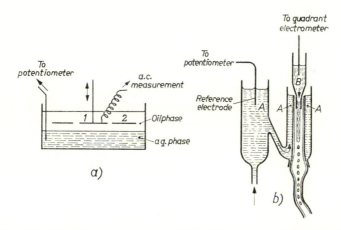

Fig. 5.3. Potential measurements. (a) Vibrating plate method (Davies and Rideal[17])—1: vibrating (gold) plate; 2: guard disc. (b) Liquid jet method (Kenrick[19]), after Randles.[22]

was measured with a quadrant electrometer. The potential of the solution A relative to each was adjusted by means of a potentiometer, to give zero deflection of the electrometer. The absence of an asymmetry potential was checked by experiments in which A and B were the same liquid. The main sources of error were: lack of insulation of the reservoir supplying the jet, slowly leaking electrostatic charges, a too rapid stream of the jet liquid, and an evaporation of substances dissolved in one of the liquids followed by adsorption on the opposite surface. A reproducibility of 1 mV was reported.

A third method employs an ionizing electrode.[23] Here the air between electrode and surface is ionized, i.e. made conducting by means of a radioactive agent. The variations of the surface potential were then recorded by the electrode equipment.

For mixed adsorbed surfactants the problem of surface composition arises. To solve this problem, Corkill et al.[24] used a tracer method: a

Geiger counter was placed at a reproducible distance from the surface and C^{14} or S^{35} tagged surfactants were used.

Measurements of the interfacial (surface-) tension γ and of the interfacial (surface-) pressure π can be carried out in a number of ways, which can be divided in two groups: first, γ is determined as a function of the bulk concentration of the dissolved surfactants. Second, an equation of state is directly measured by a simultaneous measurement of π and of the interfacial (surface-) concentration of the surfactant Γ. This is the two-dimensional equivalent of pressure–volume measurements of gases, vapours (and liquids). It should be kept in mind that, whereas in the three-dimensional case the total number of particles involved can be held constant, the particles in the two-dimensional case can escape to the bulk phase after contraction of the interfacial area and can reappear after expansion. Therefore, it is only when the dissolution or adsorption processes are slow or/and when the solubility is low that the second group of measurements gives reliable results. Descriptions of the experimental methods have been given in many of the references so far given. We may add references to the work described by Guastalla[25] and Ter Minassian-Saraga.[26]

A large number of π–A ($A = \Gamma^{-1}$) curves for insoluble surfactants was, already at an early date,[4] analysed by means of the following equation:

$$(\pi - \pi_0)(A - A_0) = kT. \qquad (5.2.1)$$

This equation shows some resemblance with the Van der Waals equation of state, the difference being that the Van der Waals "a" term is replaced by the constant $-\pi_0$. By analogy with the Van der Waals equation, A_0 is the area occupied by the surfactant molecules when highly compressed. Langmuir[27] gave the following physical interpretation of the constant $-\pi_0$: the hydrocarbon tails of the surfactant molecules form essentially a thin layer of a hydrocarbon liquid whose component molecules are free to move in relation to one another just as in ordinary liquids, except that one end of each molecule is constrained to stay in contact with the aqueous phase because of the hyrophilic group attached to it. Therefore the constant $-\pi_0$ may be compared with the surface tension of the hydrocarbon/air surface. Equation (5.2.1) will be valid only when the monolayer is compressed in such a way that a more or less coherent hydrocarbon surface is formed. When this is not the case, i.e. when the surface concentration is low, it will be more appropriate to replace the constant $-\pi_0$ by the Van der Waals term $+a\Gamma^2$. In the following (where attention is mainly directed to ionized surfactants) we stress the importance of the contribution of electrical terms to the surface pressure and rarely make use of eq. (5.2.1).

5.2.2. *The χ-potential and the ψ₀-potential*

The change of the potential drop ΔV arising from the formation of a monolayer can be divided into a change of the χ-potential, $\Delta\chi$, and a change of the ψ-potential, $\Delta\psi_0$,

$$\Delta V = \Delta\chi + \Delta\psi_0. \qquad (5.2.2)$$

Three possible contributions to $\Delta\chi$ may be distinguished: that of a permanent dipole of the hydrophilic head of the surfactant molecule, that which is due to a reorientation of the water molecules in the immediate vicinity of the hydrophilic heads, and finally a contribution arising from a dipole moment in the hydrophobic tail of the surfactant

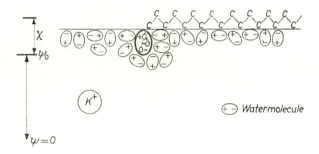

Fig. 5.4. Water surface with adsorbed ionized surfactant molecule, and potentials.

molecule. It is difficult to distinguish between these three contributions. According to Frumkin[28] and Adam[29,4] the reorientation effect of the water molecules is rather small because $\Delta\chi$ may have either sign and may be large as 1 V. Therefore the original orientation (of water molecules at the clean water surface) is probably weak.[28] Frumkin[28] found that the adsorption of lepidin (4-methyl-quinoline) at the air/water interface led to a negative shift of 0·8 V. In contrast for ω-bromohexadecanoic acid (a C_{16} acid with Br substituted in the terminal hydrocarbon group) there was a shift in the positive direction of about 0·8 V while for perfluorodecanoic acid the (positive) shift was as much as 1 V. Because he disregarded the latter observations (i.e. positive shift), Kamiènski (see ref. 21)) concluded that there was at the free water surface an orientation of the water molecules, which led to a χ-potential of about 1 V, the positive hydrogen part of the molecules being turned to the water phase.

Bernett *et al.*,[30] comparing the properties of adsorbed C_{16} and C_{18} fatty acids and fatty alcohols, and their ω-monohalogenated derivatives, calculated values of the dipole moments ranging from 0.7 D ($C^+ - F^-$) via 0.85 D ($C^+ - Cl^-$) to about 0.95 D ($C^+ - Br^-$ and $C^+ - I^-$) for the components of the dipole moments of the terminal halogenated groups which were assumed to be perpendicular to the plane of the interface. A value of 0.3 D was used for the contribution of the $C^+ - H^-$ dipole. It was assumed that the halogen substitution left the other contributions to $\Delta\chi$ unaltered. Langmuir and Schaefer[31] found that polyvalent ions had a condensing effect on adsorbed monolayers of the higher fatty acids. Therefore an aqueous substrate containing 5×10^{-5} M $Th(NO_3)_4$ at pH $= 3.4$ was used. This was particularly useful for the halogenated surfactants, because π–A curves showed that halogenation led to an increased repulsion between the adsorbed molecules. Since this could not be explained by steric hindrance, Bernett $et\ al.$ ascribed the effect to a repulsion between adjacent C^+-$(Hal)^-$ dipoles. The dipole moment associated with a C^+ Hal^{-1} group deduced from these experiments is probably much smaller than that of an isolated dipole. This was assumed to be caused by mutual induction effects between neighbouring dipoles. Another possibility is that, since the C^+-$(Hal)^-$ groups are placed at the ends of rather long chains, the perpendicular components of the moments are much smaller than expected on the basis of the assumption of a stretched molecule.

With regard to the change of the ψ_0 potential, we note first that electrolyte solutions (e.g. of KCl, NaCl and so on) usually have a slightly higher surface tension than water. This points to a negative adsorption of the ions. The relative increase is about 0.3% per added mole of ionic substance per litre. A negative ionic adsorption can be understood on the basis of electro static theory. The ionic charges are in a medium with a high dielectric constant, ε_h, whereas at the other side of the interface the relative dielectric constant, ε_1, is low. Under these circumstances the ions tend to keep away from the interface and to stay in the medium with ε_h. This tendency is represented by an image force f_i

$$f_i = \frac{e\,e'}{\varepsilon_h(2r)^2} \qquad (5.2.3)$$

where

$$e' = \frac{\varepsilon_h - \varepsilon_1}{\varepsilon_h + \varepsilon_1}. \qquad (5.2.4)$$

Onsager and Samaras[32] have given a theory, based on this image force repulsion, of the distribution of ions near an interface. Image forces alone would not lead to a change of a potential drop across the interface. Nevertheless, negative shifts of about 0.1 V have been found. See

Randles' review.[22b] The sign of ΔV indicates that anions come closer to the interface than do cations.

Kamiènski studied interesting systems which were intermediate between those of small ions and those containing surfactants. He considered acids and bases with a low ionization constant and containing fairly small hydrophobic groups. The hydrophilic character could here be modified by varying the pH. Thus, weak bases at high pH were

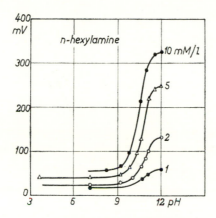

FIG. 5.5. Shift of ΔV as a result of the adsorption of n-hexylamine at the water surface, at various pH-values. No shift below pH = 9, the amine molecules being completely dissolved (from Kamiènski[21c]).

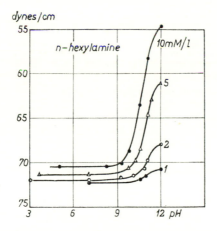

FIG. 5.6. Surface tension of aqueous surface with adsorbed n-hexylamine at various pH-values. No shift below pH = 9. (1 dyne/cm = 10^{-3} N m^{-1}) (Kamiènski[21c]).

H

adsorbed, at least to some extent, and a shift of the potential V, to be interpreted as a shift $\Delta\psi_0$, was produced. No adsorption was taking place at a low pH where the completely ionized salt was formed and ΔV was close to zero.

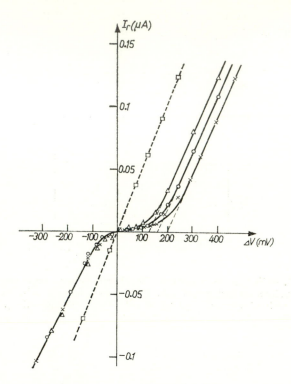

Fig. 5.7. Current–voltage curves showing overpotentials which are ascribed to electroadsorption (Guastalla[25b]). Aq. solution: 0·001 M KCl. Nitrobenzene solution:

△	Dodecyl trimethyl ammonium picrate			
○	Tetradecyl ,,	,,	,,	5×10^{-4} M
×	Hexadecyl ,,	,,	,,	
□	,, ,,	,,	chloride	

Another way to "enforce" an adsorption is by using a potential difference applied across the interface between an aqueous electrolyte and an ionic solution in oil. After application of the potential difference (V) the interfacial tension decreases, which is ascribed to the adsorption of ions. This is the phenomenon of electroadsorption (Guastalla). An electric current (i_r) flows and the i_r–V curves show the characteristics of over-potential curves provided the two adjacent inmiscible solutions have no ions in common. Figure 5.7 gives an example. The shape of the $i_r - V$ curves point to a non-zero activation energy for the process of electron transfer, accompanied by a prolonged time of residence of the ions at the interface. In contrast to the adsorption phenomena observed by Kamieński *et al.* electroadsorption is a non-equilibrium phenomenon.

5.2.3. *The Stern layer and the Gouy layer. Equations of state*

Just as in the case of hydrophobic colloids we may apply the Stern–Gouy model of the charge distribution in the double layer. However, it is probable that the plane $x = \delta$ of the position of the specifically adsorbed gegen-ions is located very close to the plane $x = 0$ where the ionized head groups of the surfactant molecules are located. The reason is that the potential drop across the layer can, in this way, be made less drastic and that especially at high coverage the accommodation of a rather close-packed surface is facilitated. Levine *et al.*[33] were the first to give a correct treatment of the problem. They arrived at the conclusion that δ could be negative, of the order of -1 Å to -2 Å, i.e. they found it probable that the gegen-ions protruded somewhat beyond the plane of location of the head groups in the direction of the non-aqueous phase, i.e. in the "frustrated region".

The work of Corkill *et al.*[24] is also relevant here. They found that for mixed anionic and cationic surfactants of equal chain length, a maximum interaction occurred at a 1 to 1 cation/anion ratio.

In the following we mainly consider the role of the bulk concentration of (small) ions and shall give some simple equations for the adsorption. We identify the inner and the outer Helmholtz planes and locate them at $x = \delta$. Denoting the charge of a surfactant ion as $z_i e$, and assuming that only one type of ion, with charge $z_j e$, is adsorbed at the plane $x = \delta$, we have

$$\sigma = z_i e \Gamma \alpha \qquad (5.2.5)$$

where Γ is the number per unit area of ionized and non-ionized surfactant molecules, and α is the degree of ionization at the surface. Furthermore,

$$\sigma_{\text{Stern}} = z_j e \frac{\Gamma_0}{1 + \dfrac{N}{n} \exp \dfrac{\varphi_j + z_j e \psi_\delta}{kT}} \qquad (5.2.6)$$

where Γ_0 is the number per unit area of the available adsorption sites for the gegen-ions, N is the number per unit volume of the aqueous phase of the "sites" (see Stern's theory, Chapter 4) for the gegen-ions j, n is the bulk concentration (number per unit volume), φ_j is the adsorption potential of the gegen-ions j and ψ_δ is the electrostatic potential at $x = \delta$.

For the charge in the Gouy layer we have

$$\sigma_{\text{Gouy}} = -\sqrt{\left(\frac{2}{\pi}\, \varepsilon n k T\right)}\, \sinh \frac{e\psi_\delta}{2kT}. \qquad (5.2.7)$$

We have assumed a 1–1 electrolyte with n cations and n anions per unit volume. One of the sorts of these ions is designated as j.

Electroneutrality requires

$$\sigma + \sigma_{\text{Stern}} + \sigma_{\text{Gouy}} = 0. \qquad (5.2.8)$$

A simple equation can now be derived as follows: for sufficiently small n, eq. (5.2.6) can be written in the form of a Boltzmann equation. That means that σ_{Stern} must be rather small and that ψ_0 and ψ_δ must lie close together. Furthermore, assume that $|\psi_\delta|$ is large compared with $2kT/e$. Then eq. (5.2.7) can be written in exponential form. A large $|\psi_\delta|$ value points to a large Γ value. We assume that we are here in the saturation region, i.e. that Γ hardly increases when more surfactant is added. The Γ value in the saturation region will also hardly be affected when n is varied.

For complete ionization ($\alpha = 1$) one can write

$$-e\Gamma + e\Gamma_0\, \frac{n}{N}\, \exp \frac{-(\varphi_j + e\psi_\delta)}{kT} + \sqrt{\left(\frac{1}{2\pi}\, \varepsilon n k T\right)}\, \exp \frac{-e\psi_\delta}{2kT} = 0. \quad (5.2.9)$$

In this equation we have taken $z_i = -1$ and $z_j = +1$. That means, that ψ_0 and ψ_δ are negative and that σ_{Gouy} is positive. Equation (5.2.9) is obeyed when $n \exp -e\psi_\delta/kT$ (or $n \exp +e\psi_\delta/kT$ for $z_j = -1$) is constant with respect to varying n. Thus, curves relating $\log n$ and ψ_δ should give straight lines with a slope of 59 mV per decade. It is unlikely that the χ-potential varies as a result of variations of n, although this possibility can not be ruled out. Assuming χ to be constant, measurements of ΔV as a function of n can be interpreted as indicating curves of ψ_δ versus n. Figure 5.8 gives an example.

The non-saturated region will now be considered. Since we stress the electrical effects we make the simple approximation that in the absence of charges the adsorbed particles show ideal behaviour. The surface pressure π of Γ adsorbed partly ionized particles per unit area is

$$\pi = \Gamma k T + \int_0^{\psi_0} \sigma\, d\psi_0 = \Gamma k T - F_{\text{el}} \qquad (5.2.10)$$

where we have denoted the integral by $-F_{el}$. The value of the integral depends on the distribution of the gegen-ions. We briefly consider four cases.

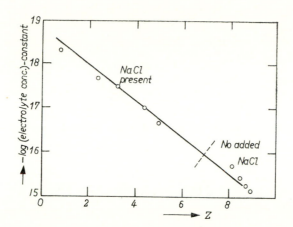

Fig. 5.8. Electrostatic potential against electrolyte concentration. The "constant" is equal to $+\log \Gamma + 15 \cdot 4(A)^{-\frac{1}{2}}$, where Γ is the number per cm² of adsorbed molecules and A is the molecular area in Å². The line has the theoretical Nernst slope.

(a) SWAMPING ELECTROLYTE. NO STERN LAYER

Here $n \gg n_i$, i.e. the salt concentration (e.g. NaCl, $n = n_{Na} = n_{Cl-}$) is much larger than the bulk concentration of the surfactant, n_i. For large $|\psi_\delta|$, σ_{Gouy}, and therefore σ, is proportional to

$$\exp \frac{e|\psi_\delta|}{2kT}.$$

Then it is easy to see that (see also Section 2.3), identifying ψ_δ and ψ_0:

$$-F_{el} = 2\alpha\Gamma kT \qquad (5.2.11)$$

and

$$\pi = (2\alpha + 1)\Gamma kT. \qquad (5.2.12)$$

For small potentials, $|\psi_\delta| < 25\,\text{mV}$ we have (see also Section 2.3)

$$-F_{el} = \tfrac{1}{2}\sigma\psi_\delta = (\alpha\Gamma)^2 \frac{\kappa}{4n}\,kT. \qquad (5.2.13)$$

This contribution can be considered as a virial term. The small potential approximation is, in a 10^{-1} N solution, valid for $(\alpha\Gamma) < 5 \times 10^6$ cm⁻².

(b) NO SALT ADDED. COMPLETE IONIZATION.
NO STERN LAYER

Now $n \approx n_i$ and this means that there are special relationships between Γ, ψ_0 and n_i. Let us, in spite of the fact that we are discussing the non-saturated region, assume that the high potential approximation for σ_{Gouy} applies and let us neglect n with respect to n_i. Then, for $z_i = -1$ and $z_j = +1$

$$\sigma_{\text{Gouy}} = +\sqrt{\left(\frac{1}{2\pi}\,\varepsilon n_i kT\right)} \exp \frac{-e\psi_\delta}{2kT} \quad (-\psi_\delta \gg 2kT/e), \quad (5.2.14)$$

i.e. those cations, present in the solution serve to participate in the Gouy layer. Concerning the surfactant ions, assume a Boltzmann relationship as follows:

$$\Gamma = \frac{\Gamma_{\text{surf}}}{N_0}\,n_i \exp \frac{-(\varphi_i + z_i\,e\psi_0)}{kT} \quad (5.2.15)$$

where $(\Gamma_{\text{surf}}/N_0)$ is a normalization factor with dimension of length, and where φ_i is the specific adsorption energy. In the absence of a Stern layer the requirement of electrical neutrality is $(z_i = -1)$:

$$\sigma_{\text{Gouy}} = -\sigma = e\Gamma. \quad (5.2.16)$$

If we insert eq. (5.2.14) and eq. (5.2.15) into eq. (5.2.16) it is seen that n_i is proportional to $\exp -3\,e\psi_0/kT$. We again identify ψ_0 and ψ_δ. For $-F_{\text{el}}$ one finds

$$-F_{\text{el}} = \tfrac{1}{2}\Gamma kT \quad (5.2.17)$$

and for the surface pressure

$$\pi = \tfrac{3}{2}\Gamma kT. \quad (5.2.18)$$

The pertinent adsorption equation is

$$\Gamma = A^{\frac{1}{3}}n_i^{\frac{2}{3}} \quad (5.2.19)$$

where the dimensionless quantity A is

$$A = \frac{\Gamma_{\text{surf}}}{N_0} \cdot \frac{\varepsilon kT}{2\pi\,e^2} \cdot \exp \frac{-\varphi_i}{kT}. \quad (5.2.20)$$

Davies found experimental evidence[17,18] for this adsorption equation. It is easily seen that for swamping electrolyte the adsorption equation would have been

$$\Gamma = (An)^{\frac{1}{2}}n_i^{\frac{1}{2}} \quad (5.2.21)$$

but according to Davies there is no experimental evidence confirming this law. The equations of state considered so far are special cases of the Davies equation in which the corrections for the finite size of the adsorbed ions have been omitted.

Goddard *et al.*[33a] found surface pressures for spread palmitic-, stearic- and arachidic acid layers on aqueous electrolytes, which were lower than those predicted on the basis of a simple double layer theory as outlined by Davies. They invoked as a possible explanation the discreteness-of-charge or fluctuation effect (see Chapters 3 and 4).

(c) WEAKLY IONIZED ACIDS HA. CONSTANT Γ

Here molecules HA and ions A^- are assumed to be adsorbed. For Γ one has

$$\Gamma = (1-\alpha)\Gamma_{HA} + \alpha\Gamma_{A^-}. \qquad (5.2.22)$$

This case has been extensively treated by Payens.[34] The degree of ionization, α, was given by

$$\frac{1}{1-\alpha} = n_{H^+}^{-1} \exp \frac{-\Delta\mu_0 + e\psi_0}{kT} \qquad (5.2.23)$$

where $-\Delta\mu_0$ is given by standard chemical potentials as follows:

$$-\Delta\mu_0 = (\mu_{HA})_0 - (\mu_{A^-})_0 - (\mu_{H^+})_0 = kT \ln K \qquad (5.2.24)$$

K being the equilibrium constant.

It is not *a priori* certain that the standard chemical potentials have the same value in the bulk and at the interface. Payens found for stearylphosphonic acid, adsorbed at the petroleum ether/water interface, that $\Delta\mu_0(\text{bulk}) - \Delta\mu_0(\text{interface})$ was of the order of $-3kT$. Thus the effect of the interface was to decrease the value of the equilibrium constant. This effect was ascribed to a difference of hydration energies in the bulk and at the interface.

Owing to the charges of the adsorbed ions A^-, a term $\exp e\psi_0/kT$ is present in eq. (5.2.23). A new equilibrium constant K' may be used:

$$K' = K \exp \frac{-z_i e\psi_0}{kT}. \qquad (5.2.25)$$

We have reintroduced z_i because we want to stress the fact that z_i and ψ_0 have the same sign. Consequently $z_i\psi_0$ is always positive (see Payens) and $K' < K$. This is understandable because the dissociation leads to an extra repulsion between the adsorbed particles. Concerning the Gouy layer, Payens considered the case of a mixture of the acid HA and the salt NaCl in water. For n one has

$$n = n_{H^+} + n_{Na^+}. \tag{5.2.26}$$

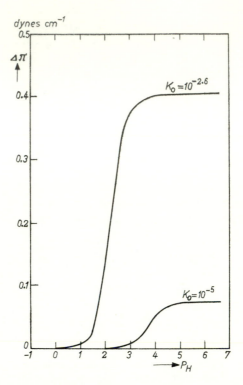

FIG. 5.9. Increase of surface pressure $\Delta\pi$ against pH of a subphase of weakly ionized monolayers with two values of K_0. Molecular area 100 Å². (1 dyne/cm $= 10^{-3}$ N m^{-1}) (from Payens[34]).

Without a Stern layer the charge in the Gouy layer can be written as $e\alpha\Gamma$ and we have the equation (ψ_0 and ψ_δ are identified)

$$\frac{-e\Gamma}{1+n_{H^+}K'} = \sqrt{\left(\frac{2}{\pi}\,\varepsilon(n_{H^+} + n_{Na^+})\right)} \sinh\frac{e\psi_0}{2kT}. \tag{5.2.27}$$

Curves of the interfacial pressure versus pH (Fig. 5.9) or versus pNaCl ($= -\log c_{NaCl}$) (Fig. 5.10) demonstrate the role of a varying degree of ionization α. Figure 5.9 shows that as the pH is increased, α increases and with it the value of $-F_{el}$. Figure 5.10 shows a maximum. The reason (Payens) is that the addition of NaCl has two effects: first, owing to an enhanced shielding of the A$^-$ ions by the added small ions, the ionization is facilitated and the interface charge is increased. Secondly, the addition of the small ions leads to an increase of the Gouy capacity, i.e. to a lowering of $|\psi_0|$ if σ is assumed constant. The two effects work in opposite directions. The increase in $\Delta\pi$ with c_{NaCl} at low NaCl concentration was confirmed experimentally, but no maximum

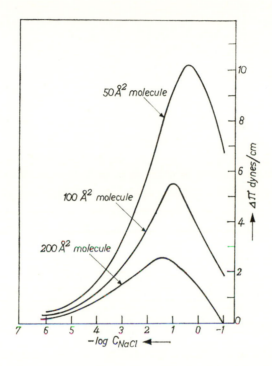

FIG. 5.10. Increase of surface pressure $\Delta\pi$ against log $(c_{NaCl})^{-1}$ ($= p_{NaCl}$) for three values of surface coverage. For all curves the same value 4·004 is taken for c_{H^+}/K_0 (1 dyne/cm $= 10^{-3}$ N m^{-1}) (Payens[34]).

was found, the value of the surface pressure continuously increasing with addition of NaCl. This behaviour points to an enhanced adsorption, the total amount of adsorbed particles being larger than assumed in the theory given here. If ions had been adsorbed in the Stern layer instead of in the Gouy layer, the double-layer capacity would have been higher than assumed in the theory and this would not have explained the effect. A theoretical basis for the interpretation of an increased adsorption must probably be sought in a further study of the precise location, and the energies involved, of the adsorbed particles.

5.3. Micelles

5.3.1. *The kinetic micelle. Experimental methods*

We consider here mainly charged micelles in which the number, n, of (ionized) monomers is about 50 or less. In these cases the micelles have a spherical shape. Micellar solutions then provide model systems for testing theoretical expressions for such phenomena as viscosity, electrophoresis or light scattering. Larger micelles have a non-spherical, often rod-like, shape and are briefly discussed below.

Three regions can be distinguished in each micelle :[35]

(a) a hydrocarbon core with a density, d_c, which is approximately equal to that of the liquid hydrocarbon. In most cases $d_c \approx$ 0·8 g cm^{-3};

(b) a Stern layer, thickness δ, which contains a fraction $(1-f)$ of the counter ions;

(c) a Gouy layer surrounding the micelle. We assume a shear surface placed between Gouy and Stern layer.

The hydrocarbon core forms, together with the Stern layer, the "kinetic micelle" and has a radius $(r+\delta)$, r being the radius of the hydrocarbon core (about the length of a hydrocarbon chain). Its charge is $+zenf$, where ze is the charge of a monomer ion while the Gouy layer has a charge $-zenf$.

The most important parameters are n, f, δ, and of course the critical micelle concentration (cmc). A number of ways are available for the determination of the cmc:[11] (1) Solubilization, preferably of a dye. There is a sharp rise of the solubility after the detergent concentration is increased beyond the cmc. (2) Osmosis. The osmotic coefficient decreases rapidly after the cmc has been reached. (3) Conductance. The same is true for the equivalent conductance. The conductance method can also be used for the determination of the micellar charge. (4) Surface tension. This tends to be constant at detergent concentrations above the cmc. (See Section 5.1.)

We discuss briefly three other methods which give more information, in particular for spherical micelles, concerning the four parameters just mentioned. These methods are those using: (4) viscosity, (5) electrophoresis and (6) light scattering.

VISCOSITY

The intrinsic viscosity of a rigid charged (kinetic) micelle is

$$\left[\eta\right] = \frac{5}{2}\frac{S}{d}\left(1+b_2\left(\frac{e\zeta}{kT}\right)^2\right) \tag{5.3.1}$$

where d is the density of the micelle, S is a shape factor, $S = 1$ for spheres, $S > 1$ otherwise; ζ is the potential at the shear surface (the zeta potential). The correction factor containing the coefficient b_2 describes the electroviscous effect. Booth[36] found

$$b_2 = \frac{\varepsilon kT}{\eta_0 \, e^2 u} f(\kappa a) \qquad (5.3.2)$$

where η_0 is the viscosity of the solution in the absence of micelles, ε the dielectric constant and u the ionic mobility. We have simplified Booth's expression in that it is assumed here that all the ions have the same mobility. Finally $f(\kappa a)$ in which $a = r + \delta$ describes the electroviscous effect as a function of κa. For large κa the electroviscous effect is small, as expected. Thus Booth found that $f(\kappa a) \approx 0.003$ for $\kappa a = 3$ whereas for $\kappa a = 0.3$, $f(\kappa a) \approx 0.008$. With $\varepsilon = 80$, $kT = 4 \times 10^{-14}$ erg, $\eta_0 = 10^{-2}$ poise and $w = 10^8$–10^9 cm^2/erg sec the value of $(\varepsilon kT/\eta_0 \, e^2 w)$ is of the order of 10. Measurements (sodium dodecyl sulphate micelles[35]) have indicated that the density d was smaller than d_c, the core density. Since $d_\delta < 1$ g cm^{-3} it seems improbable that the "kinetic micelle" contains a thick layer of fixed water molecules, i.e. the shear surface must be placed close to the hydrophilic heads. Stigter[35] found δ-values of 4–5 Å, the Stern layer containing only hydrated ions and ionic groups with open space between them. Stigter carried out his analysis after subtracting the electroviscous contribution from $[\eta]$. The remaining part, $[\eta]_0$, appeared to be independent of the salt concentration as should be expected. He found $[\eta]_0 = 5/2d \approx 3\frac{1}{2}$.

ELECTROPHORESIS

When a rigid non-conducting, charged, spherical particle, immersed in a liquid electrolyte, is brought under the influence of an electric field, it obtains an electrophoretic velocity, v, which we write

$$v = \frac{\varepsilon kT}{6\pi\eta_0 \, e} \, E f(\kappa a, y) \qquad y = \frac{e\zeta}{kT} \qquad (5.3.3)$$

where E is the field strength and $f(\kappa a, y)$ is the electrophoretic mobility in dimensionless units.

For small κa ($\kappa a \lesssim 0.1$), $f(\kappa a, y) \to y$, and eq. (5.3.3) reduces to the Hückel equation

$$v = (\varepsilon\zeta/6\pi\eta_0)E. \qquad (5.3.3')$$

An elementary consideration would provide v as the result of the compensation of two forces, an electric force $f_1 = QE$ where Q is the particle charge, and a hydrodynamic force $f_2 = -6\pi\eta_0 av$ (Stokes' law).

This simple calculation would give too large a value of v, because the effect of the ions in the Gouy layer is ignored. The Gouy layer has a charge $-Q$ and it tends to drive back the particle with a force[37] $f_3 = (\varepsilon \zeta a - Q)E$. This is the electrophoretic retardation. The condition $f_1 + f_2 + f_3 = 0$ leads to Hückel's equation.

For large κa ($\kappa a \gtrsim 100$), $f(\kappa a, y) \rightarrow (3/2) y$ and eq. (5.3.3) reduces to Smoluchowski's equation

$$v = (\varepsilon \zeta / 4\pi \eta_0)E. \tag{5.3.3''}$$

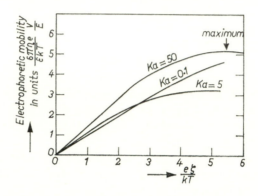

FIG. 5.11. Electrophoretic mobility as a function of the zeta potential for three values of κa. (After calculations of Wiersema.[40])

The difference from Hückel's equation, a factor $3/2$, is due to the fact that Hückel assumed the lines of force in the liquid not to be deformed by the moving particle whereas Smoluchowski let them run parallel to the particle surface. Therefore the electrophoretic retardation is, in Smoluchowski's case, less important than in Hückel's case.

Overbeek,[38] Booth[39] and most recently Wiersema[40] incorporated a fourth force in the calculation of v. This force arises from the deformation of the ionic atmosphere around a particle when the particle moves. It tends to restore the equilibrium situation in which the centres of the spherical particle and of the surrounding atmosphere coincide. In electrolytes a temporary deformation of an ionic atmosphere has a typical relaxation time $\tau = (w\kappa^2 kT)^{-1}$. A similar tendency to restoration of the equilibrium situation is found in the deformed ionic atmosphere around a moving particle. This tendency is called the relaxation effect. For low y, combined with low κa (where Hückel's model is valid), the relaxation effect is small and it is also small for high κa where τ becomes small. The calculation of v, emerging from the condition $f_1 + f_2 + f_3 + f_4 = 0$, is very complicated. Wiersema carried out numerical computations based on the approach of Overbeek and obtained results for widely varying values of κa and y. Figure 5.11 gives some examples. The relaxation effect is particularly important for $\kappa a = 5$ (although less important than would be inferred from Overbeek's calculations which were only valid for $y < 1$). In Fig. 5.11 there is a weak maximum for $\kappa a = 50$ at about $y = 5\frac{1}{2}$. Now we have seen in Chapter 4 that AgI particles may have an electrophoretic mobility which shows a maximum at a given value of the surface potential. Levine *et al.* interpreted such a maximum along the lines of a discreteness-of-charge theory. Wiersema pointed out, however, that the relaxation effect may also explain such a maximum.

Electrophoresis experiments (and conductance measurements) led for a number of cases (dodecyl ammonium chloride, sodium dodecyl sulphate[35]) to f-values of about $\frac{1}{2}$. That means that about half of the counter ions are present in the Stern layer.

LIGHT SCATTERING

Debye[41] derived the following equation, valid for two-component systems (e.g. some micellar solutions, or a colloid and a solvent),

$$\frac{Hg}{\tau - \tau_0} = \frac{1}{RT}\frac{d\pi}{dg} \tag{5.3.4}$$

where τ is the turbidity† (dimension length^{-1}), τ_0 the turbidity of the solvent, g is the concentration in g cm^{-3} ($g = cM$) of the dissolved component, π is the osmotic pressure against the solvent, and

$$H = \frac{32\pi^2 n^2 (dn/dg)^2}{3\lambda_0^4 N} \tag{5.3.5}$$

in which n is the refractive index of the solution measured at a wavelength (in vacuum) λ_0. Debye arrived at eq. (5.3.4) by combining Rayleigh's scattering equation for gases with Einstein's fluctuation theory. Typical values for τ^{-1} in air are of the order of 100 km, whereas in liquid systems $\tau^{-1} \approx 1$ km or less. This means that the intensity of a light beam is reduced to $1/e$ of its original value, I_{pr}, after transmitting 100 km through air or 1 km through a liquid. The scattered light is often observed in a direction normal to that of the transmitting beam at a given distance r. Denoting the observed light intensity in this

---

† There should be no confusion between the use of τ for turbidity here and its use elsewhere for relaxation time. In both cases the symbol "τ" is the established notation.

direction by I_{90} the turbidity is approximately given by (see Doty and Steiner[42]) $I_{90}r^2/I_{pr}$.

For the osmotic pressure one writes (see Chapter 2)

$$\pi = \frac{g}{M} RT + B_2 RT g^2. \tag{5.3.6}$$

where B_2 is the second osmotic virial coefficient and M the molar mass. Thus eq. (5.3.4) becomes

$$\frac{Hg}{\tau - \tau_0} = \frac{1}{M} + 2B_2g. \tag{5.3.7}$$

Measurements of the refractive index as a function of the concentration g, together with turbidity measurements, give the values of M and B_2 which is most easily seen by plotting $Hg/(\tau - \tau_0)$ against g (Debye plot).

Zernike[43] extended Debye's treatment to multicomponent systems. This extension is pertinent to the case of charged micelles because here at least three components are present: detergent (1), salt (2) and solvent (3). It was shown that in that case the refractive index increment (dn/dg in eq. (5.3.5)) must be measured at constant μ_2. However, the experiments usually give values of the increment at constant concentrations g_2. This leads to a modification of the process used for obtaining the molar mass.[44] Denoting the true molar mass of a micelle by M_1 and the apparent molar mass by M_1^* (this is measured at constant concentration) then the modification is expressed as follows:

$$M_1^* = M_1 \left[1 + \frac{\left(\frac{\partial n}{\partial g_1}\right)_{g_2} \left(\frac{\partial g_2}{\partial g_1}\right)_{\mu}}{\left(\frac{\partial n}{\partial g_2}\right)_{g_1}} \right]^2 \tag{5.3.8}$$

in which $g_i = c_i M_i (i = 1, 2)$. Usually

$$\left(\frac{\partial g_2}{\partial g_1}\right)_{\mu_2} = \frac{M_2}{M_1} \left(\frac{\partial c_2}{\partial c_1}\right)_{\mu_2}$$

is negative and therefore $M_1 > M_1^*$, the correction amounting to 10–25%.

For a three-component system such as Na-lauryl sulphate (1), NaCl (2), water (3) we may write

$$c_2 = c_1 n_2{}^s + c_2{}^0 \tag{5.3.9}$$

where $n_2{}^s$ is the number of NaCl molecules adsorbed per detergent micelle of molecular weight M_1, and $c_2{}^0$ is the NaCl concentration in the absence of detergent. For non-specific adsorption $n_2{}^s$ is a negative quantity. This is due to the charge on the micelle. For complete ionization in the absence of any specific adsorption the micellar charge is $-en_1$ where n_1 is the number of monomers in the micelle. The compensating charge is $e(n_{Na}{}^s - n_{Cl}{}^s)$. It is customary to use a parameter

$$\alpha = \frac{-n_{Cl}{}^s}{n_{Na}{}^s - n_{Cl}{}^s} .$$

In the flat plate approximation there is a low potential limit of α of $\frac{1}{2}$, whereas for high potentials $\alpha \to e^{-\frac{1}{2}|z|}$. (Here $|z| = |e\psi_0|/kT$). There naturally is a deficiency of co-ions near a charged wall. Therefore $n_{Cl}{}^s < 0$. Since $n_{Cl}{}^s = n_2{}^s$ we have

$$\left(\frac{\partial c_2}{\partial c_1}\right)_{\mu_2} = -\alpha n_1 \quad \text{or} \quad \left(\frac{\partial g_2}{\partial g_1}\right)_{\mu_2} = -\alpha \frac{M_2}{M_1} \qquad (5.3.10)$$

where M_1 is the monomer molar mass.

For incomplete dissociation of component 1, or for adsorption in the Stern layer of counter ions, there is no essential change in the treatment, because the fact remains that there must be a deficiency of the co-ions near the charged micelle. For sodium lauryl sulphate micelles αn_1 is of the order of 10.[45,46] Only in exceptional cases (HCl on half-neutralized polymethacrylic acid[44]) is there a positive instead of a negative adsorption.

When the number of monomers per micelle increases beyond 50–100, the micellar shape is no longer spherical. Information concerning the shape has been obtained by measuring the dissymmetry of the light scattered around the observation angle of 90°. For non-spherical particles the ratio of the intensities I_θ^i and $I_{180-\theta}$ ($\theta < 90°$) is[47] larger than unity. Semiquantitative information concerning the shape can also be obtained from viscosity measurements (see eq. (5.3.1)). A rod-like shape seems to be energetically the most favourable. It has been found in a number of cases, for example: n-hexadecyltrimethyl-ammonium bromide (Debye and Anacker[47]), dodecyl ammonium chloride (flexible rods, Stigter[48]).

The diameter of the rods must be expected to be about twice the length of a monomer hydrocarbon chain, the hydrophilic groups pointing to the aqueous phase. Halsey[49] pointed out that thermal agitation prevents such rods from assuming an unlimited length.

5.3.2. *Theory*

The formation process and the stability of micelles have received much attention in the past. Two main lines of approach were followed:

the mass-action approach and the two-phase approach. Statistical treatments[50] embrace these two lines.

(A) MASS-ACTION APPROACH

The micelle is considered as an "n-mer" L_n, formed through the association of n monomers L_1:

For uncharged monomers

$$nL_1 \leftrightarrows L_n. \tag{5.3.11}$$

Chemical equilibrium requires that

$$n\mu_1 = \mu_n \tag{5.3.12}$$

where we write

$$\left. \begin{aligned} \mu_1 &= \mu_{10} + kT \ln c_1, \\ \mu_n &= \mu_{n0} + kT \ln c_n, \end{aligned} \right\}, \tag{5.3.13}$$

i.e. ideal behaviour is assumed. The fraction x_n of monomers present in the micelles is proportional to nc_n. We make the simplifying assumption that the system contains only monomers and n-mers: $x_1 + x_n = 1$. Some justification of this assumption is given below. From eqs. (5.3.12) and (5.3.13) we derive

$$\left(\frac{1}{n} \mu_{n0} - \mu_{10} \right) = -kT \ln \frac{c_1}{c_n^{1/n}} . \tag{5.3.14}$$

Here $(1/n \ \mu_{n0} - \mu_{10})$ is the free energy difference per monomer involved in the micellization.

The mass-action approach explains the existence of a cmc. Upon plotting c_1 against c_n for a certain, not too small value of n ($n > 20$) it is seen (Fig. 5.12) that at $c_1 = nc_n$ an abrupt rise of c_n sets in. This value of c_1 may be identified with the cmc. Introducing

$$\ln K = (\mu_{n0} - n\mu_{10})/nkT + \frac{1}{n} \ln n \tag{5.3.15}$$

we have

$$\ln K = \frac{n-1}{n} \ln \text{cmc} \sim \ln \text{cmc} \ (n \gg 1). \tag{5.3.16}$$

Beyond the cmc it is difficult to increase further the monomer concentration. Addition of monomers merely leads to the formation of more micelles (Fig. 5.12).

For ionized micelles the treatment must be somewhat modified. Assume that L_1^- is the anion of a long-chain paraffin salt ML. Thus also cations M^+ are involved in the association reaction

$$nL_1^- + (n-p)M^+ \rightleftharpoons L_n^{-p}. \qquad (5.3.17)$$

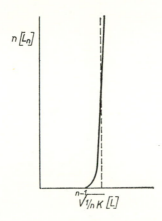

FIG. 5.12. Concentration of micelles $[L_n]$ $(=c_n)$ against monomer concentration $[L_1]$ $(=c_1)$ for $n = 20$ (From Overbeek, *Chem. Weekblad*, 1958, **54**, 687.)

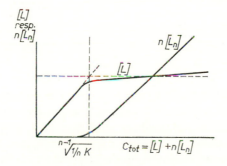

FIG. 5.13. Concentration of micelles $[L_n]$ and monomer concentration $[L_1]$ against total concentration for $n = 20$ (Overbeek, *Chem. Weekblad*, 1958, **54**, 687).

Introducing

$$\ln K = \left(\mu_{n0} - n\mu_{10} - \left(1 - \frac{p}{n}\right)\mu_{M0}\right) + \frac{1}{n}\ln n \qquad (5.3.18)$$

where, still ideal behaviour being assumed, μ_{M0} is a standard chemical potential in

$$\mu_M = \mu_{M0} + kT \ln c_M. \qquad (5.3.19)$$

Following the same procedure as for non-ionized n-mers we have

$$\ln K \cong \left(2 - \frac{p}{n}\right) \ln \text{cmc} \qquad (n \gg 1). \qquad (5.3.20)$$

Construction of Fig. 5.14 which is analogous to Fig. 5.13 for non-ionized micelles shows that there is a maximum monomer concentration just beyond the cmc. This is due to the influence of the cations M^+ which increases if more salt ML is added. It may be shown that the mass-action approach also explains the lowering of the cmc when other

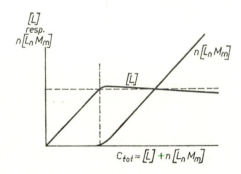

FIG. 5.14. Concentration of micelles $[L_n]$ and monomer concentration $[L_1]$ against total concentration for $n = 20$, whereas each micelle also contains monovalent cations ($m = n - p = 10$) (Overbeek, *Chem. Weekblad*, 1958, **54**, 687).

salts than ML, but containing the cation M^+, are added. Finally, according to Mukerjee,[10,51] the influence of the cations may explain why Traube's rule may not be fully obeyed for ionized micelles. Assuming that for ionized *and* non-ionized micelles the addition of one CH_2-group per monomer leads to a decrease of K of a factor 3, this decrease corresponds no longer to a decrease of the cmc of a factor 3 when the micelles are ionized. A factor 4 has sometimes been found[10] which means that $p = 0\cdot6n$, indicating that the micelles are about half-neutralized.

So far we have considered n as a given constant. We now inquire into the nature of this assumption. For this purpose consider again the association of n monomers (eq. (5.3.11)) and write

$$\frac{1}{n}\mu_{n0} - \mu_{10} = \Delta\varepsilon - T\Delta s \qquad (5.3.21)$$

where $\Delta\varepsilon$ is the energy difference and Δs the entropy difference involved in the micellization. We expect both $\Delta\varepsilon$ and Δs to be a complicated function of n. For hypothetical disc-shaped charged micelles Debye[52] assumed

$$\Delta\varepsilon - T\Delta s = a - b\sqrt{n} \qquad (5.3.22)$$

where a and b were negative constants; na was the energy gained upon micelle formation (Debye assumed mainly Van der Waals energy) and $nb\sqrt{n}$ was the electrical energy lost in putting n charges into a disc-shaped arrangement. He then considered the minimum value of $n(a - b\sqrt{n})$. The value of n so obtained was considered to indicate the size of a stable micelle. Although this procedure is not correct in principle because $\partial \ln x_n/\partial n \neq 0$ when $n(a - b\sqrt{n})$ is at its minimum value, and that the theory does not apply to non-ionized micelles, the combination of a free energy gain and a free energy loss, which are different functions of n, may be the clue to the explanation of the existence of micelles with finite n. As a semi-empirical approach we maintain a term na favouring the micellization and we introduce a term $nb'kT \ln n$, where b' is a constant, preventing the unlimited growth of the micelles. This term, although its mathematical form is rather arbitrary, arises from such influences as the thermal agitation invoked by Halsey[49] and the frustrations caused by the misfits of the molecules at the micelle/water boundaries. Hoeve and Benson[50] also used a logarithmic term but gave it a more direct, kinetic, interpretation. We write, using fractions x_1 and x_n instead of concentrations c_1 and c_n (and thereby introducing trivial constants which, however, we shall omit),

$$\mu_{n0} - n\mu_{10} = na - nb'kT \ln n = nkT \ln x_1 - kT \ln \frac{x}{n}n. \qquad (5.3.23)$$

The value of n where x_n has its maximum value is determined by the condition $\partial \ln x_n/\partial n = 0$. Neglecting terms with n^{-1} we have

$$\ln x_1 = \frac{a}{kT} - b' - b' \ln n. \qquad (5.3.24)$$

The fraction x_1 in this equation is proportional to the cmc. There is a qualitative agreement between the experimental findings and eq. (5.3.24).[53] Overbeek and Stigter[54] showed that the influence of the double layer upon the free energy of formation of the micelle is relatively small. They wrote

$$\mu_{n0} - n\mu_{10} = n(\Delta\varepsilon - T\Delta s) + F_{el} \qquad (5.3.25)$$

where F_{el} is the free energy needed for building up a double layer around the micelle and where $(\Delta\varepsilon - T\Delta s)$ is the same contribution as that used for non-ionized micelles. The contribution $F_{el}(= nf_{el})$ can be written as $A \int \psi \, d\sigma$ where $A = 4\pi\alpha^2$ is the interfacial area. For a precise calculation of F_{el} tables of Loeb *et al.*[55] are available, but qualitative estimates can rapidly be made. In contrast to Debye's concept, F_{el}, which is a generalization of Debye's $-bn\sqrt{n}$ term, is only a contribution of secondary importance. Thus, for micelles surrounded by a Gouy layer, f_{el} is typically of the order of $4\,kT$, whereas $(\Delta\varepsilon - T\Delta s)$ is typically of the order of $-13\,kT$. This value is found by a subtraction procedure: once the cmc, the micelle concentration and n are known, $\mu_{n0} - n\mu_{10}$ can be calculated, and eq. (5.3.25) gives $(\Delta\varepsilon - T\Delta s)$ after subtraction of the calculated value of F_{el}.

TABLE 5.3

(Na dodecyl sulphate and NaCl)[53]

c_{NaCl}† mol/cm³	cmc‡ mmol/cm³	n	F_{el}/nkT§	$(\Delta\varepsilon - T\Delta s)/kT$	B cm³/g	κa
0·00	8·14	57·3	4·41	−13·04	9·24	0·534
0·01	5·60	64·2	4·32	−13·37	4·10	0·780
0·03	3·13	70·8	3·91	−13·59	1·95	1·136
0·1	1·47	93·4	3·27	−13·86	0·60	3·193
0·3	0·66	123·0	2·66	−14·18	0·23	5·130

† Activity coefficients were introduced for the calculations in the table.
‡ At the cmc 2% of the detergent was assumed to be in micellar form.
§ The tables of ref. 55 were used.

Table 5.3, taken from the work of Huisman,[53] gives an overall impression of the values of the parameters pertinent to micelle systems. The values of the association number n and the second virial coefficient B were derived from light-scattering measurements. We see that, as the NaCl concentration is increased, the cmc decreases and n increases. At the same time B decreases rapidly, i.e. the micelle–micelle repulsion which is rather high in the absence of salt (small κa) is insignificant at $c_{NaCl} = 0\cdot3$ M. The quantity $(\Delta\varepsilon - T\Delta s)$ tends to be constant with respect to c_{NaCl} variations.

There are several simplifying assumptions: the micelles are spherical, the surface charge is smeared out (discreteness-of-charge effects have been considered by Stigter[56]), the ions are point charges when present in the diffuse layer, the dielectric constant in the Gouy layer is that of pure water, and dimerization of detergent molecules has been left out. Such a dimerization has been taken into consideration by Mukerjee.[10] Another, probably important, problem is that of the distribution of the ions over a Stern and a Gouy layer. It was assumed in Table 5.1 that the ionization was complete and it was implicit that no counter ions were present in the "kinetic micelle". We have seen that this may not be the case. For all these reasons the tabulated values, in particular of F_{el}, are uncertain although of the correct order of magnitude.

There is a qualitative agreement between the approach using the equilibrium reaction, eq. (5.3.17), and the approach using the free energy F_{el}, but in the latter case the behaviour of the second virial coefficient B is more understandable. The concept of Gouy layers (leading to the estimated values of F_{el}) and their interpenetration (providing for the comprehension of B) is similar to that used for the interpretation of the stability of hydrophobic colloids.

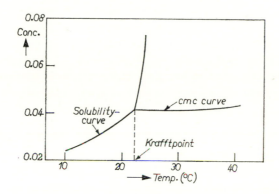

FIG. 5.15. "Phase diagram" for micelles, monomers and a solidified phase (from Dervichian[57]).

(B) TWO-PHASE APPROACH

Micelles have a number of properties in common with a phase. Thus, it is possible to construct a diagram (Fig. 5.15) which is similar to a phase diagram. Three regions of stability can be distinguished. These regions are similar to phases: a micellar phase, a monomer phase, and a crystalline phase, the "curd". The triple point is called Krafft point. By studying Krafft points of mixed detergents, Raison and Dervichian[57] found eutectic phenomena. Thus, the Krafft points of mixtures of sulphate detergents with C_{12} chains, with sulphate detergents with C_{10} or C_{14} chains, were lower than those of the separate components, the minimum being at a composition of $3C_m + 1C_{m+2}$.

For two different phases the Clausius–Clapeyron equation can be used. It reads, for the case of micelles in a detergent solution,[58-60]

$$-\frac{\Delta h}{RT^2} = \alpha \left(\frac{\partial \ln \text{cmc}}{\partial T} \right)_p \tag{5.3.26}$$

where Δh is the enthalpy change of reaction (5.3.11) and $\alpha = 1$ in "swamping electrolyte" and $\alpha = 2$ if only detergent molecules are dissolved. In other cases, for instance weak ionization, $1 < \alpha < 2$.

The Clausius–Clapeyron equation provides one way of obtaining Δh. Another way is by direct calorimetric measurement of the heat of dissolution of detergents above and below the cmc. It appears that Δh measurements are, on the whole, rather uncertain (see also Mukerjee[10]) and it seems premature to draw definite conclusions from them regarding, for instance, the entropy contribution to the micellization powers.

5.4. Thin films

Our attention is primarily directed toward films with a thickness less than about 1000 Å. These thin but stable films are called black films because light is only weakly reflected by them. The reason for selecting them for study is that they can be made understandable on the basis of the DLVO theory. Two kinds of black film can be distinguished. The "first black film" has a stable thickness which, depending on the electrolyte concentration, lies between about 60 Å and about 1000 Å. The "second black film" has a thickness of 40–50 Å depending on the chain length of the surfactant molecule used and on the specific nature of the (small) ions in the solution. Already Hooke,[61] in 1672, was acquainted with the existence of black films, or rather, with black spots ("holes") in soap films. Newton[62] observed black spots of different thicknesses in the same soap film. Many historical details can be found in the book by Mysels et al.[22] and also in the reviews by Overbeek[63] and Scheludko.[64]

Black films can be formed in a variety of ways, by using microscopic (Derjaguin et al.[65]) or macroscopic (Mysels et al.[66]) methods. Scheludko[64] gives details of most methods. One way (a macroscopic method) consists in lifting a gloss frame from a surfactant solution in which it has been submerged previously, in a vertical position with a speed which is of the order of microns per second. Some results of this method are shown in Fig. 5.16 for different electrolyte concentrations. We see that the black film thickness decreases with increasing concentration and that the speed of withdrawal is sufficiently slow to ensure equilibrium values of the thickness.

In passing we mention that upon removal with a *high* speed, films are formed whose initial thickness is of the order of 1 μm.

Mysels et al.[22] distinguished between "mobile" films and "rigid" films. Rigid films have a high surface viscosity. Just after formation, the thickness of a rigid film is about 10 μm. Thinning here is a slow process determined by the rate of drainage of the intralamellar solution.

Rigid films can be prepared from solutions of sodium dodecyl sulphate somewhat below the cmc, with lauryl alcohol added to concentrations of 10% of the sulphate concentration. Such a mixture may lead to ordered structures, either in the bulk (after prolonged standing) or at the surface. Black films can be formed by drainage of the intralamellar solution.

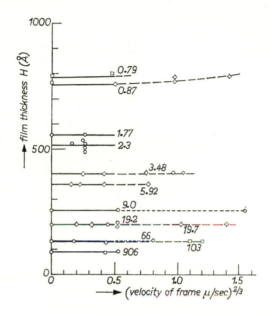

FIG. 5.16. Film thicknesses at low frame velocities are essentially constant and equal to those obtained by static drainage and shown on the vertical axis. Surfactant: sodium n–C_{12} sulphate. Salt: LiCl. The numbers at the curves are concentrations of counterions in mmol/1.

  ☐   Na n–C_{12} sulphate surfactant.
  ○   mobile film.
  ◇   rigid film (C_{12} = dodecyl = lauryl.)

(From Lijklema and Mysels.[70b])

Mobile films are the more common form. They have a low surface viscosity and are prepared from, for instance, solutions of pure sodium lauryl sulphate, of sodium alkyl benzene sulphonates, of pure dodecyl amine hydrochloride and so on, the concentrations being of the order of the cmc. According to Scheludko[64] the minimum surface concentration necessary to prevent bursting is of the order of 10^{13} molecules per cm^2. The rate of thinning is high, i.e. in the course of a minute or so, the film is completely black. The thinning mechanism has been extensively discussed by Mysels *et al.*[22] and will not be touched upon here. We merely mention that thinning does not mean a simple drainage of the intralamellar solution via the border region at the glass frame (i.e. the Plateau border) into the mother solution, but that the *whole* film is "sucked" into the border and is constantly renewed.

The "second black film" is formed when evaporation of the solvent can take place and/or when the electrolyte concentration is high, i.e. of the the the order of 1 mol/cm^3.

We now consider the stability problem of the (first) black film. We have seen in Section 5.2 that the hydrophilic heads of surfactant molecules are usually ionized and that the surface charge formed by them is largely compensated by a Gouy layer. A soap film contains two such layers opposing each other. When the intralamellar distance, or core distance is of the order of a Debye length, there is interpenetration of the two Gouy layers and we have largely the situation of two charged colloidal particles in close proximity, as described in Chapter 4. Just as in that case the overlap of two Gouy layers leads to a repulsion, the repulsion potential energy V_R being in the flat plate approximation, which is very appropriate for thin films

$$V_R = 64\kappa^{-1} \operatorname{tgh} \frac{z}{4} nkT \, e^{-2\kappa d}. \tag{5.4.1}$$

In fact, this is eq. (4.3.43), $2d$ being the distance between the planes of the hydrophilic heads.

Also, as first noted by De Vries,[67] there are Van der Waals forces which tend to make the film thinner. The formal treatment runs along the same lines as given in Chapter 4, where two thick plates of material 1 were separated by a medium of material 2 at a distance H and where the attractive potential energy V_A was given as

$$V_A = -\frac{\pi^2}{12\pi n_2^4 H^2}(\lambda_1 q_1{}^2 + \lambda_2 q_2{}^2 - \lambda_{12} q_1 q_2) \tag{5.4.2}$$

where the symbols have the same meaning as in Chapter 4. In the case considered in this section "material 1" is air. Therefore the density q_1 is almost zero. Equation (5.4.2) is strictly applicable only when "material 2" is homogeneous. However, thin films are not homogeneous. They consist of two monolayers and a solution layer. This alters somewhat the energy–distance curve, but we assume eq. (5.4.2) to be valid because the refractive index and the Hamaker–De Boer constant of water does not appreciably differ from that of hydrocarbons.

Since "material 1" has a much lower density than "material 2" the physical basis of the tendency to a thinner film as represented by V_A may not be immediately clear. However, the basis of this tendency is still the same: owing to Van der Waals forces the molecules tend to surround themselves with other molecules. This tendency can be obeyed by a local thinning combined with an increased thickness elsewhere.

Equations (5.4.1) and (5.4.2) form the basis of the DLVO theory. In addition it is possible in the thin film case, to exert a hydrostatic pressure normal to the plane of the film. The equilibrium condition becomes

$$p_R + p_A + p = 0 \qquad (5.4.3)$$

where p_R and p_A are pressures given by

$$p_R = \frac{dV_R}{dH}; \quad p_A = \frac{dV_A}{dH}$$

and where p is the external hydrostatic pressure. Derjaguin[68] introduced the notion of a "disjoining pressure" which is the pressure suffered by two solid bodies immersed in a liquid medium when they are pressed together, and exerted by the liquid layer remaining between them. In terms of eq. (5.4.3), the disjoining pressure can be put equal to $-p$. In a number of cases the Van der Waals pressure p_A can be neglected with respect to p. Thus, with $A = 10^{-20}$ J ($= 10^{-13}$ erg), $n_2 = 1 \cdot 3$ and $H = 3000$ Å the pressure p_A becomes 10 N m^{-2} ($= 10^2$ dyn cm^{-2}), whereas p-values of 10^5 N m^{-2} ($= 10^6$ dyn cm^{-2}) can be reached (see below).

Optical reflection methods are commonly used to measure the thickness H. Assuming a layer with a uniform refractive index n, and assuming the absence of secondary reflections, the ratio I_i/I_r of the intensities of an incident and a reflected monochromatic light beam is, at normal incidence, given by Rayleigh's equation:[69]

$$\frac{I_i}{I_r} = 4\left(\frac{n-1}{n+1}\right)^2 \sin^2 (2\pi n H_{app}/\lambda_{vac}) \qquad (5.4.4)$$

where λ_{vac} is the wavelength in vacuum and H_{app} is an apparent thickness which, owing to the assumption of a uniform film, is not necessarily equal to H.

Corrections for secondary reflections were already introduced by Rayleigh. Corrections for the non-uniformity of the film can also be introduced.[64] Thus, Mysels and Jones,[70a] working with dodecyl sulphate films, assumed that the film was a three-layered sandwich, the two monolayers having a refractive index of 1·45 (at $\lambda_{vac} = 5460$ Å) and a thickness of 8·5 Å. This led to a correction $H_{app} - H$ of 7·25 Å. Figure

5.17 gives some of the results of Mysels and Jones. They used a cell in which the film was held in a porous ring and in which pressures p of 10^5 N m^{-2} ($= 10^6$ dyn cm^{-2}) could be maintained. The ring was impermeable to air up to pressures determined by the pore size, but it was freely permeable to liquids. The porous ring was in contact with a capillary leading to the outside of the cell. The capillary served to

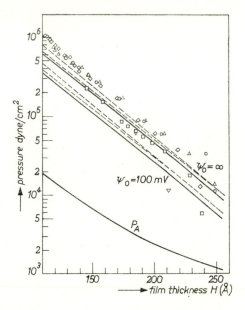

FIG. 5.17. The effect of pressure upon the thickness of films of 0·0094 M Na dodecylsulphate solutions. Van der Waals forces (pressure P_A) make a significant contribution (heavy line) to the exact or approximate calculated electrical forces (dashed lines). For H-values below 150 Å the experimental points are systematically higher than the theoretical lines (from Mysels and Jones[70a]).

supply the solution and finally to form the film. The thickness H was measured as a function of p. The lines in the figure are provided by theory (eqs. (5.4.1)–(5.4.3)). It is seen that the measured points tend to be too high in comparison with the theoretical lines. In this respect we note that eq. (5.4.1) (of 5.4.2) contains the factor $\exp(-2\kappa d)$ and not the smaller factor $\exp(-\kappa H)$. The relation between d and H is

$$H = 2(d+t) \tag{5.4.5}$$

where t is the thickness of a monolayer.

Especially for small H a precise knowledge of t is very important. When t was taken as 12·5 Å instead of 8·5 Å, corresponding to a larger double-layer repulsion, good agreement with the DLVO theory was obtained. (The optical correction increased from 7·25 Å to 10·25 Å.) Incorrect estimates of t are probably the basis of the often discussed[64] disagreement between theory and experiment.

When we pass on to consider the "second black film", the question arises as to whether a water core is still present. Derjaguin et al.[65] concluded that a water layer of a thickness equal to several molecular diameters is present.

Now it is difficult to establish a final equilibrium state for such very thin films, while in addition, as we have seen, the interpretation of the optical data is not unambiguous. Therefore the problem is not yet solved[64] but from a number of recent papers (for example, Jones et al.[71] and ref. 64) it seems probable that a water core is present of about 20 Å thickness or so. Since these films are usually obtained from concentrated electrolytes the water core will contain many ions.

In terms of the DLVO theory, the "second black film" may be related to the "first minimum", while the "first black film" corresponds to the "second minimum". It must be borne in mind, however, that the minimum film thickness cannot become zero as we have in the particle–particle model of Chapter 4. It has a minimum value of at least $2t$. This makes the minima shallower than in the original DLVO theory. When the monolayers constituting the film are uncharged, no second black stable films can be expected. Films may be stabilized in the form of a first black film, or collapse. However, experiments with neutral surface active agents[72,73] have shown that also in these cases stable first black films may be obtained. Clunie et al.[73] applied the old[74,75] method of the measurement of the conductance of the film parallel to the surface and found zeta potentials up to 100 mV for films prepared from "C_{16} sultaine" ($C_{16}H_{33} \cdot N^+(CH_3)_2 \cdot (CH_2)_3 \cdot SO_3^-$). In this method the measured conductance is compared with the conductance value which is obtained under the assumption of zero surface charge. The measured conductance is usually larger and the excess conductance is ascribed to ions which are necessary for the compensation of the surface charge (formed by preferentially adsorbed anions in the C_{16} sultaine case[73]).

In spite of all the work devoted to the exploration of the stability of thin films it is not surprising that much attention has been given to the process and mechanism of rupture. Rupture of a film may occur when for some reason a very thin region is formed. In that region V_A (see eq. (5.4.2)) tends to a very large negative value which may eventually dominate energies which tend to a restoration of a uniform thickness. The surface free energy, which is proportional to the surface area, provides such a tendency, whereas the double-layer repulsion tends to eliminate thin regions. Scheludko[76] presented a model, worked out by Vrij[77] and Vrij and Overbeek,[78] which was based on the competition between the Van der Waals energy and the surface free energy.

The Van der Waals energy V_A for a film with a uniform thickness H is given by $-A/12\pi H^2$. When the film has a non-uniform thickness, varying between $H+\frac{1}{2}\Delta H$ and $H-\frac{1}{2}\Delta H$, then V_A can be written as $-A/12\pi H^2[1+a(\Delta H/2H)]^2$ where now H is an average thickness and a is a numerical factor. A similar calculation for the non-uniformity correction of the double-layer repulsion, ΔV_R, leads to $\Delta V_R = a'V_R \cdot \frac{1}{2} \cdot (\frac{1}{2}\kappa\Delta H)^2$, where $a' = a$, provided eq. (5.4.1) is used with $\kappa\Delta H \ll 1$. When thickness variations smaller than ΔH are absent, $a = 3$. When the film surfaces are parallel (although they may be curved), then $a = 0$. For sinusoidal profiles of the surfaces with opposite maxima and minima, then according to Vrij and Overbeek[78] (who considered only V_A) $a = 3/2$. This profile arrangement was also used by these authors to calculate the increase $2\Delta A_s$ of the two surface areas of the film which had a geometric surface area of A_s. Usually $\Delta A_s/A_s$ can be written in the form $[\{1+b(\Delta H/2\lambda)^2\} - 1] = +b(\Delta H/2\lambda)^2$ where b is a geometric factor and λ the wavelength, the repeated distance along the surface with sinusoidal profile. This factor, b, has no relation to the factor a defined above. For a sinusoidal film b was found to be $\frac{1}{2}\pi^2$ and the increase of the surface free energy per unit area is equal to $\frac{1}{2}\pi^2\gamma(\Delta H/2\lambda)^2$. Here the wavelength is an important parameter, which is not so for the V_A and the V_R corrections. This is obvious. When half of the film has a thickness of $H+\frac{1}{2}\Delta H$ and the other half has a thickness of $H-\frac{1}{2}\Delta H$, the correction $\Delta V_A = aV_A(\Delta H/2H)^2$ irrespective of the distribution of the regions of large and small thickness. Thus, a "chess-board" distribution with many "white blocks" having a small thickness and many "black blocks" having a large thickness (this situation corresponds to a small λ) leads to the same values of ΔV_A and ΔV_R as a film consisting of only one large "white block" and only one large "black block".

The Van der Waals forces tend to a rupture of the film whereas the double-layer interaction and the surface tension tend to stabilize it. When the opposing tendencies are equal, the following condition is obeyed:

$$b\gamma\left(\frac{\Delta H}{\lambda_{\text{crit}}}\right)^2 + a\left\{V_A\left(\frac{\Delta H}{2H}\right)^2 + \frac{1}{6}V_R(\frac{1}{2}\kappa\Delta H)^2\right\} = 0. \qquad (5.4.6)$$

Here λ_{crit} is a critical wavelength; for $\lambda > \lambda_{\text{crit}}$ the tendency to rupture dominates, for $\lambda < \lambda_{\text{crit}}$ the fluctuations tend to die out although their spontaneous formation prevents ΔH from becoming zero.

Typically λ_{crit} may become of the order of 1 micron [e.g. with $H = 100$ Å, $\gamma = 10^{-2}\,N\,m^{-1}\,(=10\,\text{dyn cm}^{-1})$, $A = 10^{-19}\,J\,(=10^{-12}\,\text{erg})$ and ignoring the double-layer contribution]. In that case fluctuations with a wavelength larger than 1 micron lead to rupture. Time effects are here important: the larger the wavelength, the more time is needed for the necessary displacements of the (viscous) film material. Therefore the wavelength of the fluctuation which leads most rapidly to thinning is only slightly (a factor $\sqrt{2}$) larger than λ_{crit}.

External drainage may also be a cause for thinning, and the combination of the two thinning processes has provided a method to check the theory. Thus, assuming a circular film with radius r, and taking the suction pressure (which leads to drainage) the same for all r-values then, according to the theory of Vrij and Overbeek, in which for that matter the double layer plays a subordinate role, the film thickness at the moment of bursting was proportional to $r^{2/7}$, in agreement with experiments by Exerowa and Kolarov.[79,64] Surface elasticity was ignored in the theory. Whereas the density of bulk elastic waves, or Debye waves, is proportional to λ^{-3}, the density of surface elastic waves tends to be proportional to r^{-2}.[80] It is improbable[78] that surface elasticity is an important factor in the process of the breaking of thin films.

5.5. Membranes

5.5.1. *Introductory remarks*

The films of the preceding section can be considered as examples of membranes, but the membranes which are usually discussed have a more complicated structure.

In this section we broadly divide membranes into two classes: biological membranes, including artificial model membranes, whose structure may show affiliation with that of the films of the preceding section (see for biological membranes and their models, subsection 5.5.3), and membranes used for industrial (e.g. desalting of sea water) or laboratory (e.g. purification of colloidal solutions) purposes.

The interest in membranes mainly derives from their different permeability for different species: perm selectivity or ion selectivity. Owing to different permeabilities, electrical rectification effects may be found with membranes. Figure 5.18, from the work by Teorell,[81] gives an illustration. Data for this figure were obtained by considering a "multi-membrane" consisting here of four sheaths. The ions were forced through the multi-membrane by means of an electrostatic potential difference. The sheaths were separated afterwards. According to Teorell most biological tissues show electrical rectification effects. Among the classical examples of semipermeable membranes are collodion films, pig's bladder and parchment paper. These membranes were and are used for the purification of sols in the process of dialysis. The membranes are impermeable to colloidal particles and large non-electrolyte molecules, and permeable to water and (small) ions, especially (in the case of the classical membranes) to cations. It is difficult, though not impossible, to force anions through these membranes, because the membrane material usually carries fixed negative charges. Therefore the equilibrium anionic concentration in the membrane, and the anionic

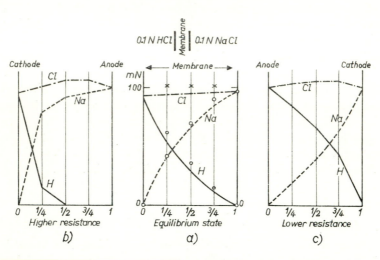

FIG. 5.18. Rectification shown by multimembrane consisting of four sheaths (from Teorell[81]).

permeability are low. This explanation of a low permeability is in agreement with the classical view on this point: one of the most important factors determining the permeability of a certain species is its solubility in the membrane material. Salt contaminations in sols can be removed by using a membrane which separates the sol from pure water, but owing to the ion selectivity the rate of removing cations is faster than that of anions. Since electroneutrality should be preserved, hydrogen ions diffuse into the sol-compartment thus making it more acid. At present there are refined experimental techniques which partly overcome this difficulty and which speed up the whole process of dialysis. Thus, the ions may be accelerated by an electric field in the process of electrodialysis.

Other examples of ion-selective membranes are porous porcelain plates in which $Cu_2Fe(CN)_6$ is formed. A modern version of the $Cu_2(CN)_6$

membrane is the "Hirsch" membrane.[82] This is a film of polymer material separating a sulphate solution from a barium salt solution. $BaSO_4$ crystals are formed on the film. The crystals are positively charged at the Ba^{++} side and negatively charged at the SO_4^{--} side. The distance between the tiny crystals is of the order of a Debye length or less. Therefore the Ba^{++} concentration between positively charged crystals is very low, whereas the SO_4^{--} concentration is very low between negatively charged crystals. This means that the membrane is effectively impermeable to these and other (polyvalent) ions. The passage of monovalent ions is somewhat easier. When $Ba(OH)_2$ and H_2SO_4 solutions are used to form the membrane, a Donnan potential difference of somewhat less than $60\Delta pH$ mV is found at room temperature.

Another group of membranes is that of thin (e.g. 0·05 mm) crystalline plates of ionic conducting material such as AgBr, separating two solutions. Hall and Bruner[83] measured the resistance of the platelets as a function of the Ag^+ concentration. At the c_{pzc} the resistivity should be the same as that of a large crystal since no concentration gradients of the mobile Ag^+ ions inside the AgBr are expected. In this way Hall and Bruner found values for the c_{pzc} which were, at room temperature, much higher than those found for the classical sols[84] ($p_{Ag} \sim 5\frac{1}{2}$) but within reasonable agreement with estimates made by Ottewill and Woodbridge[85] for their monodisperse AgBr sols. For high temperatures the c_{pzc} approaches the "classical" value. This fact lends support to the view[85] that lattice imperfections are present in the sol particles of classical sol but not, or to a lesser extent, in the monodisperse sols. At high temperatures these imperfections lose their significance.

Finally, a very important group of membranes is that of the ion-exchange membranes, made of ion-exchange resins. These resins, first prepared by Adam and Holmes,[86] are sometimes highly ion selective. One way to make ion-exchange resins is by polycondensation of aromatic acids or aromatic bases with formaldehyde or other aldehydes. The polycondensates are three-dimensional networks, sufficiently open to allow the penetration of electrolyte solutions. The networks carry fixed negative charges or fixed positive charges. The fixed-charge density depends on the value of the pH of the solution. The electroneutrality condition requires the presence of gegen-ions, cations in an acid resin and anions in a base resin. There are also (few) ions of the same sign as that of the fixed charges. These are the co-ions. Gegen-ions and co-ions are mobile and can be replaced by other ions. This replacement is called ion-exchange. Ion-exchange resins have precursers in nature: certain clay minerals,[87] zeolites, have ion-exchange properties. Products, similar to zeolites, can also be synthesized ("Permutit").

There is a rapid progress in this field. Thus, Loeb and Sourirajan[88] found a procedure for making membranes based on cellulose acetate, which can be used to desalinate sea water. Probably owing to narrow pores these membranes are permeable to water molecules, but impermeable to (hydrated) ions.

5.5.2. *Potentials*

There is a remarkable resemblance between a semiconductor and an ion-exchange resin. In both cases there are a number of fixed charges and a number of freely moving charges. In the semiconductor, for instance Ge or Si, the fixed charges are provided by ionized impurity atoms, which occupy lattice positions, and the free charges are conduction electrons and holes. It must immediately be added that there are also important differences. Thus the recombination and the formation of conduction electrons and holes usually plays a more important role than the comparable phenomenon of ion association and dissociation. Furthermore, whereas there are numerous sorts of different ions there are, in electronic semiconductors, only two sorts of mobile charge carriers: the conduction electron and its counterpart, the hole. Nevertheless, the electrostatic aspects of both systems have much in common. Since more calculations seem to have been carried out for semiconductors an (elementary) treatment will be postponed to Chapter 6. There also is a resemblance between charged long chain polymers in solution, and ion-exchange resins. The latter may be described as solid solutions of polyelectrolytes in water. One property they have in common is that of swelling and shrinking when the pH is appropriately altered. The fixed charges, being of like sign, repel each other. Swelling and shrinking are explained respectively by an increase and a decrease of the fixed-charge density. The effects of swelling and shrinking are diminished when many free ions are available for screening. Hermans and Overbeek,[89] and independently Künzle and Katchalsky,[90] worked out the theory for charged long-chain molecules in ionic solutions. These, and similar problems are dealt with in Morawetz's book.[91]

We shall now give simple potential equations pertinent to the case of a membrane separating two different electrolyte solutions. We ignore the microstructure (pores) of the membranes and consider the root mean square distance between the fixed charges as small compared to the Debye length.

Figure 5.19 illustrates the membrane system with which we are concerned. The potential difference between two points far in the solutions I and II is the membrane potential E_M. It is not possible to measure E_M directly because any measurement requires two electrodes, one in each solution, and this introduces new potentials.[92] It is assumed that the membrane is a planar sheet which contains fixed charges but in which no space charge is present. Therefore, in Fig. 5.19 the potential-distance curve inside the membrane is a straight line. The boundaries of the membrane with the surrounding solutions are denoted as α and ω. We write E_M as a sum of four potential differences:

$$E_M = E_D{}^\alpha + E_D{}^\omega + E_{\text{diff}} + E_{\text{liq}} \qquad (5.5.1)$$

where $E_D{}^\alpha$ and $E_D{}^\omega$ are two Donnan potentials. These arise from the presence of fixed charges in the membrane leading to an unequal

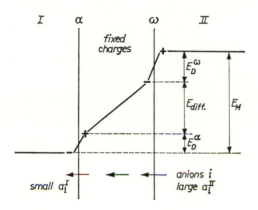

FIG. 5.19. Potentials across membrane, limited by boundaries α and ω.

distribution of permeable ions at the boundaries α and ω. For any permeable ion i one can write

$$E_D{}^\alpha = \frac{kT}{z_i\,e}\ln\frac{a_i{}^{\text{I}}}{a_i{}^\alpha} \quad\text{and}\quad E_D{}^\omega = -\frac{kT}{z_i\,e}\ln\frac{a_i{}^{\text{II}}}{a_i{}^\omega} \qquad (5.5.2)$$

where $a_i{}^{\text{I}}$ and $a_i{}^{\text{II}}$ are the activities of ion i far in the solutions I and II and where $a_i{}^\alpha$ and $a_i{}^\omega$ are the activities just inside the membrane. The case for easily permeating anions ($z_i < 0$) is shown in Fig. 5.19, for the case in which $a_i{}^{\text{II}} > a_i{}^{\text{I}}$.

Furthermore, E_{liq} is a potential drop which may be present in the solutions near the boundaries, because stirring may not be effective. This has been pointed out by Helfferich.[93] In biological systems E_{liq} may be of importance but it will not be considered here.

The diffusion potential E_{diff} is caused by the fact that ions of different kind diffuse with different velocities with respect to a given reference frame in the same homogeneous medium. As a general rule

J

different diffusion velocities should lead to the formation of space charges but in Fig. 5.19 we have assumed that charge effects are only found at the boundaries.

A well-known expression for E_{diff}, obtained by using irreversible thermodynamics[95] or by a quasi-thermodynamic method,[96] reads

$$E_{\text{diff}} = E^b - E^a = -\frac{kT}{e} \int_a^b \sum_n t_i \, d \ln a_i. \qquad (5.5.3)$$

One-dimensional diffusion, say in the x-direction of the frame of reference, is considered here, $x = a$ and $x = b$ representing two planes in the medium where the potentials are E^a and E^b, a_i is the activity of component i, n is the number of diffusing components, and t_i is the transference number which, following Scatchard,[94] we define as the number of moles of component i transferred in the direction of positive current for one Faraday of charge. For t_i one has

$$t_i = \frac{u_i' c_i}{\sum_n u_j' z_j c_j} \qquad (5.5.4)$$

where u_i' and u_j' are mobilities, c_i and c_j are molalities and z_j is the valency of component j.

When a component is uncharged, it may have a non-zero mobility u'. In that case it contributes to E_{diff}. According to Staverman et al.[97] water transport across membranes in aqueous solutions may lead to a contribution of about 4%.

Transport numbers T_i should be distinguished from transference numbers. The transport number is the fraction of current carried by component i.[96] It is equal to $z_i t_i$ so that $\sum T_i = 1$. Transport numbers are positive or zero. Transference numbers of anions are counted negative in this scheme of definitions. Often the quantity $u_i = z_i u_i'$ is called the mobility, for instance in semiconductors where the mobilities of conduction electrons and holes are both counted positive.

Thus

$$T_i = z_i t_i = \frac{u_i c_i}{\sum_n u_j c_j} \; ; \quad u_i = z_i u_i' \; ; \quad \sum_n T_i = 1. \qquad (5.5.5)$$

It is customary to introduce a diffusion coefficient D_i as follows:

$$D_i = \frac{kT}{e} u_i. \qquad (5.5.6)$$

This is the Einstein relation. It is convenient to express D_i in cm^2 per sec, and u_i or u_i' in cm^2 per V sec.

The simplest application of eq. (5.5.3) is to the diffusion of a salt CA with activities equal to concentrations. Then

$$E_{\text{diff}} = -(kT/e)D \ln \frac{c_{\text{CA}}{}^b}{c_{\text{CA}}{}^a}; \quad D = \frac{u_{\text{C}} + u_{\text{A}}}{u_{\text{C}} - u_{\text{A}}}. \tag{5.5.7}$$

This equation is due to Nernst (see his textbook[98]). In semiconductors one has a similar equation, the Dember equation,[99] leading to the Dember potential. Here the "salt" consists of conduction electrons and holes, created by shining light on the crystal. The mobility of conduction electrons is usually higher than that of holes.

The membrane potential can also be written as

$$E_M = -kT/e \int_{\text{I}}^{\text{II}} \sum_n t_i \, d \ln a_i \tag{5.5.8}$$

where now the integration limits are replaced to planes in the solutions I and II. In this way the Donnan potentials are incorporated in the integral. This is permitted because the mobility of the fixed charges in the membrane is zero and because for all the mobile charges we have the same value of $E_D{}^\alpha$ and $E_D{}^\omega$.

We give expressions for E_M for three special cases:

(a) The membrane has a high positive charge (i.e. a large number of fixed positive charges). The contribution of co-ions is neglected. We consider one type of gegen-ion A and assume that $t_A = 1$ at all positions. Then, with $z_A = -1$:

$$E_M = +kT/e \ln \frac{a_A{}^{\text{II}}}{a_A{}^{\text{I}}}. \tag{5.5.9}$$

The membrane acts as an electrode, which is reversible to A. For a negative membrane with gegen-ions C one has ($z_C = +1$):

$$E_M = -kT/e \ln \frac{a_C{}^{\text{II}}}{a_C{}^{\text{I}}}. \tag{5.5.10}$$

"Membrane electrodes" can be made for a number of gegen-ions but they are not very selective.

(b) The membrane, having a, say, negative charge, is permselective with respect to a number k of univalent gegen-ions. The influence of co-ions is neglected. It is assumed that u_j/f_j (where $f_j = a_j/c_j$ is the activity coefficient of the jth component) has the same value everywhere in the membrane. The diffusion potential becomes

$$E_{\text{diff}} = -kT/e \int_\alpha^\omega \sum_{j=1}^k \frac{(u_j/f_j)a_j}{\sum (u_j/f_j)a_j} \, d \ln a_j = kT/e \ln \frac{(\sum (u_j/f_j)a_j)^\alpha}{(\sum (u_j/f_j)a_j)^\omega} . \tag{5.5.11}$$

Assume now that at the α-side of the membrane, in solution I and around the interface, only ions 1 are present and at the ω-side only ions 2 ($z_1 = z_2 = -1$). Then eq. (5.5.11) becomes

$$E_{\text{diff}} = -kT/e \ln \frac{(u_2/f_2)a_2{}^{\omega}}{(u_1/f_1)a_1{}^{\alpha}}. \tag{5.5.12}$$

The membrane potential for this case is the biionic potential (BIP). The BIP is obtained by adding two Donnan potentials to eq. (5.5.12)

$$\text{BIP} = -kT/e \ln \frac{(u_2/f_2)a_2{}^{\text{II}}}{(u_1/f_1)a_1{}^{\text{I}}} \tag{5.5.13}$$

in which it is assumed that the uniformity of mobilities and activity coefficients is extended to solutions I and II.

Biionic potentials can be measured by placing both in solution I and in solution II an electrode which is reversible to the co-ions. Thus, if solution I is a NaCl solution and solution II is a KCl solution, two AgCl electrodes can be used. The measurements are carried out at numerous values of $c_1{}^{\text{I}}$ and $c_2{}^{\text{II}}$ and the results are usually given as two plots in one figure. One plot gives the BIP as a function of $\ln c_1{}^{\text{I}}$ at constant $c_2{}^{\text{II}}$ and the other plot gives the BIP as a function of $\ln c_2{}^{\text{II}}$ at constant $c_1{}^{\text{I}}$. At the point of intersection the biionic potential is zero. By this method the transport ratio $u_2 f_1/u_1 f_2$ in the membrane can be found and appears to be of the same order as the value found in diffusion experiments carried out with the same membrane. The absolute values of the slope of the BIP-log conc. curves is somewhat less than 59 mV per concentration decade. Bergsma and Staverman[97] ascribed this difference, and the difference of the transport ratios found by potential measurements and by diffusion measurements, to water transport which had previously been neglected. That means that in eq. (5.5.11) a term

$$-kT/e \int t_w \, d \ln a_w$$

has been neglected. Scatchard[94] has given an analysis of the problem.

(c) The theory of Teorell and Meyer and Sievers (TMS).[100] Here the charge of the membrane appears explicitly in the expression of the membrane potential. Assume that the membrane separates two solutions of a univalent salt CA with concentrations c^{I} and c^{II}. The concentration of the fixed charges in the membrane is A. These charges may be univalent positive or negative. Considering ideal solutions one has

$$c^{\text{I2}} = c_{\text{A}}{}^{\alpha}c_{\text{C}}{}^{\alpha} = (c_{\text{C}}{}^{\alpha} - A)c_{\text{C}}{}^{\alpha}, \tag{5.5.14}$$

and

$$c^{II2} = c_A{}^\omega c_C{}^\omega = (c_C{}^\omega - A)c_C{}^\omega; \qquad (5.5.15)$$

or

$$c_C{}^\alpha = \tfrac{1}{2}(A + \surd(A^2 + 4c^{I2})), \qquad (5.5.16)$$

$$c_C{}^\omega = \tfrac{1}{2}(A + \surd(A^2 + 4c^{II2})). \qquad (5.5.17)$$

In the TMS treatment the diffusion potential is given by the Henderson equation (as explained in the book of MacInnes[96]). It is assumed that at a given position in the membrane the cation concentration is given by $c_C{}^\alpha - y(c_C{}^\alpha - c_C{}^\omega)$ where y varies linearly from 0 as at $x = \alpha$ to 1 at $x = \omega$. The anion concentration is given by $c_A{}^\alpha - y(c_A{}^\alpha - c_A{}^\omega)$. The parameter y is the only variable. Now

$$E_{\text{diff}} = DTk/e \ln \frac{u_C c_C{}^\omega - u_A c_A{}^\omega}{u_C c_C{}^\alpha - u_A c_A{}^\alpha}. \qquad (5.5.18)$$

In order to obtain the membrane potential it is convenient to write for the Donnan potentials

$$E_D = kT/e \ln \frac{c_C{}^\omega c^I}{c_C{}^\alpha c^{II}}. \qquad (5.5.19)$$

After application of eqs. (5.5.16) and (5.5.17) E_M becomes

$$E_M = kT/e\left\{ D \ln \frac{DA + \surd(A^2 + 4c^{II2})}{DA + \surd(A + 4c^{I2})} + \ln \frac{c^I(\surd(A^2 + 4c^{II2}) + A)}{c^{II}(\surd(A^2 + 4c^I)^2 + A)} \right\}.$$

$$(5.5.20)$$

Kedem and Katchalsky,[101] applying the methods of irreversible thermodynamics to composite membranes, confirmed the results obtained by Staverman[97] in 1952 for single membranes and gave numerical examples based on the TMS model. For a "mosaic" membrane, consisting of positive and negative sections, there was a salt diffusion which could become much larger than expected on the basis of the separate cation and anion permeabilities. The effective diffusion constant of the salt can, under favourable circumstances, be higher than the diffusion constant of free diffusion. In another example the membrane was assumed to be formed by two sheets, a and b, placed in series, a being a negative membrane and b being comparatively loose, and uncharged. The presence of sheet b, which itself does not show up an electro-osmotic flow, may have an important effect upon the electro-osmotic flow of the composite membrane. This was ascribed to salt accumulation (or depletion, depending on the characteristics of sheet a) between the sheets.

There is another approach to diffusion problems of charged species. The starting point of that approach is a set of Nernst[102] and Planck[103] equations. The Nernst–Planck equation of a charged component i reads:

$$\phi_i = -D_i\left(\text{grad } c_i + z_i c_i \frac{e}{RT} \text{grad } \psi + c_i f_i\right) + c_i v \qquad (5.5.21)$$

where ϕ_i $(=c_i v_i, v_i$ is velocity of component i) is the flux of the ith component, v is the translational velocity of the centre of gravity of the system and f_i is an external force acting on one mole of component i. In the original Nernst–Planck equations f_i, and v are taken zero. The bracketed part of eq. (5.5.21) represents, in molar units the force acting on the particles. We met this part in Chapter 2. Teorell,[104] Helfferich[105] and Schlögl[106] have extensively studied sets of Nernst–Planck equations and have calculated flow and concentration profiles and current densities as a function of the potential in the membrane. In view of the mathematical difficulties to be overcome, the problems studied were highly simplified. Thus, ideal ionic solutions were often considered and space charges in the membrane were assumed absent. The mobilities of the ions were taken as uniform. On the whole, the same insights were gained as obtained earlier, starting with eq. (5.5.2). In addition, rectification[107] could be explained and very interesting results were obtained by Schlögl, who explained the phenomena of anomalous osmosis and incongruent salt flow. These phenomena can only take place when the membrane is charged. There is positive anomalous osmosis when the solvent flow is (temporarily) higher than that across a neutral membrane, other parameters remaining the same (the osmotic pressure difference may be temporarily higher than in the case of a neutral membrane). There is negative anomalous osmosis when the solvent flow is directed such that the osmotic pressure difference is temporarily "too low". There is incongruent salt flow when salt moves from the more dilute toward the more concentrated solution. This can happen only in the case of positive anomalous osmosis. All these phenomena can be considered as a consequence of an electro-osmotic flow (the flow of a charged liquid due to an electric field). Electro-osmosis is usually demonstrated by means of a system containing capillaries or pores, but strictly speaking, this is not necessary. In anticipation of the discussions in the next subsection we mention here that Teorell constructed a membrane model in which an oscillating liquid flow and oscillating potentials could be obtained. A crucial factor here was the possibility of electro-osmotic flow across the membrane.

5.5.3. *Biological membranes*

Electrical processes in biological tissues are probably largely governed by the properties of membranes. These are thin ($\lesssim 100$ Å) layers consisting of lipids (e.g. lecithin) and proteins, enclosing a cell (plasma membrane) or enclosing part of a cell (endoplasmic reticulum membranes, nuclear membranes, mitochondrial membranes). We shall occupy ourselves only with plasma membranes because the data pertaining to double layers are derived from experiments carried out with plasma membranes (or cell-surface layers).

Around the turn of the century, Bernstein[108] and Overton[109] (see also the book by Fleckenstein[110] and the review paper by Robertson[15]) discussed bioelectric phenomena and connected these phenomena with the permselectivity of cell membranes with respect to K^+ and Na^+ ions. According to Bernstein, membranes were permselective with respect to cations, K^+ ions having the highest permeability. Since the K^+ concentration inside the cell was higher than outside, a potential difference was set up across the membrane. Specifying the case of a nerve membrane, excitation led to an increased permeability of other cations such as Na^+. This effect resulted in a depolarization, the total cation concentration inside and outside the nerve cells being of the same order of magnitude. Overton discussed the enhancement of the muscle excitability which was known to take place after increasing the external Na^+ ion concentration and he considered the possibility of a Na^+/K^+ exchange (Na^+ going inward, K^+ going outward) during the excitation, restoration by the reversed process taking place afterwards. The importance of a Na^+/K^+ balance was further stressed by Overton's experiments in 1904, which showed that an increased extracellular K^+-concentration had a paralysing influence. After decreasing the K^+-concentration, or increasing the Na^+-concentration, the excitability returned.

The membrane potential envisaged by Bernstein can be measured when the cells are sufficiently wide to permit the introduction of an electrode. This is the case for giant squid axons as first pointed out by Young.[111] These axons have a width of about 0·5 mm. Cole[112] and Hodgkin and Huxley[113] found for the membrane potential in the resting state (the resting potential) a value of about -50 mV (negative inside). This was measured with two Pt electrodes placed at either side of the membrane. Upon excitation the potential value changed sign and became of the order of $+50$ mV (positive inside).

Excitation can be generated by an electric pulse across the membrane of 1 msec in duration and an intensity of 1 mA/cm². The excitation region is propagated with a velocity of about 10 m/sec along the nerve

fibre and its intensity remains constant; it is an all-or-none effect. The
resistivity in the tangential direction is much lower than in the transverse direction. However, in the excited region there is a sharp fall of
the transverse resistance to about 5 Ω/cm^2. On the whole,[114] nerve and
muscle fibres behave as if they have a cable-like structure, the membrane which surrounds a low resistance core having a capacitance of
about 1 $\mu F/cm^2$. The membrane shows, after excitation, a damped
oscillatory behaviour, in particular when a constant current is applied.
Hodgkin and Huxley found, by varying the Na^+ and K^+ ion concentrations, that the potential difference across the membrane, V, could
be given as:

$$V = -kT/e \ln \frac{(K^+ + bNa^+)_{ins}}{(K^+ + bNa^+)_{outs}} \qquad (5.5.22)$$

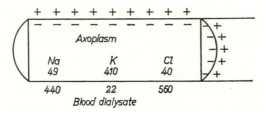

FIG. 5.20. Membrane of an axon with double layer: concentrations in
mmol/l. of K^+, Na^+ and Cl^- ions. Reproduced by Teorell[81] from *The
Neurophysiological Basis of Mind* by Eccles, Oxford, 1953.

where the subscripts "ins" and "outs" indicate inside and outside
concentrations; b is the ratio of the mobilities in the membrane of the
K^+ and the Na^+ ions, in the resting state $b \ll 1$ and in the excited state
$b \gg 1$. According to Tasaki *et al.*[115] eq. (5.5.22) is an over-simplification
because also other cations, such as Ca^{++}, are important for the determination of V. The role of Ca^{++} ions was already stressed by Brink.[116]

Tasaki *et al.* performed experiments as illustrated in Fig. 5.21.
Giant axons, immersed in aqueous solution of known composition, were
mounted on a platform and perfused with a second aqueous solution at
a rate of 15 mm^3 per minute. The axon length was of the order of
10–20 mm and the diameter 0·5–1 mm. The inside of the axon was
cleaned by this method, but according to Nachmansohn[117] it is probable
that acetylcholine ($CH_3COOCH_2CH_2N(CH_3)_3OH$) still adheres to the
membrane. Acetylcholine is a physiologically active substance which is
known to be very important in the mechanism of transfer of a nerve
signal to muscle. The reference electrode (M) in the axon was a micropipette (outer diameter 0·1 mm) filled with a 1 M NH_4Cl solution. The

axon was brought in the excited state through a pair of Pt electrodes (St in the figure). Propagation of nerve impulses across the perfused zone was monitored through another pair of Pt electrodes (Re). A standard perfusing liquid was used, containing 0·25 M of K_2SO_4 and 0·5 M of sucrose per litre of double-distilled water, with its pH adjusted to 7·3 by adding a small amount of K_2HPO_4. Perfused and unperfused axons showed the same value of the resting potential between Re and the reference electrode (-55 mV± 5 mV) and of the action potential ($+110$ mV± 10 mV). The action potential is the peak potential reached in the excited state. It was remarkable to note that in other perfusing liquids, without K^+ ions, normal action potentials were observed. This

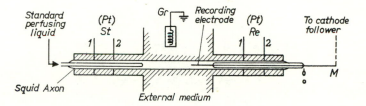

FIG. 5.21. Schematic illustration of experimental arrangement employed for intracellular perfusion. St: stimulating electrodes. Re: extracellular recording electrodes. M: intracellular micropipette electrode. Gr. electrode placed in solution and connected to earth. Temperature 23°C (from Tasaki *et al.*[115]).

fact may, according to Tasaki, be attributed to the existence of an unstirred layer adjacent to the membrane. When the perfusing liquid was an isotonic NaCl solution, excitability was lost in 1 minute. Addition of Ca^{++} ions or of Mg^{++} ions to the surrounding medium restored the excitability. According to Tasaki, action potentials can be observed in K^+-free media and also in Na^+-free media. After stating that these results do not support any known hypothesis, it was suggested that in the resting state Ca^{++} ions are bound at the outer membrane surface. According to Lehninger[117] it is possible that binding has taken place by acid groups, which are constituents of a special group of neuronal lipids, the gangliosides.† The gangliosides, which comprise about 5% of the total amount of neuronal lipids, extend relatively far into the intercellular space. The acid groups can also bind other ions, and protamine (a basic protein of low molecular weight, predominantly arginine). Protamine, added to the extracellular solution, blocks the

† Here R″ of the *general lipid formula* has long-branched polar head groups composed of oligosaccharides containing one or more residues of sialic acid, a generic term referring to N-acetyl or N-glycolyl derivatives of the nine-carbon sugar derivative neuraminic acid, which is negatively charged at pH = 7.

electrical excitability while the addition of gangliosides leads to its restoration. The question has arisen whether data obtained for very few membranes, such as the membrane enveloping the squid axon, must be considered as representative for all membranes. In the past there was a tendency to answer this question in the affirmative, but more recently, when an overwhelming number of highly specific membrane functions have been revealed, this is no longer the case. There is a considerable discussion in the literature as to the general validity of Robertson's "unit membrane" concept.[15] The "unit membrane" concept has its roots in the bi-molecular or paucimolecular leaflet model of Danielli and Davson[118] (Fig. 5.22). Danielli and Davson[118] proposed that a membrane

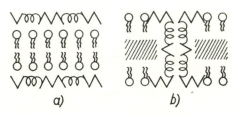

FIG. 5.22. (a) Membrane model according to Danielli and Davson.[118] (b) Membrane model proposed by Lenard and Singer.[136]

consists of a lipid core covered on each side with a monomolecular film of protein. The lipid core was assumed to be a bimolecular leaflet which could be compared with an "inverted soap film", the polar ends pointing outward instead of inward. In this way the model is connected with the views developed by Adam, Rideal and Langmuir as given in the preceding sections. The high ohmic resistivity in the resting state (low ionic permeability) had already indicated that the core is hydrophobic. The Danielli–Davson model improved, or rather replaced, that of Gorter and Grendel.[119] These workers extracted the lipid from a known amount of red blood cells. The extracted lipid was spread as a a monolayer and from a force-area curve the area of the monomolecular film was obtained. From the known total area of the red blood cells it was concluded that the cell membrane contained a bimolecular lipid leaflet. However, the surface tension of a membrane in its natural environment was found by Cole[120] and by Harvey and Schipiro[121] to be of the order of 0·1–0·2 dyne/cm. Since this value was much lower than that found for monolayers Danielli and Davson inferred that the bimolecular lipid leaflet contained a considerable amount of adsorbed species, which was assumed to be protein. Support for an ordered layer in the membrane was provided by Schmitt et al.[122] on the basis of optical (birefringence) measurements. Cohen et al.[123] found that, following excitation, the birefringence of crab nerve and of squid giant axon changed in such a way as to indicate a greater degree of order.

The unit-membrane concept implies that the Danielli–Davson model is to be regarded as a general molecular membrane model. The concept is based largely on data obtained by electron microscope and X-ray studies, which were originally used by Robertson to indicate that the repeating unit of the membrane system surrounding vertebrate nerve fibres or axons is a tripartite layer. Typically the tripartite layer is $(20 + 35 + 20)$ Å thick, the spacing between two tripartite layers being 100–150 Å. At least two tripartite layers are present around the axons. This is so because these axons are embedded over their whole length in tubular cells: the mesaxons or Schwann cells. In a number of cases the axon fills up the core of the mesaxon, in other cases the mesaxon is "wrapped" many (say five) times around the axon. If so the membrane system consists of many (say eleven) membrane units at regular distances. Such a thick membrane system is a myelin sheath.

Myelin is reminiscent of the multilayers obtained by the Langmuir–Blodgett[124] technique. In this technique a monolayer of, for instance, barium stearate on water is deposited on a glass surface in the following way: a glass plate is placed in a vertical position in the water phase and is lifted whilst being held in that orientation. The stearate molecules adhere to the glass and form a coherent monolayer with the hydrophobic tails pointing away from the glass. The following step is a reversal of the direction of the movement of the vertical glass plate. This results in the formation of a second layer on top of the first, the hydrophobic tails of the second layer turned toward those of the first layer. This procedure can be repeated many times. Kuhn et al.[125] have elaborated on this method by introducing organic dye molecules in the original monolayers. Thus, when two different kinds of dye molecules, one absorbing u.v. light and emitting blue light and the other absorbing blue light and emitting yellow light, were introduced in subsequent monolayers, the light intensity, I_d, of the yellow light emission is determined by the distance between the chromophore groups in the subsequent layers. It turned out that the intensity–distance relationship was exactly that, which was predicted on the basis of the assumption that the two different kinds of dye molecules were present in different monolayers and that no "diffusion" from one monolayer to another was taking place. Thus, the multilayer arrangement was found to be a very stable one.† Concerning the problem whether irradiation of u.v. light,

---

† That dye molecules were indeed held fixed inside one monolayer was confirmed by den Engelsen[125a] who applied the optical method of ellipsometry to these layers and found that the optical constants parallel, and normal to the plane of the monolayer (these constants were different) did not change in the course of at least a week.

leading to excitation of dye molecules, results in the transfer of photons or of electrons, it was found that the nature of the solid material, which served as the substrate, was of importance, AgBr favouring photon transmission and CdS favouring electron transmission.[126] Kuhn's results may be relevant to the case of myelinated axons although it is of course dangerous to consider a multilayer system obtained by the Langmuir–Blodgett technique as a model for a myelin sheath.

The unit membrane concept has been challenged, in particular after Stoeckenius[127] and Luzzatti and Husson[128] published their work (in 1962) on phospholipid–water systems. Luzzatti and Husson applied an X-ray diffraction technique and found a number of lamellar phases at water concentrations higher than 4% and at physiological temperatures. Here water formed sheets between the lipid leaflets. When the water concentration was decreased and/or the temperature was increased, an hexagonal phase was formed with (probably) water cylinders in the lipid bulk, the lipid molecules having turned the hydrophilic groups toward the water. The role of ionic charges in these transitions is not very clear.

Stoeckenius obtained results similar to those of Luzzatti and Husson by using electron-microscope techniques. Thus in both cases the distance between the parallel and hexagonally arranged water cylinders in the hexagonal phospholipid–water phase appeared to be 42–45 Å. Such a common result was considered largely to eliminate the doubts which might be raised as to a possible annihilation of delicate structures during the fixation and staining procedure in the electron-microscope techniques.† The findings of Luzzatti and Husson stressed the possibility that the lamellar structures such as proposed in the unit membrane concept are not the only relevant structures. Thus Sjöstrand[131] presented electron micrographs showing globular substructures (globules of 60–70 Å in diameter) in mitochondrial mem-

---

† For instance, a sample is fixed with 1–5% glutaraldehyde, post-fixed and stained with 1% OsO_4 or with 1% $KMnO_4$, dehydrated by extraction with a series of aqueous ethanol solutions, washed in propelene oxide, soaked in epoxy monomers and polymerized at 60°C for 24–48 hours.[129] The heavy Os and Mn atoms act as "stains" because they interfere with the electron beams from the electron microscope. Two other techniques are the freeze-drying technique and the negative staining technique. In the first case the sample is, in the presence of glycerol, rapidly brought to liquid nitrogen temperature (−196°C) and in vacuum a thin slice is cut with a cold knife. The slice is brought to −100°C and a thin layer of ice is sublimed off. The surface is shadowed with a heavy metal and a carbon replica is made. Close examination of this technique by Branton et al. (J. G. Davy and D. Branton, Science 1970, 160, 1216) showed that it is not without problems. In the second case, membranes are dispersed in a 1–2% of sodium-phosphotungstate solution. A small drop of the dispersion so obtained is placed on a collodion-carbon support film of a grid and almost dried with filter paper. Phosphotungstate avoids the hydrophobic regions and forms structureless layers in the aqueous regions. This is "negative staining".

branes from the mouse kidney. Glauert and Lucy[132] argued, on the basis
of their electron micrographs of structures obtained by negatively
stained preparations of lecithin and cholesterol, that the lipids in
biological membranes may be arranged in the form of adjacent globular
micelles with 40 Å diameter, held together in leaflets by protein.
Permeability was assumed possible in the open (or rather protein-
filled) space between the globules.

The unit membrane concept was particularly criticized by Korn.[133]
Thus, the older data of Gorter and Grendel were considered to be
inaccurate, the lipid/membrane area ratio probably being considerably
lower than 2, e.g. 1·3. Furthermore, phospholipids, "building bricks"
for membranes, are capable of forming bilayers with a very low interfacial tension. This eliminates the necessity to assume adsorbed
(protein) layers. Another point was that most of the electron-microscope work has been performed with myelin, but myelin has the unusual
low protein/lipid ratio of 0·25 and contains the unusual high cholesterol
percentage of 25. Myelin is metabolically rather inactive. It is "essentially an inert lipid shield"[133] whose role is mainly that of an electrical
resistor. Finally, although it is true, according to Korn, that electron
micrographs often show up layered structures, the physical significance
of these layers is doubtful. The reason is our lack of knowledge of the
chemical interactions between the staining agent (e.g. OsO_4) and the
lipids and proteins in question. Korn noted that even after removal of
the lipid, triple-layered membranes have been observed in osmium-
fixed material.[134]

However, according to Chapman[130] results obtained by the freeze-
drying technique, which involves no chemical treatment until the
replica is formed, broadly support the unit membrane concept. Recently
optical rotatory dispersion (ORD), circular dichroism (CD)[135] and
nuclear magnetic resonance[130] (nmr) measurements have been carried
out and their interpretation has contributed to the discussion of the
unit membrane concept. Molecules containing a non-symmetrical
arrangement of atoms are optically active, the activity (m) being a
function of the wavelength (λ). ORD measurements are given as m vs. λ
curves. The same molecules have the property of absorbing left and
right circularly polarized light to different extents (circular dichroism).
The difference is also wavelength-dependent. Nuclear magnetic
resonance measurements lead to information concerning the freedom
which certain atomic groups have in a molecular structure.

Lenard and Singer[136] have carried out ORD and CD measurements
between $\lambda = 180$ mμ and $\lambda = 250$ mμ with the protein part of red-
blood-cell membranes, which were obtained via an ultracentrifuge
method. The intact membranes were dispersed in various solvents,

aqueous PO_4^{3-} buffer, 2-chloro ethanol, etc. The ORD and CD spectra of these membranes were totally different from those of the random coil form of the protein poly-L-lysine but there was a striking similarity with the spectra of the α-helical form. These results pointed to the presence of a protein layer, which was (partly, say about 30%) in a structured, helical, form. The small but systematic difference between the spectra of the membranes and of the separate helices were ascribed to a slight perturbation of the helical protein in the membrane due to neighbouring lipid or protein molecules. Furthermore, Lenard and Singer pointed out that the hydrophobic interaction was not maximized in the Danielli–Davson and unit membrane models, large hydrophobic parts of the proteins being exposed to the aqueous phase. They proposed a new membrane model in which protein-lined forces connect the two faces of the membrane (Fig. 5.22b). This model accounts for hydrophobic bonding and for the permeation of hydrophilic substances. Already Haydon and Taylor[137] and Glauert and Lucy[132] argued that the addition of lipids such as cholesterol to phospholipid bilayers had a decreasing effect upon the stability of the bilayer. Nuclear magnetic resonance measurements by Chapman[130] supported the view that hydrophobic bonding is important. He found that the lipid hydrocarbon chain was not as free as in the isolated lipid and that the molecular movements of membrane proteins were rather limited. Addition of materials which were known to disrupt hydrophobic bonds were, however, found to increase the freedom of movement.

Starting from the unit membrane concept, techniques were developed for making inverted soap films separating two aqueous phases, the aim being to compare the electrical and the permeability properties of the artificial systems with those of biological membranes. In fact, this line of work had already been initiated by Langmuir and Waugh.[138] Black lipid films were made by Mueller et al.,[139] by Haydon et al.[140] and they are now beginning to be widely studied. One technique[139–40] is to introduce the lipid solution onto an opening (diameter about 1 mm) in a thin plate of a hydrophobic support, e.g. Teflon, placed in an aqueous electrolyte. After drainage of the apolar solvent a bimolecular membrane remained, the hydrophobic parts of the molecular layers facing each other. Van den Berg[142] modified this technique and was able to obtain lipid membranes with a surface area of about 1 cm².

The capacitance of these membranes was of the order of 1 $\mu F/cm^2$. Hanai et al.,[140] using purified egg lecithin, investigated the frequency dependence of the capacitance. They found one relaxation time, the low-frequency value of the capacitance being independent of the electrolyte concentration. They concluded that only the hydrocarbon part was involved in the dispersion.

White[141] studied voltage- and temperature effects. The capacitance C_m of certain lipid bilayer membranes increased according to a law $C_m = C_0 + BV^2$ where V, the applied voltage, i.e. the potential difference between the two sides of the membrane, was of the order of 100 mV, and where β was a constant, equal to 10^{-6} μF cm^{-2} (mV)$^{-2}$ at 20°C, whereas $\beta \cong 0$ when the temperature was 15°C or lower. Concerning the temperature effect, the C_0 values decreased from 0·55 μF cm^{-2} at

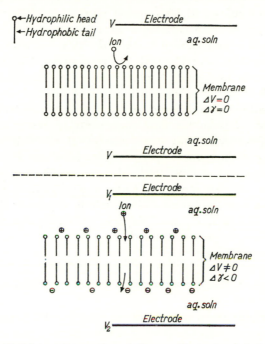

Fig. 5.23. Model membrane (inverted soap film) placed between two electrodes and showing the effect of a potential difference.

20°C to 0·45 μF cm^{-2} at 35°C. White interpreted these effects in terms of thickness variations.

At $V > 100$ mV the films were usually disrupted. The existence of this break-down voltage may be explained as follows:

The applied potential difference introduces a negative contribution $\Delta\gamma_e$ to the interfacial tension of the membrane, which is assuming that the membrane is a flat condenser:

$$\Delta\gamma_e = -\tfrac{1}{2}\sigma V. \qquad (5.5.23)$$

The charge σ is present in the form of adsorbed ions. If their density is 10^{12} per cm^2, which is a reasonable estimate in view of the capacitance value of about 1 μF/cm^2 and possible values of V, then, with V being of the order of 100 mV ($=4$ kT/e) a representative value of $\Delta\gamma_e$ is minus 4×10^{12} kT per cm^2 or 0·16 dyn/cm. When minus $\Delta\gamma_e$ exceeds the value of the interfacial tension of the membrane the membrane is no longer stable. The charge density at thin regions is higher than elsewhere, because the capacitance is higher at thin regions. Therefore bursting may start at thin regions. The membrane resistivity was high: 10^7–10^8 Ω/cm. Finkelstein and Cass[143] studied the permeability of a number of molecules and ions across phospholipid membranes. They found that the membrane core behaved as expected from bulk properties of a hydrocarbon and that it governed the permeation process. The water permeability was studied by comparing the results of two different methods, an osmotic method and a tracer method. In the first, osmotic, method the membrane separated two aqueous solutions of an impermeant solute with different concentrations in each compartment. Water permeated from the low concentration reservoir to the high concentration reservoir and measurement of the rate of permeation led to a value of the osmotic permeability constant P_f. In the second method the membrane separated solutions with equal concentrations of the solute but here some of the water molecules in one reservoir were marked so that its rate of permeation could be followed. Here a diffusion permeability constant P_d was obtained. When pores through which water can diffuse are present then[144]

$$P_f > P_d \quad \text{(pores present)}, \qquad (5.5.24)$$

but when water traverses by means of a solution–diffusion–dissolution sequence (i.e. without pores), then $P_f = P_d$. In the lipid membranes studied by Finkelstein and Cass $P_f = P_d = 10^{-3}$ cm/sec and the conclusion was that aqueous pores were absent.

A most important observation, first made by Mueller and Rudin,[145] was that the addition of very small amounts (e.g. 10^{-7} g/ml) of macrocyclic antibiotics such as valinomycin† or alamethicin† caused the membrane resistivity to decrease drastically, up to a factor 10^7. For alamethicin the resistivity was no longer ohmic. For instance, the steady state conductance λ at 30 mV and in a 0·1 M NaCl concentration was 10^{-4} Ω^{-1} cm^{-2} for an alamethicine concentration (c_{ala}) of $\frac{1}{2} \times 10^{-5}$ g/cm^3. At 50 mV, λ was increased to about 2×10^{-3} Ω^{-1} cm^{-2}.

Sixth-power relationships were found to be characteristic for the behaviour of alamethicin. Thus, upon varying c_{ala}, λ varied, at constant salt concentration c_{NaCl} and constant applied potential ΔV, with c_{ala}^6; at constant c_{ala} and constant ΔV, it varied with c_{NaCl}^6 and at constant c_{ala} and constant c_{NaCl} but varying ΔV, it varied according to a $(\Delta V)^6$ law, or thereabouts.

These results indicated that complexes of six molecules of alamethicin co-operated in the conduction mechanism. The suggestion was made[144] that these ring-shaped molecules were joined in such a way as to form a "channel" which was sufficiently wide for the passage of a metal ion. Analogous experiments with valinomycin showed a much better intracationic (K^+/Na^+) selectivity than for alamethicin. For valinomycin the permeability for K^+ was 400 times greater than for Na^+, while for alamethicin the permeabilities were equal. This was ascribed by Mueller and Rudin to a larger ring diameter (13 Å) for alamethicin than for valinomycin (7 Å). Finkelstein and Cass produced evidence for the hypothesis that the cyclic dipepsides (nystatin† in their case) led to the formation of pores or channels across the membrane. They found that P_f was much larger than P_d when nystatin was present, $P_f = 25 \times 10^{-3}$ cm/sec, which was a fifty-fold increase compared with a nystatin-free membrane. Furthermore, whereas the nystatin-free membrane was impermeable to small hydrophilic solutes, nystatin

---

† Valinomycin

$$\left[\text{D-Val} \quad \text{L-Lact} \quad \text{L-Val} \quad \text{D-Oxisoval}\right]_3$$

three units forming a ring. Ring diameter about 7 Å.[145] Shemyakin and his school have investigated the formation of κ-complexes with valiomycin in solution.[145a]

Alamethicin contains the amino-acids $(GluN)_2$, $(Glu)_1$, $(Pro)_2$, $(Gly)_1$, $(Ala)_2$, (methyl Ala)$_8$, $(Val)_2$ and $(Leu)_1$. Mueller and Rudin[145] constructed ring with diameter of 13 Å.

Nystatin

$C_{46-47} H_{73-75} O_{18} N$

allowed them to permeate according to their size; the smaller the size of the solute molecules, the higher was the permeability. In the presence of alamethicin Mueller and Rudin created a negative resistance region by introducing an ionic gradient (different concentration at either side of the membrane) or by introducing protamine. Suitable combinations of alamethicin and protamine concentrations produced bistable, oscillating membrane potentials, which resembled the time course of action potentials. The physical mechanism behind these oscillations may be elucidated when first a comparison is made with the oscillatory behaviour of Teorell's large-scale membrane models.[81,145] In its simplest form Teorell's model was as follows:

A porous membrane separates two solutions. The height of these solutions is different so that there is a hydrostatic pressure difference ΔP acting on the membrane. The two solutions have different ionic concentrations. Thus there is an ionic gradient in the pores of the membrane. There is a reversible electrode in each solution. We assume that there is only one pore and that there are fixed charges on the pore surface such that there is a zeta potential, ζ, at the hydrodynamic slipping plane near the pore/solution surface. Furthermore, we assume that there is an electric current flowing between the two electrodes. Ions are the charge carriers. There is an almost unidirectional flow of ions through the pore, their sign being opposite to that of the fixed charges on the pore surface. The ionic flow leads to an electro-osmotic flow of the liquid through the pore. We write for the quantity of liquid, V_{el}, transported per unit time through the pore

$$V_{el} = \frac{\varepsilon j}{4\pi\eta} \frac{\zeta}{\lambda} \qquad (5.5.25)$$

where j is the electric current, which is taken constant in what follows, η is the viscosity of the liquid and λ its specific conductivity. Finally ε is its relative dielectric constant.

Equation (5.5.25) is closely related to eq. (5.3.3) which describes the electrophoresis of a charged particle. As mentioned in the Introduction, electrophoresis, electro osmosis and other electrokinetic phenomena, such as that of streaming potential, can be placed on a common basis by using the methods of irreversible thermodynamics, the electrokinetic phenomena being known separately for a long time.

The pore has also the function of a leakage. If the pore radius, measured from the slipping plane, is r, than the Poiseuille flow, V_P, is

$$V_P = \frac{\pi r^4}{8\eta l} \Delta P \qquad (5.5.26)$$

where l is the length of the pore.

Both the electro-osmotic flow and the Poiseuille flow have the effect of changing the hydrostatic pressure difference ΔP:

$$V_{el} + V_P = -\alpha \frac{d\Delta P}{dt} \tag{5.5.27}$$

where α is a positive constant depending on the surface areas of the liquid reservoirs separated by the porous membrane.

Because there is an ionic gradient, the values of ζ and λ are not uniform along the length of the pore. The values in eq. (5.5.25) are average values and, for one thing, depend on ΔP. Even when ΔP is held constant there is a continuous change in concentrations in and near the pore. It is now assumed that the following relation is generally valid:

$$\frac{d(\zeta/\lambda)}{dt} = \beta(\Delta P - \Delta P_0) \tag{5.5.28}$$

where ΔP_0 is the hydrostatic pressure difference at which a stationary concentration distribution is ensured and where β is a constant, which is positive when the liquid flow is in the direction from the solution with low ionic concentration to that with a high concentration. The combination of eqs. (5.5.25)–(5.5.28) leads to the following differential equation:

$$x'' + Ax' + \omega^2 \dot{x} = 0 \tag{5.5.29}$$

where

$$x = \Delta P - \Delta P_0; \quad A = \frac{\pi r^4}{8\eta l \alpha}; \quad \omega^2 = \frac{\varepsilon j}{4\pi\eta} \frac{\beta}{\alpha}$$

and has the following solution:

$$x = c_1 e^{-s_1 t} + c_2 e^{-s_2 t} \tag{5.5.30}$$

where $s_{1,2} = \frac{1}{2}A \pm \frac{1}{2}\sqrt{(A^2 - 4\omega^2)}$ and $c_{1,2}$ are constants.

The solution indicates a damped oscillation for $4\omega^2 > A^2$. It is essential that ω^2, and therefore β, is positive. For small r and/or large l, so, that damping is negligible, the oscillation frequency is given by ω. Also V_{el} and V_P may oscillate. In that case also E, the potential difference between the two ends of the pore, must oscillate. This is understandable because E is the streaming potential. A streaming potential is the result of the flow of a charged liquid through the pore. At constant j an oscillating streaming potential must lead to an oscillating potential difference between the two reversible electrodes.

Teorell explained the oscillatory behaviour by assuming special properties of the pore conductance (in our treatment eq. (5.5.28)). The similarity of the model with that of Mueller and Rudin implied that also in that case the existence of pores with special electrical properties was essential. These properties were presumed to be due to the presence of alamethicin, protamine, and a ionic gradient. The alamethicin might have fixed charges at the probably hydrophilic inside part, its hydrophobic outside part serving to attach the pore in the hydrophobic membrane core. The presence of valinomycin instead of alamethicin resulted in an ohmic character of the conductance. So far no oscillations seem to have been found with valinomycin. Peculiar electrical effects have been found with a number of ill-defined metabolically active substances (termed EIM: excitability inducing materials). A drastic decrease (about 100-fold) of the membrane resistivity was reported by Van den Berg[147] resulting from the addition to the membrane system of a combination of protein derived from ox red blood cells together with Na-polyphosphates. Pore formation might have taken place here too.

Returning to biological membranes, the similarity of their behaviour to the findings reported above suggests that biological membranes might also contain pores. In view of the numerous experimental facts concerning transport through biological membranes[148] it is difficult to give a firm statement. However, if pores are present, their density is probably low; an estimate of about 13 per μ^2 has been suggested for lobster nerve membrane. This was measured by adding a known amount of tetrodoxin (mol. wt. 319) and assuming that each molecule blocks one pore,[149] by means of the occupation of a cationic site, which is assumed to be situated near the "entrance" of a pore. We have seen that a biological membrane contains a structured protein layer. If "pores" are real entities in these membranes they are probably connected with these structured surface layers. It seems too early to give a detailed description of the exceedingly complicated phenomenon of membrane excitation. Therefore we conclude this section with some remarks, which may have a general character, although it is realized that the extent of this field of research may make, and indeed has made, the value of such remarks a matter of controversy.

The fact that only few cationic sites are responsible for the all-or-none effect of excitation was used by Lehninger[117] as a basis for his hypothesis of a lipid–protein superstructure in which structured water layers adsorbed between protein and lipid[150] play a role. Upon excitation a "phase-change" of this superstructure will take place. Arguments in favour of such a transient phase change are the observations that the birefringence and also the light-scattering properties[123] change somewhat during the excitation. Calorimetric measurements point in the same direction. Hill[151] and later Abbott et al.[152] measured a heat production of $37 \cdot 7$ μJ g^{-1} ($= 9$ μcal g^{-1}) during excitation, followed by a heat absorption of $29 \cdot 3$ $\mu J\,g^{-1}$ ($=7\,\mu cal\,g^{-1}$). This figure seemed fairly representative for a number of nerve membranes.[153] The calorimetric effects point to a temporary entropy decrease. Finally, the fact that the potential drop across the membrane decreased from about 50 mV to zero and eventually changed its sign means that a "natural" negative

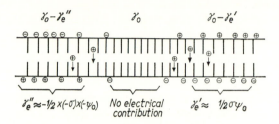

FIG. 5.24. Membrane showing lateral effects of a sign reversal of potential difference across the membrane.

contribution $\Delta\gamma_{el}$ of about $10^{-4}\,\mathrm{Nm^{-1}}$ ($=0 \cdot 1$ dyn cm^{-1}) (see eq. (5.5.23)) to the interfacial tension, which is present in the resting state, is lost during the excitation. A temporary contraction must therefore take place in the excited part of the membrane.

REFERENCES

1. J. W. GIBBS, *Collected Works*, Vol. I, Yale U.P., p. 275, 1948.
2. K. J. MYSELS, K. SHINODA and S. FRANKEL, *Soap Films*, Pergamon Press, 1959.
3. A. POCKELS, *Nature*, 1891, **43**, 437; 1892, **46**, 418; 1893, **48**, 152; 1894, **50**, 223.
4. N. K. ADAM, *The Physics and Chemistry of Surfaces*, Oxford, 1930.
5. E. K. RIDEAL, *Surface Chemistry*, Cambridge, 1930; see also ref. 17.
6. I. LANGMUIR, *J. Am. Chem. Soc.*, 1916, **38**, 2221; 1917, **39**, 1848.
7. W. D. HARKINS, *The Physical Chemistry of Surface Films*, New York, 1952.
8. G. S. HARTLEY, *Aqueous Solutions of Paraffin Chain Salts*, Paris, 1936.
9. C. P. ROE and P. D. BRASS, *J. Am. Chem. Soc.*, 1954, **76**, 4703.
10. P. MUKERJEE, *J. Phys. Chem.*, 1962, **66**, 1375.
11. K. SHINODA, T. NAKAGAWA, B-I. TAMAMUSHI and T. ISEMURA, *Colloidal Surfactants*, Academic Press, 1963.
12. F. VAN VOORST VADER, *Trans. Faraday Soc.*, 1960, **56**, 110.
13. H. KLING and W. LANGE, *Proc. IInd Intern. Congr. of Surface Act.*, **I**, 295, Butterworth, 1958.
13a. M. VAN DEN TEMPEL, *Rec. Trav. Chim. Pays-Bas*, 1953, **72**, 419.
14. J. TRAUBE, *Liebig's Ann.*, 1891, **265**, 27.
15. J. D. ROBERTSON, *Proc. Symp. Biophysics and Physiology of Biological Transport*, Frascati 1965, Springer 1967, p. 218, reviews the history of the "unit-membrane" concept.

16. W. A. ZISMAN, *Rev. Sci. Instr.,* 1932, **3**, 367; M. K. BERNETT and W. A. ZISMAN, *J. Phys. Chem.,* 1963, **67**, 1534.
17. J. T. DAVIES and E. K. RIDEAL, *Interfacial Phenomena*, Academic Press, 1963.
17a. J. T. DAVIES, *Z. Elektrochemie*, 1951, **55**, 559; *Nature,* 1951, **167**, 193.
18. J. MINGINS, F. G. R. ZOBEL, B. A. PETHICA and C. SMART, *Proc. Roy. Soc. London* A, 1971, **324**, 99.
19. F. B. KENRICK, *Z. Physik. Chem.,* 1896, **19**, 625.
20. A. N. FRUMKIN, (a) *Z. Physik. Chem.,* 1924, **109**, 34; (b) *Electrochimica Acta*, 1960, **2**, 351.
21. B. KAMIÈNSKI (a) *Proc. IInd Intern. Congr. of Surface Act.,* 1958, **III**, 103; see also his papers in: (b) *the Proceedings of the IIId* (1960) and: (c) of the *IVth Congr. of Surface Act* (held 1964, Brussels, and published 1967, Vol. **2**, pp. 211 and 225)
22. J. E. B. RANDLES, (a) *Trans. Faraday Soc.,* 1956, **52**, 1573; (b) *Adv. Electrochem. and Electrochem. Eng.,* 1962, **3**, 1.
23. J. GUYOT, *Ann. Phys.,* [11] 1924, **2**, 506.
24. J. M. CORKILL, J. F. GOODMAN, S. P. HARROLD and J. R. TATE, *Trans. Faraday Soc.,* 1967, **63**, 247.
25. (a) J. GUASTALLA, *J. Coll. Sci.,* 1956, **11**, 623; (b) C. GAVACH, T. MLODNIKA and J. GUASTALLA, *C. R. Acad. Sci. Paris*, 1968, **266C**, 1196.
26. L. TER MINASSIAN-SARAGA, *J. Coll. Sci.,* 1956, **11**, 398.
27. I. LANGMUIR, *J. Chem. Phys.,* 1933, **1**, 756; *J. Franklin Inst.,* 1934, **218**, 143.
28. A. N. FRUMKIN, *Electrochimica Acta*, 1960, **2**, 351.
29. N. K. ADAM and J. HARDING, *Proc. Roy. Soc.* A, 1932, **138**, 411.
30. M. K. BERNETT, N. L. JARVIS and W. A. ZISMAN, *J. Phys. Chem.,* 1964, **68**, 3520.
31. I. LANGMUIR and V. J. SCHAEFER, *J. Am. Chem. Soc.,* 1936, **58**, 284; 1937, **59**, 2400.
32. L. ONSAGER and N. M. I. SAMARAS, *J. Chem. Phys.,* 1934, **2**, 528.
33. S. LEVINE, G. M. BELL and B. A. PETHICA, *J. Chem. Phys.,* 1964, **40**, 2304.
33a. E. D. GODDARD, S. R. SMITH and L. H. LANDER, *Proc. IVth Intern. Congr. Surface Act,* Brussels, 1964, Gordon & Breach, 1967, **2**, 199.
34. TH. A. J. PAYENS, *Philips Res. Repts.,* 1955, **10**, 425.
35. D. STIGTER, *J. Coll. Interface Sci.,* 1967, **23**, 379.
36. F. BOOTH, *Proc. Roy. Soc.* A, 1950, **203**, 533.
37. E. HÜCKEL, *Physik. Z.,* 1924, **25**, 204; A. J. RUTGERS and J. TH. G. OVERBEEK, *Z. physik. Chem.* A, 1936, **177**, 33.
38. J. TH. G. OVERBEEK, Thesis, Utrecht, 1941; *Kolloidchem. Beihefte,* 1943, **54**, 287; see also *Philips Res. Repts.*, 1946, **1**, 315.
39. F. BOOTH, *Nature,* 1948, **161**, 83; *Proc. Roy. Soc.* A, 1950, **203**, 514.
40. P. H. WIERSEMA, A. L. LOEB and J. TH. G. OVERBEEK, *J. Coll. Interface Sci.,* 1966, **22**, 78; P. H. WIERSEMA, Thesis Utrecht, 1964.
41. P. DEBYE, *J. Appl. Phys.,* 1944, **15**, 338; *J. Phys. Colloid Chem.,* 1947, **51**, 18. (b) B. CHU, *Molecular Forces*, Interscience, 1967, Chap. 6; (c) J. TH. G. OVERBEEK, A. VRIJ and H. F. HUISMAN, *I.C.E.S.* (1963) 320, Pergamon Press.
42. P. DOTY and R. F. STEINER, *J. Chem. Phys.,* 1950, **18**, 1211.
43. F. ZERNIKE, *Arch. Néerl. Sci.,* 1918, **IIIA**, 74; thesis Amsterdam, 1915; see also ref. 41 (b) and (c).
44. A. VRIJ and J. TH. G. OVERBEEK, *J. Coll. Sci.,* 1962, **17**, 570.

45. K. J. MYSELS, *J. Coll. Sci.,* 1955, **10**, 507.
46. D. STIGTER, *Rec. Trav. Chim. Pays-Bas,* 1954, **73**, 593.
47. P. DEBYE and E. W. ANACKER, *J. Phys. and Colloid Chem.,* 1951, **55**, 644.
48. D. STIGTER, *J. Phys. Chem.,* 1966, **70**, 1323.
49. G. D. HALSEY, *J. Phys. and Colloid Chem.,* 1953, **57**, 87.
50. R. H. ARANOW, *J. Phys. Chem.,* 1963, **67**, 556; C. A. J. HOEVE and G. C. BENSON, *J. Phys. Chem.,* 1957, **61**, 1149.
51. P. MUKERJEE, *J. Phys. Chem.,* 1962, **66**, 1375.
52. P. DEBYE, *Ann. N.Y. Acad. Sci.,* 1949, **51**, 575.
53. H. F. HUISMAN, thesis Utrecht, 1964, p. 41; *Proc. Koninklijke Nederland. Akad. Wetenschap.* 1964, **B67**, 367, 476, 388, 407; in part, p. 406.
54. J. TH. G. OVERBEEK and D. STIGTER, *Rec. Trav. Chim. Pays-Bas,* 1956, **75**, 1263.
55. A. LOEB, P. H. WIERSEMA and J. TH. G. OVERBEEK, *The Electrical Double Layer Around a Spherical Colloid Particle,* M.I.T. Press, 1961.
56. D. STIGTER, *J. Phys. Chem.,* 1964, **68**, 3603.
57. D. G. DERVICHIAN, *Proc. IIIrd Intern. Congr. Surface Act.,* 1960, **1**, 182.
58. E. MATIJEVIČ and B. A. PETHICA, *Trans. Faraday Soc.,* 1958, **54**, 587; D. G. HALL and B. A. PETHICA, in Non-ionic Surfactants, I, chap. 16, (Ed. M. J. SCHICK), Marcel Dekker, 1967.
59. G. STAINSBY and A. E. ALEXANDER, *Trans. Faraday Soc.,* 1950, **46**, 587.
60. P. WHITE and G. C. BENSON, *Trans. Faraday Soc.,* 1959, **55**, 1025.
61. R. HOOKE, *Comm. to the Royal Society,* March 28, 1672; ref. 63.
62. I. NEWTON, *Opticks* 1704, Book II, Part I, obs. 17–21; ref. 63.
63. J. TH. G. OVERBEEK, *J. Phys. Chem.,* 1960, **64**, 1178; *Proc. IVth Intern. Congr. Surface Act,* 1964, **2**, 19, Gordon & Breach, 1967.
64. A. SCHELUDKO, *Adv. Coll. and Interface Sci.,* 1967, **1**, 391.
65. B. V. DERJAGUIN and A. S. TITIJEVSKAJA, *Koll. Zh.,* 1953, **15**, 416; *Proc. IInd Intern. Congr. Surface Act.,* 1957, **1**, 211; *Discussions Faraday Soc.,* 1954, **18**, 24; *J. Coll. Sci.,* 1964, **19**, 113.
66. J. LIJKLEMA, P. C. SCHOLTEN and K. MYSELS, *J. Phys. Chem.,* 1965, **69**, 116.
67. A. J. DE VRIES, *Rec. Trav. Chim. Pays-Bas,* 1958, **77**, 383.
68. B. V. DERJAGUIN and M. M. KUSSAKOV, *Acta Physicochim. U.R.S.S.,* 1939, **10**, 25.
69. LORD RAYLEIGH, *Proc. Roy. Soc. A,* 1936, **156**, 343.
70. (a) K. J. MYSELS and M. N. JONES, *Discussions Faraday Soc.,* 1966, **42**, 42; (b) J. LIJKLEMA and K. J. MYSELS, *J. Am. Chem. Soc.,* 1965, **87**, 2539.
71. M. N. JONES, K. J. MYSELS and P. SCHOLTEN, *Trans. Faraday Soc.,* 1966, **62**, 1336.
72. E. M. DUYVIS, thesis Utrecht, 1962.
73. J. S. CLUNIE, J. M. CORKILL, J. F. GOODMAN and C. P. OGDEN, *Trans. Faraday Soc.,* 1967, **63**, 505.
74. A. W. REINHOLD and A. W. RÜCKER, *Phil. Trans. Roy. Soc.,* 1883, **174**, 6435.
75. J. J. BIKERMAN, *Z. Physik. Chem. A,* 1933, **163**, 378.
76. A. SCHELUDKO, *Proc. Kon. Ned. Akad. Wetenschap B,* 1962, **65**, 65.
77. A. VRIJ, *Discussions Faraday Soc.,* 1966, **42**, 23.
78. A. VRIJ and J. TH. G. OVERBEEK, *J. Am. Chem. Soc.,* 1968, **90**, 3074.
79. D. EXEROWA and T. KOLAROW, *Ann. Univ. Sofia, Fac. Chim.,* 1964/5, **59**, 207; see also ref. 64.
80. J. FRENKEL, *Theory of Liquids,* Oxford U.P., London, 1955, chap. 6.
81. T. TEORELL, *Discussions Faraday Soc.,* 1956, **21**, 9.

82. E. P. Honig, J. H. Th. Hengst and P. Hirsch-Ayalon, *Ber. Bunseng. für Physik. Chemie*, 1968, **72**, 1231; P. Hirsch-Ayalon, *Rec. Trav. Chim. Pays-Bas*, 1956, **75**, 1065.

83. J. E. Hall and L. J. Bruner, *J. Chem. Phys.*, 1969, **50**, 1596.

84. G. H. Jonker and H. R. Kruyt, *Discussions Faraday Soc.*, 1954, **18**, 170.

85. R. H. Ottewill and R. F. Woodbridge, *J. Coll. Sci.*, 1964, **19**, 606.

86. B. A. Adam and E. C. Holmes, *J. Soc. Chem. Ind. (London)*, 1935, **54**, IT; 1935, EP 450308.

87. e.g. H. Van Olphen, *An Introduction to Clay Colloid Chemistry*, Interscience, 1963.

88. S. Loeb and S. Sourirajan, *Adv. Chem. Ser.*, 1963, **38**, 117; J. Th. G. Overbeek, *Verslagen Kon. Ned. Akad. Wetenschap*, 1968, **77**, 44.

89. J. J. Hermans and J. Th. G. Overbeek, *Rec. trav. chim. Pays-Bas*, 1948, **67**, 761; *Bull. Soc. Chim. Belges*, 1948, **57**, 154.

90. W. Kuhn, O. Künzle and A. Katchalsky, *Bull. Soc. Chim. Belges*, 1948, **57**, 421.

91. H. Morawetz, *Macromolecules in Solution*, Interscience, 1966 (Volume 21 of the series *High Polymers*).

92. J. Th. G. Overbeek, *J. Coll. Sci.*, 1953, **8**, 593.

93. F. Helfferich, *Ionenaustauscher*, I Verlag Chemie, Weinheim, 1959.

94. G. Scatchard, *J. Am. Chem. Soc.*, 1953, **75**, 2883; see also contribution in Cohn and Edsall's *Proteins, Amino Acids and Peptides*, Reinhold, 1943.

95. F. Bergsma and Ch. A. Kruissink, Fortschritte Hochpolymeren-Forschung. *Adv. Pol. Sci.*, 1961, **2**, 307.

96. D. A. MacInnes, *The Principles of Electrochemistry*, chap. 13, Reinhold, 1939.

97. J. W. Lorimer, Miss E. I. Boterenbrood and J. J. Hermans, *Disc. Faraday Soc.*, 1956, **21**, 141 (water contribution 4%); F. Bergsma and A. J. Staverman, *Disc. Faraday Soc.*, 1956, **21**, 61 (bi-ionic potentials); A. J. Staverman, *Trans. Faraday Soc.*, 1952, **48**, 176 (classical paper).

98. W. Nernst, *Theoretische Chemie*, 11th–15th ed., Stuttgart, 1926.

99. H. Dember, *Physik. Z.*, 1931, **32**, 554, 856.

100. T. Teorell, *Proc. Natl. Acad. Sci.*, 1935, **21**, 152; K. H. Meyer and J. F. Sievers, *Helv. Chim. Acta*, 1936, **19**, 649, 665, 987; 1937, **20**, 634.

101. O. Kedem and A. Katchalsky, *Trans. Faraday Soc.*, 1963, **59**, 1918, 1931, 1941.

102. W. Nernst, *Z. Physik. Chemie*, 1884, **2**, 613; 1889, **4**, 129.

103. M. Planck, *Ann. d. Chemie*, 1890, **39**, 161.

104. T. Teorell, *Z. Elektrochemie*, 1952, **55**, 460; *Proc. biophysics*, 1951, **3**, 305.

105. F. Helfferich, *Discussions Faraday Soc.*, 1956, **21**, 83; ref. 93.

106. R. Schlögl, *Stofftransport durch Membranen*, Darmstadt, 1963; *Z. physik. Chemie* N.F., 1954, **1**, 305; 1955, **3**, 73 (osmosis).

107. M. Schödel, *Z. physik. Chemie* N.F., 1955, **5**, 372 (rectification).

108. J. Bernstein, *Pflüger's Arch.*, 1902, **92**, 521; 1910, **131**, 589.

109. E. Overton, *Pflüger's Arch.*, 1902, **92**, 346; 1904, **105**, 176.

110. A. Fleckenstein, *Der Kalium-Natrium Austausch*, Springer, 1955.

111. J. Z. Young, *Cold Spring Harbor Symposium*, 1936, **4**, 1.

112. K. S. Cole and H. J. Curtis, *J. gen. Physiol.*, 1939, **22**, 37, 649; K. S. Cole and A. L. Hodgkin, *ibid.*, p. 671; K. S. Cole and R. F. Baker, *ibid.*, 1941, **24**, 771; K. S. Cole, *ibid.*, 1941, **25**, 29.

113. A. L. Hodgkin and A. F. Huxley, *J. Physiol.*, 1952, **116**, 449, 473, 497; 1952, **117**, 500.

114. R. D. Keynes and R. H. Adrian, *Disc. Faraday Soc.,* 1956, **21**, 265.
115. I. Tasaki, A. Watanabe and S. Tanaka, *Proc. Natl. Acad. Sci.,* 1962, **48**, 1177; I. Tasaki and M. Shimamura, *ibid.,* 1571.
116. F. Brink, *Pharmacol. Rev.,* 1954, **4**, 243.
117. A. L. Lehninger, *Proc. Natl. Acad. Sci.,* 1968, **60**, 1069; for acetylcholine: e.g. Nachmansohn, *Ann. New York Acad. Sci.,* 1966, **137**, 877.
118. J. F. Danielli and H. A. Davson, *J. Cell. Comp. Physiol.,* 1935, **5**, 495; *The Permeability of Natural Membranes,* Cambridge U.P., 1952.
119. E. Gorter and R. Grendel, *J. Exper. Med.,* 1925, **41**, 439.
120. K. S. Cole, *J. Cell. Comp. Physiol.,* 1932, **1**, 1.
121. E. N. Harvey and H. Schipiro, *J. Cell. Comp. Physiol.,* 1934, **5**, 255.
122. F. O. Schmitt and R. S. Bear, *Biol. Rev.,* 1939, **14**, 27.
123. L. B. Cohen and R. D. Keynes and B. Hille, *Nature,* 1968, **218**, 438.
124. I. Langmuir, *J. Am. Chem. Soc.,* 1934, **56**, 495; K. B. Blodgett, *ibid.,* 1935, **57**, 1007; K. B. Blodgett and I. Langmuir, *Phys. Rev.,* 1937, **51**, 964.
125. H. Bücher, K. H. Drexhage, M. Fleck, H. Kuhn, D. Möbius, F. P. Schäfer, J. Sondermann, W. Sperling, P. Tillmann and J. Wiegand, *Molecular Crystals,* 1967, **2**, 199.
125a. D. den Engelsen, *J. Opt. Soc. Am.,* 1971, **61**, 1460.
126. H. Bücher, H. Kuhn, B. Mann, D. Möbius, L. von Szentpály and P. Tillmann, *Photographic Sci. and Eng.,* 1967, **11**, 233.
127. W. Stoeckenius, *J. Cell Biol.,* 1962, **12**, 221.
128. V. Luzzatti and F. Husson, *J. Cell Biol.,* 1962, **12**, 207.
129. E. D. Korn, *J. Gen. Physiol.,* 1968, **52**, 257.
130. D. Chapman, *Science J.,* 1968, no. 3, 55.
131. F. S. Sjöstrand, *Proc. Symposium Biophysics and Physiology of Biol. Transport,* Frascati, 1965, Springer, 1967, p. 248; in *The Membranes* Vol. 4 of the series *Ultrastructure in Biological Systems,* Academic Press, 1968, p. 151.
132. A. M. Glauert and J. A. Lucy, in *The Membranes,* Vol. 4 of the series *Ultrastructure in Biological Systems,* Academic Press, 1968, p. 1.
133. E. D. Korn, *Science,* 1966, **153**, 1491, and ref. 129.
134. S. Fleischer, B. Fleischer and W. Stoeckenius, *Fed. Proc.,* 1965, **24**, 296.
135. D. F. H. Wallach, *Proc. Natl. Acad. Sci.,* 1966, **56**, 1613.
136. J. Lenard and S. J. Singer, *Proc. Natl. Acad. Sci.,* 1966, **56**, 1828.
137. D. A. Haydon and J. Taylor, *Proc. Roy. Soc. A,* 1963, **281**, 377.
138. I. Langmuir and D. F. Waugh, *J. Gen. Physiol.,* 1938, **21**, 745.
139. P. Mueller and D. O. Rudin, H. I. Tien and W. C. Westcott, *Nature,* 1962, **194**, 979; *J. Phys. Chem.,* 1963, **67**, 534.
140. T. Hanai, D. A. Haydon and J. Taylor, *Proc. Roy. Soc. A,* 1964, **281**, 377.
141. S. H. White, *Biophys. J.,* 1971, **10**, 1127.
142. H. J. Van den Berg, *J. Mol. Biol.,* 1965, **12**, 290.
143. A. Finkelstein and A. Cass, *J. Gen. Physiol.,* 1968, **52**, 145s.
144. E. Robbins and A. Mauro, *J. Gen. Physiol.,* 1960, **43**, 523.
145. P. Mueller and D. O. Rudin, *Biochem. Biophys. Res. Comm.,* 1967, **26**, 398; *Nature,* 1968, **217**, 713.
145a. V. T. Ivanov, I. A. Laine, N. D. Abdulaev, L. B. Senyavina, E. M. Popov, Yu. A. Ovchinnikov and M. M. Shemyakin, *Biochem. Biophys. Res. Comm.* 1969, **34**, 803.
146. T. Teorell, *Ann. New York Acad. Sci.,* 1966, **127**, 950.

147. H. J. Van den Berg, *Adv. in Chem. Series* (1968), **84,** 99.
148. E. Heinz, *Ann. Rev. Physiol.,* 1967, **29,** 21.
149. J. W. Moore, T. Marahashi and T. I. Shaw, *J. Physiol.,* 1967, **188,** 99.
150. F. A. Vandenheuvel, *Proc. Symposium Biophysics and Physiology of Biol. Transport,* Frascati, 1965, Springer, 1967, p. 188.
151. A. V. Hill, *Chemical Wave Transmission in Nerve,* Cambridge U.P., 1952, and e.g. paper by Danielli in *Surface Phenomena in Chemistry and Biology,* Pergamon Press, 1957, p. 246.
152. B. C. Abbott, A. V. Hill and J. V. Howarth, *Proc. Roy. Soc.* B, 1958, **148,** 149.
153. B. C. Abbott, J. V. Howarth and J. M. Ritchie, *J. Physiol.,* 1965, **178,** 368.

CHAPTER 6

SEMICONDUCTOR SURFACES

6.1. Introduction

6.1.1. *Types of conduction*

Crystalline solids can, as far as their electrical properties are concerned, be broadly divided into three classes:

1. *Metals*

2. *Semiconductors*. We distinguish between three groups:

   (a) Electronic semiconductors, e.g. Ge, Si. The behaviour of the electrons in these semiconductors, and in metals, has formed an important chapter in solid state physics.[1,2]

   (b) Ionic semiconductors, e.g. numerous halides, including silver halides. Materials, showing a very high (10^{-1}–10^{-2} Ω^{-1} cm^{-1}) ionic conductivity, such as XAg_4I_5 (X alkali metal)[3] and Ag_3SBr[4] have recently been found.

   (c) Mixed semiconductors, e.g. Ag_2S. Faraday[5,6] discovered typical semiconductor properties in his experiments with Ag_2S. Thus he found that the electrical conductance increased when the temperature was raised. The properties of ionic and mixed semiconductors also form an important chapter of solid state chemistry.[7]

3. *Insulators*

Wagner[8] developed experimental techniques, based on earlier observations by Tubandt[9] to distinguish between the electronic and the ionic contributions to the total conductance. They used the following facts. The interfaces between (solid) Ag and (solid) AgI, and between (solid) AgI and (solid) Ag_2S are non-polarizable in the sense of Chapter 4, Ag^+ ions being capable of passing across the interfaces.

Furthermore, the Pt/(solid) AgI interface is polarizable in the sense of Chapter 3. Neither ions nor electrons are capable of passing across the interface under normal laboratory conditions. Finally only electrons can pass across the Pt/Ag_2S interface. Wagner constructed galvanic cells based on the behaviour of these and similar interfaces for investigations of ionic and electronic conduction.

We shall be mostly concerned with electronic semiconductors. These have been widely studied, in particular after the advent of the transistor in 1948–9. Also the *surfaces* of these semiconductors have received much attention, and are discussed in textbooks such as those of Many et al[10]. and of Frankl,[11] and also in articles such as the one by Green.[12] There are two reasons for this interest. The first is that semiconductor surface studies have proved very important in the manufacture of transistors and other semiconductor devices. The second is that the mobilities and the concentrations of conduction electrons and of holes have convenient values which can be varied and controlled with great precision. A pure Ge crystal at 0°C has a specific resistance of 60 Ωcm. It then contains 3×10^{13} conduction electrons and holes per cm^3. These are formed through an "ionization" reaction:

$$\text{Ge} \leftrightarrows \text{Ge}^+ + e^- \quad \leftarrow \text{recombination}$$

or   (6.1.1)

$$\text{O} \leftrightarrows h^+ + e^- \qquad \rightarrow \text{ionization, formation of hole–electron pair}$$

where h^+ (an ionized Ge atom minus a neutral Ge atom) denotes a hole and where e^- denotes an electron. The mobility of a conduction electron at room temperature in Ge is 3900 cm^2/V sec, and that of a hole is 1900 cm^2/V sec. These large mobilities should be contrasted with the mobilities of ions in aqueous solutions or in ionic semiconductors, which are a factor 10^8 or more lower. The large conductivity values in XAg_4I_5 and in Ag_3SBr are explained by assuming an almost random distribution of the Ag^+ ions through the lattice, the energy barriers between two neighbouring positions being only small. In other ionic semiconductors a vacancy mechanism is invoked. Some ions are removed from their lattice positions by thermal agitation. This leads to Schottky or Frenkel disorder. It is relatively easy for ions on lattice positions neighbouring a vacancy to move and to occupy a vacant lattice site. Conduction electrons and holes in germanium and silicon can be considered as almost free.

On the whole the ionic conductance depends more strongly on temperature than does the electronic conductance. Usually (at room temperature) the values of the ionic conductivity are much smaller than those of the electronic conductivity, although the concentration of mobile ions may be much larger than that of the electrons.

6.1.2. *Conduction electrons and holes*

We assume a crystal which initially is expanded in such a way that the interatomic distance is very large. In this case the atoms can be considered as isolated. Upon a decrease of the interatomic distance the wave functions of the outer electrons of these atoms overlap and as a result the originally identical energy levels pertaining to these electrons spread out and form energy bands. Figure 6.3(a) represents the general situation of diamond and diamond-type crystals. In this figure the $2s$ and $2p$ levels of carbon become bands. These are indicated by the shaded areas. Eventually, when the interatomic distance has become sufficiently small, these bands cross. Before band crossing (large interatomic distance) the $2s$ band is capable of holding two electrons per atom. After crossing the lower band can hold four electrons per atom. For diamond-type crystals we are at some value at the left-hand side of the crossing point of the bands. The highest energy value of the lower band (the valence band) is denoted as E_v and the lowest value of the upper (conduction) band is E_c. The energy values between E_c and E_v form the forbidden energy gap. The number of discrete levels in the energy bands is very high and, as will be pointed out in Subsection 6.1.2, the energy distribution can be considered as quasi-continuous. In equilibrium some of the electrons of the valence band are excited and have energy values higher than E_c. These are the conduction electrons. The excitation of these electrons gives rise to the formation of holes according to eq. (6.1.1). They have energies lower than E_v. Owing to the quasi-continuous nature of the bands, both conduction electrons and holes can be considered as free, the only effect of the presence of semiconductor atoms of the crystal being that the electron mass m_e must be replaced by an effective mass $m_e{}^*$. This is an anisotropic quantity. An electron in free space has an energy E, given by

$$E = \frac{1}{2m_e} (p_x{}^2 + p_y{}^2 + p_z{}^2).$$ (6.1.2)

For free electrons we have the de Broglie expressions

$$p_x = \hbar k_x; \quad p_y = \hbar k_y; \quad p_z = \hbar k_z$$ (6.1.3)

where the quantities k_x, k_y and k_z are components of the wave vector in the x-, y-, and z-directions.

It is customary to make use of the concept of k-space when the energy of free electrons, of electrons in metals, and of electrons in electronic semiconductors, are discussed. It is easily seen that, in k-space, the energy is a surface and that for free electrons each E-value is represented by a spherical surface. As a general rule, this is no longer the case in metals and electronic semiconductors, where surfaces of constant energy may have a much more complicated shape, depending on symmetry properties and other characteristics of the crystal. Solid state theory tells us that in Ge and Si crystals energy surfaces for small k-values are ellipsoids. For Ge the shape of each ellipsoid is such that in two directions an effective mass of $0 \cdot 082\ m_e$, and in the third direction an effective mass of $1 \cdot 59\ m_e$ must be assigned to the electrons. For Si these figures are $0 \cdot 19\ m_e$ and $0 \cdot 98\ m_e$. However, for statistical purposes it is often sufficient to consider an average effective mass, which we denote as $m_e{}^*$. For Ge, $m_e{}^* = 0 \cdot 55\ m_e$ and for Si, $m_e{}^* = 1 \cdot 08\ m_e$. For large k-values the situation is more complicated (see also below).

With all these simplifications in mind we write for an electron in the conduction band

$$E = E_c + \frac{\hbar^2}{2m_e{}^*}\,(k_x{}^2 + k_y{}^2 + k_z{}^2) \qquad (6.1.4)$$

which we only consider approximately valid for low k-values.

In principle the electrons in the conduction band can occupy only discrete energy levels. In a macroscopic crystal with sides L_x, L_y and L_z and lattice spacings a, b and c elementary solid state theory shows that the allowed energy states are given by k-values within the following limits:

$$-\frac{\pi}{a} < k_x \leqslant \frac{\pi}{a}; \quad -\frac{\pi}{b} < k_y \leqslant \frac{\pi}{b}; \quad -\frac{\pi}{c} < k_z \leqslant \frac{\pi}{c} \qquad (6.1.5)$$

the distances between allowed k-values being given by

$$\Delta k_x = \frac{2\pi}{L_x}; \quad \Delta k_y = \frac{2\pi}{L_y}; \quad \Delta k_z = \frac{2\pi}{L_z}. \qquad (6.1.6)$$

For each value of k there are two possible spin states. Since a, b and c are of the order of angstrom units, whereas L_x, L_y and L_z are of the order of cm, the number of allowed energy states is very large and it is usually permissible to consider the allowed energies as quasi-continuous.

The reasoning for holes is analogous. The overlapping wave functions for valence electrons lead to a valence band with an upper level denoted as E_v. When no "ionization" has taken place all the states in the valence band are occupied and no conduction is possible. Upon "ionization" some states, just below E_v, are vacant. This leads to the possibility of conduction in the valence band. A hole can now be conceived of as a mobile particle with a positive charge and a positive effective mass $m_h{}^*$. For free holes in the crystal under consideration we write

$$E = E_v - \frac{\hbar^2}{2m_h{}^*}\,(k_x{}^2 + k_y{}^2 + k_z{}^2). \qquad (6.1.7)$$

That the band theory is actually more complicated than suggested by the eqs. (6.1.4) and (6.1.7) is illustrated by Fig. 6.1. In Fig. 6.1(a) a direct transition of an electron of the valence band to the conduction band is shown. We assume that it follows the simple scheme outlined above. In a direct transition the k-value remains constant. In Fig. 6.1(b) the more complicated case of an indirect transition is shown. Here the energy vs. k curve is not a simple parabola, but it shows two minima. In an indirect transition the electron momentum is not conserved. This

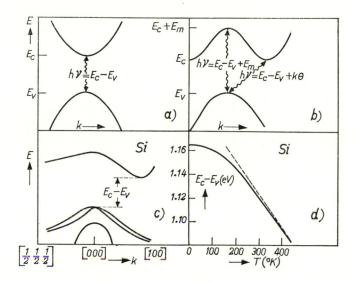

FIG. 6.1. (a) A semiconductor for which the valence band minimum and the conduction band maximum are at the same k-value. A direct transition is shown. It is the least energetic optical transition. (b) A semiconductor for which the valence band minimum and the conduction band maximum are not at the same k-value. The least energetic optical transition is an indirect transition (non-vertical arrow). This involves the emission or absorption of a phonon with energy $k\theta$. (c) Actual situation for Si. (d) Temperature dependence of intrinsic energy gap in Si (from Blakemore[1b]).

is due to an interaction with the lattice, lattice phonons being emitted or absorbed simultaneously with the electron transition. In Fig. 6.1(c) the actual situation for Si is depicted. The energy is given as a function of the wave vector along one of the cubic axes and along a cube diagonal. Here indirect transitions are possible. The valence band structure is rather complex. There is a twofold degeneracy at the extremum (the top of the valence band). From the two energy vs. k-curves we see that there are two sorts of holes: heavy holes ($m_h{}^* = 0.53\ m_e$, upper curve) and light holes ($m_h{}^* = 0.16\ m_e$), the fraction of light holes being found by theory as 0·14. For Ge these numbers are: heavy holes 0·36 m_e, light holes 0·043 m_e and the fraction of light holes: 0·04. Band theoretical calculations also provided a third, split-off band. Figure 6.1(d) gives the temperature dependence of the band gap $(E_c - E_v)$ as inferred from optical transition measurements. For Si and for Ge this is given as

$$(E_c - E_v)\text{eV} \cong 0\cdot66 + 2\tfrac{1}{2} \times 10^{-4}\,(300 - T) \quad \text{Ge};$$

$$(E_c - E_v)\text{eV} \cong 1\cdot12 + 2\tfrac{1}{2} \times 10^{-4}\,(300 - T) \quad \text{Si}.$$

$$(6.1.8)$$

An important method of obtaining information concerning the carrier concentration and of showing that the concept of a hole, i.e. of a mobile charge carrier with a positive charge, corresponds to a physical reality, is the measurement of the Hall effect. Imagine a rectangular bar, cross-section 1 cm², of semiconducting material, the length being in the x-direction. An electrostatic potential gradient in the x-direction causes a current J_x. Assuming that only one type of mobile charge carrier, charge ze, concentration n per cm³, is present, we write

$$J_x = zenv_x \tag{6.1.9}$$

where v_x is the velocity in the x-direction.

Assume a magnetic field in the z-direction, with field strength H_z. In that case there is a Lorentz force $(ze/c)v_x \wedge H_z$ in a direction normal to the x–z-plane. We assume that the deflection of the charge carriers due to the Lorentz force is small. When the charge carriers move (without the application of the magnetic field) from the right to the left, and the direction of H_z is upward, then, in the case of electrons, the charge carriers are driven toward the observer by the Lorentz force. They create an electric potential-difference across the bar in the y-direction. The electric field, the Hall field, has a field strength E_y. After the Hall field is built up, the force upon the charge carriers in the y-direction is zero

$$f_y = 0 = \frac{ze}{c}\, v_x \wedge H_z + zeE_y. \tag{6.1.10}$$

The Hall coefficient R_H is

$$R_H = \frac{E_y}{J_x \wedge H_z} = \frac{1}{zenc}. \tag{6.1.11}$$

Since E_y, J_x and H_z are known or can be measured, n can be found. Also the sign of the charge carriers is known, z being -1 for electrons and $+1$ for holes. For mixed conductance, e.g. for conductance by both electrons *and* holes, the analysis is more complicated. For small H_z-values the result is

$$R_H = \frac{\sigma_e^2 R_e + \sigma_p^2 R_p}{(\sigma_e + \sigma_p)^2} \qquad (6.1.12)$$

where $\sigma_e = n\,eu_e$ and $\sigma_p = p\,eu_p$ are the conductivities of the electrons and holes (u_e and u_p are the mobilities) and where R_e and R_p are respectively $-(nec)^{-1}$ and $(pec)^{-1}$. When also light holes are present, the equations are still more complicated (Frankl's book gives further details).[11]

We shall consider electrons and holes as components in the sense of Chapter 2. In equilibrium we have the following relation between the electrochemical potentials $\tilde{\mu}_e$ (electrons) and $\tilde{\mu}_h$ (holes):

$$\tilde{\mu}_e + \tilde{\mu}_h = 0. \qquad (6.1.13)$$

This is found by inspection of eq. (6.1.1). The electrochemical potential of the electrons is called the Fermi energy and denoted as E_F. Thus

$$\begin{aligned} \tilde{\mu}_e &= E_F, \\ \tilde{\mu}_h &= -E_F. \end{aligned} \qquad (6.1.14)$$

In equilibrium, E_F has a uniform value through the whole crystal, i.e. a curve of E_F versus the distance x from the surface is a straight line parallel to the x-axis which we take as the abscissa. When E_c and E_v are plotted in the same way, assuming a free particle approximation for conduction electrons and holes, two horizontal straight lines are obtained, one (for E_c) above, and one below the E_F–x curve. The exact relative position of the three lines will be discussed in Subsection 6.1.4. The often used band picture is obtained when numerous horizontal lines above E_c and below E_v, representing allowed electron states, are drawn. These lines are often omitted or indicated by shading. Whereas in equilibrium the E_F–x curve is a straight horizontal line, this need not be so for the E_c–x and the E_v–x curves. These are usually bent (but remaining parallel to each other) in the neighbourhood of the surface. This band bending will be discussed in Subsection 6.1.4 and in Section 6.2.

6.1.3. *States in the forbidden zone. Surface states. Double layers*

Impurities and flaws (dislocations) in the crystal structure provide for states in the "forbidden" zone which were not accounted for in the previous subsection. There we considered a crystal which was pure and which had an ideal, i.e. completely ordered, structure.

K

As a simple example, consider a pentavalent foreign atom D on a lattice position in the Ge- or Si-lattice. Four electrons are sufficient for saturation of the chemical bonds, the fifth electron may become a conduction electron according to an "ionization" reaction:

$$D \leftrightarrows D^+ + e^-. \qquad (6.1.15)$$

The impurity atom D is a donor atom. When ionized it is empty and positively charged. When non-ionized, the electron under consideration occupies an energy state with a value, E_d, which is in the forbidden zone and which is close to E_c when the "ionization" of D is easy. On the other hand, the presence of a trivalent impurity atom A on a lattice position leads to the formation of a hole according to

$$A \leftrightarrows A^- + h^+. \qquad (6.1.16)$$

The impurity atom A is an acceptor atom. When ionized it is occupied by an electron captured at an energy E_a which, in the forbidden zone, is usually close to E_v.

Concerning the conductivity, three cases are distinguished:

1. *Intrinsic conductivity*

In a pure, ideal, crystal the conductivity is determined by the crystal properties alone. Denoting the concentration of the conduction electrons by n, and that of the holes by p, one has, according to eq. (6.1.1),

$$n = p \quad \text{(intrinsic conductivity).} \qquad (6.1.17)$$

2. *n-type conductivity*

When there are excess donor impurities or dislocations, then

$$n > p \quad \text{(\textit{n}-type conductivity. Donors).} \qquad (6.1.18)$$

3. *p-type conductivity*

When there are excess acceptor impurities or dislocations, then:

$$n < p \quad \text{(\textit{p}-type conductivity. Acceptors).} \qquad (6.1.19)$$

Impurity atoms, to concentrations ranging from 10^{14} to 10^{18} per cm^3, can be purposely built in on lattice positions, for example by diffusion at elevated temperatures. They usually dominate the effects of the dislocations in the bulk of the crystal. The diffusion process is for various cases reproducible to within 1%. Other processes are also available in semiconductor technology and all these processes are of basic importance in the electronic industry.

The largest "flaw" in the crystal is the surface. Surface atoms usually have an affinity for electrons and (or) holes, which is different from that of bulk atoms. This means that there are surface states with energies in the forbidden zone. These states may be either donor states or acceptor states, or both. This behaviour of the surface atoms leads to the existence of a surface charge σ, which may be positive or negative,

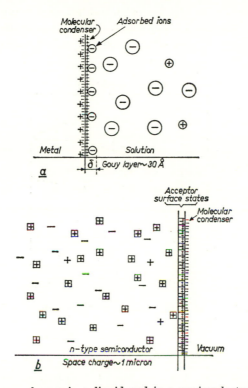

FIG. 6.2. Space charges in a liquid and in a semiconductor compared. (a) + and − signs in circles: mobile ions. (b) + signs in squares: ionized donor atoms at fixed positions.

or, in exceptional cases, zero. The surface charge is often of the order of 10^{11}–10^{12} charges per cm^2. It can be modified through chemisorption. The surface charge is compensated by a space charge bordering the surface, the depth of the space charge region being of the order of a Debye length. As we shall see in Section 6.2, the Debye length can equally well be defined in semiconductors as in liquid electrolytes. (Figure 6.2 illustrates both liquid and solid diffused layers.) This charge system constitutes the double-layer system with which we will be concerned in this chapter. In the present theoretical development one can broadly distinguish two lines: theories based on the nearly free electron (NFE) model, and secondly, theories based on the linear combinations of atomic orbitals (LCAO) model. Tamm and Shockley initiated calculations based on the NFE model, Mark and Levine introduced LCAO-type considerations. Tamm[13] predicted in 1932 the existence of surface states. He considered a periodic potential for the electrons, a potential barrier being placed between neighbouring atoms. He furthermore assumed a deformation of the potential at the outermost atoms. Using Kronig–Penney model calculations he solved the

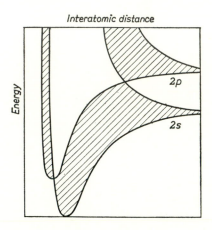

FIG. 6.3(a). A plot of energy against interatomic distance for diamond and diamond-type crystals. (Slater.[1f])

Schrödinger equation for this problem with the condition that the wave function be continuous at the surface. Tamm arrived at the conclusion that, depending on the potential deformation, one state per atom at the surface may exist. For a surface state the wave function decays at both sides of the surface. According to one recent estimate[14] the wave function penetrates less than 1 angstrom in vacuum and about 20 angstrom units in the interior of the crystal.

In the discussion that followed Tamm's work, the paper by Shockley[15] has an outstanding position. Shockley considered in detail a linear set of atoms with initially a very large interatomic distance. He also discussed, more briefly, the three-dimensional case. Upon decreasing the interatomic distance (a in Fig. 6.3(b), which is taken from Shockley's one-dimensional model) the allowed energies for the electrons formed bands as usual. A potential deformation as envisaged by Tamm might have led to a separate level in the forbidden gap. However, Shockley made the point that in the forbidden zone where band crossing has taken place, energy levels may also be found in the forbidden zone *without* the introduction of a potential deformation at the surface. These are Shockley states. Thus, in this view, there is reason to distinguish between Tamm and Shockley states. Koutecky[16] interpreted Shockley states as due to unsaturated valencies at the

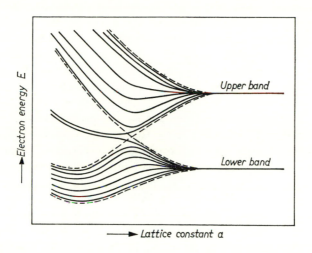

Fig. 6.3(b). Surface states according to Shockley.[15] The dotted lines enclose the energy bands. The two remaining solid lines in the forbidden gap, at sufficiently small values of the lattice constant, indicate energy levels at the surface. These are the surface states.

surface atoms or "dangling bonds" (viz. Frankl's book) but this interpretation has been challenged. In Tamm's and in Shockley's model, when extended to three dimensions, the crystals were assumed to be ideally periodic right up to the surface. In a number of cases this is certainly not the situation of minimum free energy and in these cases *ab initio* calculations, if these were possible at all, should have shown this. A lack of crystal periodicity near the surface would have led to a Tamm-type deformed potential. On the other hand, displacements of surface atoms could have been predicted on the basis of such a potential deformation. In the light of such considerations it is difficult to give general statements and therefore a rather intuitive approach might be helpful: a surface atom may be looked upon as being intermediate between a gas atom and an atom placed in the interior of a crystal. When the bands are uncrossed this means that the allowed energy levels at the surface are found *in* the bands. That means that without the introduction of a potential deformation at the surface there are no surface states. On the other hand, when the bands are crossed, the allowed energy levels at the surface *may* fall in the forbidden gap depending on the "*a*-value" reflecting best the physical situation of the surface atoms. Neighbouring surface atoms may be characterized by different "*a*-values". This possibility will be discussed in Subsection 6.4.1.

For two other cases, which are frequently discussed, the intuitive approach may be relevant: surfaces of ionic crystals, and surfaces of *d*-band metals. Levine and Mark[17] introduced the concept of surface states on ionic crystals M^+X^-. In the gas phase the atoms M and X are stable, but in the interior of the crystal the ions M^+ and X^- are stable. At the surface their ionic character will be weakened. This simple reasoning leads one to conceive of the surface anions as donors $X^- \leftrightarrows X + e^-$. The surface cations act as acceptors: $M^+ + e^- \leftrightarrows M$. The surface must therefore contain 10^{14}–10^{15} states, donors and acceptors, per cm^2. These views may be contrasted to those of Mead.[18] Mead followed Shockley's reasoning and considered an ionic crystal as a crystal in which no band crossing had taken place, saturation at the surface being established owing precisely to the strong ionic character of the crystal. Assuming a crystal like CdS, the ionic character can be weakened when S is replaced by Se. In that case Shockley states must distinguish themselves more strongly. Both Mark and Mead presented experimental results supporting their theories. These experiments are mentioned in Section 6.3. Their interpretations may not be mutually exclusive.

Concerning *d*-band metals, surface states have been postulated by Pendry, Forstmann and Heine[19-21] as a result of solid-state theory in an attempt to explain photo emission data obtained with Ni and Cu samples.[22] Later work on tungsten, in which the energy distribution of field emitted electrons was measured,[23] also pointed to the existence of surface states.

In transition elements the 3*d*-shells are characterized by energies which lie close to those of the 4*s*-shells. In the Cu-atom the energy of the $(3d)^9(4s)^2$ configuration has only a slightly higher energy than the $(3d)^{10}(4s)^1$ configuration, the difference being about $1\frac{1}{2}$ electron volts. Thus, *d*-electrons can easily be excited to become electrons in the *s*-shell. At this point it is instructive to apply the description of *d*-band metals as given by Mott[24] in 1949. When less than ten 3*d* electrons per atom are present, a hole is formed in the *d*-band. For metallic nickel it is generally assumed that there are 0·6 electrons per atom in the *s*-band and 9·4 electrons per atom in the *d*-band. The formation of these hole–

electron pairs requires, in a metal, little energy because the field between electrons and holes is screened. According to Pendry, Forstmann and Heine, at the surface the balance between excited and non-excited electrons will be different from that in the bulk in such a way that the hole concentration is decreased. Screening here may be more difficult and is absent at the vacuum side. But it is dangerous to push a simple model too far. We refer the reader to some recent review articles for the present status in this field.[25-27]

6.1.4. *Fermi distribution*

Electrons and holes obey the laws of Fermi or Fermi–Dirac statistics. The most elegant way to derive expressions for the Fermi distribution function is to apply the grand partition function.[28] For a one-component (electrons) system the grand partition function is

$$\Xi = \sum_n e^{-nE_F/kT} Z_n \qquad (6.1.20)$$

where Z_n is the partition function of a subsystem containing n electrons

$$Z_c = \sum_i e^{-n_i E_i/kT} \qquad n = \sum n_i \qquad (6.1.21)$$

where n_i is the number of electrons in the ith state and E_i is the energy of an electron in that state.

According to Pauli's principle each state can be occupied by one electron at the most. This means that n_i can be either zero or one. It is essential, for the purpose of obtaining Fermi distribution functions, to consider subsystems with zero particles, "a barren subject of study" according to Gibbs.[29]

We consider here four special cases. For a full treatment see textbooks such as those of Tolman,[30] of Landau and Lifshitz,[28] of Blakemore[1c] and of Hill.[31]

1. An empty ionized donor state (i.e. a subsystem with $n_i = 0$) and the same state (energy E_d) in which one electron is present (i.e. a subsystem with $n_i = 1$). Then

$$\Xi = 1 + e^{(E_F - E_d)/kT}. \qquad (6.1.22)$$

The probability f_d that the state is occupied is

$$f_d = \frac{e^{(E_F - E_d)/kT}}{1 + e^{(E_F - E_d)/kT}} = \frac{1}{1 + e^{-(E_F - E_d)/kT}} \qquad (6.1.23)$$

which is the Fermi distribution function for the occupation of a donor state with energy E_d. We shall use this equation for surface donor states.

2. An empty ionized donor state ($n_i = 0$), the same state containing an electron with given spin, say in the positive direction ($n_i = 1$), and the same state containing an electron with negative spin ($n_i = 1$). This case has been discussed by Landsberg[32] and by Guggenheim.[33] Now:

$$\Xi = 1 + 2e^{(E_F - E_d)/kT} \tag{6.1.24}$$

Thus the fact that the state can accommodate an electron with positive *or* negative spin leads to a factor 2 in the equation.

The probability that the state is occupied is now

$$f_d = \frac{1}{1 + \frac{1}{2}e^{-(E_F - E_d)/kT}}. \tag{6.1.25}$$

Only when $(E_F - E_d)$ is known with a precision higher than $kT \ln 2$ is it necessary to consider the new factor $\frac{1}{2}$. This is rarely the case for surface states.

3. An empty (neutral) acceptor state containing one electron (energy E_1) with positive spin ($n_i = 1$), the same state containing an electron (energy E_1) with negative spin ($n_i = 1$), and an occupied state containing two electrons ($n_i = 2$, energy per electron E_2). We have

$$\Xi = 2e^{(E_F - E_1)/kT} + e^{2(E_F - E_2)/kT} \tag{6.1.26}$$

The probability f_a that the state is occupied is

$$f_a = (\Xi)^{-1} e^{2(E_F - E_2)/kT} = \frac{1}{1 + 2e^{-(E_F - E_a)/kT}} \tag{6.1.27}$$

where $E_a = 2E_2 - E_1$ is the energy of the acceptor state.

Again the factor 2 is relevant when the value of $(E_F - E_a)$ is known with a precision higher than $kT \ln 2$.

4. There are numerous (N) states, all having energies in a narrow range ΔE between E and $E + \Delta E$. The number of the available electrons, n, is large, but the value of E is such that only few states are occupied:

$$n \ll N. \tag{6.1.28}$$

This case is relevant to conduction electrons in a semiconductor or in a metal. Each state can accommodate one electron at the most. The number, g, of distinguishable combinations of n electrons out of N states is

$$g = \frac{N!}{n!(N-n)!} 2^n. \tag{6.1.29}$$

In view of eq. (6.1.28) this becomes

$$g = \frac{N^n}{n!}\, 2^n. \tag{6.1.30}$$

Apart from the spin factor 2^n this is the same expression as the one used for the calculation of the thermodynamic functions of a classical ideal gas. We adopt a well-known result for a classical ideal gas, and write

$$N = (2\pi m_e{}^* kT/h^2)^{-\frac{3}{2}} \approx 1{\cdot}25 \times 10^{19}\sqrt{(m_e{}^*/m_e)}\ \mathrm{cm}^{-3} \tag{6.1.31}$$

in which we have used for our "classical" electron gas the effective mass $m_e{}^*$ instead of the mass m_e.

In most practical cases n does not rise above 10^{18} cm^{-3} or so and the electron gas can be considered as a non-degenerate gas.

For the electrochemical potential we can now use the expression

$$\tilde{\mu}_e = E_F = E_c + kT \ln \frac{n}{2N} \quad (n \ll N). \tag{6.1.32}$$

For the holes we arrive in an analogous way at

$$\tilde{\mu}_h = -E_F = -E_v + kT \ln \frac{p}{2N} \quad (p \ll N), \tag{6.1.33}$$

where we have assumed only heavy holes with $m_h{}^* = m_e{}^*$.

The product $n \times p$ becomes

$$n \times p = 4N^2\, e^{-(E_c - E_v)/kT}. \tag{6.1.34}$$

In the case of intrinsic conductivity we have $n = p = n_i$ and n_i, the intrinsic concentration of conduction electrons and holes, is

$$n_i = 2N\, e^{-(E_c - E_v)/2kT}. \tag{6.1.35}$$

It is easy to see that in this case $(E_c - E_F) = (E_F - E_v)$ so that in the band picture the E_F–x curve is a line half-way between E_c and E_v. For n-type conductivity, E_F is in the upper half and for p-type conductivity, E_F is in the lower half of the gap. The product $n \times p$ is independent of the donor and acceptor concentration.

6.2. The space charge. The barrier potential

6.2.1. *The space charge. Three approximations*

Owing to the existence of charged surface states there is at the semiconductor surface a surface charge, σ, and a potential ψ_0, the barrier potential. The electrostatic potential in the bulk is taken zero. In the space charge region it is, as usual, denoted as ψ. The difference between the space charge in a solid semiconductor and the Gouy layer in a liquid electrolyte is found in the existence of fixed charges in the semi-conductor: ionized donor and (or) acceptor atoms. These fixed charges contribute to the charge density ρ and thereby affect the properties of other quantities such as the free energy and the capacity of the double layer.

In the space charge region we write, instead of eqs. (6.1.32) and (6.1.33),

$$E_F = E_{c0} + kT \ln \frac{n}{2N} - e\psi \qquad (6.2.1)$$

and

$$-E_F = -E_{v0} + kT \ln \frac{p}{2N} + e\psi \qquad (6.2.2)$$

where we have written $E_{c0} - e\psi$ instead of E_c, and $-E_{v0} + e\psi$ instead of $-E_v$, E_{c0} and E_{v0} being the bulk values of the energies of the bottom of the conduction band and the top of the valence band respectively. It is seen that $n \times p$ is still given by eq. (6.1.34). The effect of the contribution $e\psi$ is that the bands are curved near the interface. Figure 6.4 gives an illustration. If the bands are curved upward, the electrostatic potential is curved downward.

Concerning the description of the concentrations of conduction electrons and holes, a parameter φ is introduced[34] in such a way that for the bulk concentrations we write

$$n = n_i\, e^{\varphi/kT} \quad \text{and} \quad p = n_i\, e^{-\varphi/kT} \quad \text{(bulk)}. \qquad (6.2.3)$$

Since there is electrical neutrality in the bulk, the effect of the donors and the acceptors must be to provide a neutralizing charge of $-e(p-n)$ per unit volume. This charge is produced by fixed ionized donor and acceptor atoms and it is assumed that this part of the charge density is independent of the value of ψ.

Inside the space charge region we have

$$n = n_i\, e^{(\varphi + e\psi)/kT} \quad \text{and} \quad p = n_i\, e^{-(\varphi + e\psi)/kT} \quad \text{(space charge)}. \qquad (6.2.4)$$

With respect to the value of $e\psi$ compared to that of φ we have three cases:

1. φ and $e\psi$ have the same sign. Then the excess carriers (conduction electrons when φ is positive, holes when φ is negative) have a higher concentration near the surface than in the bulk. The space charge layer is an *accumulation* layer.

2. φ and $e\psi$ have opposite signs, but the absolute value of $e\psi$ is smaller than that of φ. The total carrier concentration is less than in the bulk. When this is the case at all positions in the space charge, one has a *depletion* layer.

3. φ and $e\psi$ have opposite signs but now $|e\psi| > |\varphi|$. In that case there is an *inversion* layer, because there is an inversion of the conductivity type near the surface compared to that in the bulk.

The charge density ρ becomes:

$$\rho = -en_i[e^{(\varphi+e\psi)/kT} - e^{-(\varphi+e\psi)/kT} - e^{\varphi/kT} + e^{-\varphi/kT}] \quad (6.2.5)$$

cond. electrons   holes   donors and acceptors

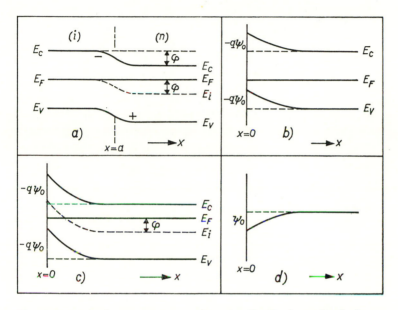

FIG. 6.4. Band picture; $x = 0$: surface. (a) Band scheme intrinsic to n-type. Double layer indicated by $-$ and $+$. (b) Band scheme at surface, intrinsic conductivity. (c) Band scheme at surface. Inversion layer. (d) Electrostatic potential vs. distance.

This expression must be inserted into the Poisson equation. The total charge in the space charge region, σ_{sc}, is found by integration and the same integration procedure can be followed as in Chapter 2 for the calculation of the charge in the Gouy layer. We find

$$\sigma_{sc} = \pm \frac{\varepsilon kT}{4\pi eL} \sqrt{2}[\cosh(u+z) - \cosh u - z \sinh u]^{\frac{1}{2}} \quad (6.2.6)$$

where $z = e\psi_0/kT$ ($+$ sign in eq. (6.2.6) when $z < 0$); $u = \varphi/kT$ and

$$L = \sqrt{\left(\frac{\varepsilon kT}{8\pi n_i e^2}\right)} \quad \text{(Debye–Hückel length)}. \quad (6.2.7)$$

The charge in the space charge region is often compensated by the surface charge σ. In that case we have

$$\sigma = -\sigma_{\text{sc}}. \tag{6.2.8}$$

Figure 6.5 gives curves, calculated by Kingston and Neustadter,[34] of σ_{sc} as a function of φ and ψ_0. The case $\varphi = 0$ corresponds to the case calculated in Chapter 2.

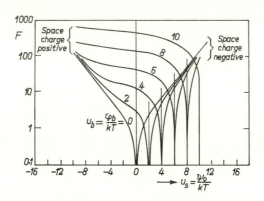

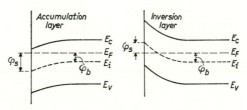

FIG. 6.5. Space charge as a function of ψ_0 for various values of φ. (Kingston and Neustadter,[34] whose notation is used: $u_b = u$; $\varphi_s = (u+z)kT$.)

$$F = [\cosh(u+z) - \cosh u - z \sinh u]^{\frac{1}{2}}$$

The bracketed part of eq. (6.2.6) can be written as

$$\frac{z^2}{2!} \cosh u + \frac{z^3}{3!} \sinh u + \frac{z^4}{4!} \cosh u + \ldots$$

or as $(\cosh z - 1)\cosh u + (\sinh z - z)\sinh u$. It is sometimes convenient to rewrite eq. (6.2.6) as

$$\sigma_{\text{sc}} = \sigma_{u=0}\sqrt{(\cosh u)}\left[1 + \frac{\sinh z - z}{\cosh z - 1}\tanh u\right]^{\frac{1}{2}} \tag{6.2.9}$$

where

$$\sigma_{u=0} = -\frac{\varepsilon kT}{2\pi\, eL}\sinh\frac{z}{2}. \tag{6.2.10}$$

For small z we have

$$\sigma_{\text{sc}} = -\frac{\varepsilon \psi_0}{4\pi L} \sqrt{(\cosh u)} \left[1 + \frac{z}{3} \sinh u\right]^{\frac{1}{2}}. \tag{6.2.11}$$

We have applied here Debye–Hückel approximations. Such approximation must lead to a linear relationship between σ and ψ_0. Therefore it is doubtful whether the term $z/3 \sinh u$ should be retained, unless its use is justified by a complete theory.

For $|z| \gg 1$ one has

$$\sigma_{\text{sc}} = +\sqrt{\left(\frac{1}{2\pi} \varepsilon n_i \, kT\right)} e^{\frac{1}{2}|z| + \frac{1}{2}u}. \tag{6.2.12}$$

A third approximation can be obtained when u is large. For n-type material (large positive u) eq. (6.2.6) becomes

$$\sigma_{\text{sc}} = +\frac{\varepsilon kT}{4\pi \, eL} e^{\frac{1}{2}u} [e^z - z - 1]^{\frac{1}{2}}. \tag{6.2.13}$$

This equation can also be obtained from:

$$\rho = -eN_D(e^y - 1) \tag{6.2.14}$$

where N_D is the donor concentration, $N_D = n_i \, e^u$.

In this expression for the charge density the minority charge carriers are ignored.

The Mott[35]–Schottky[36] approximation obtains when z is strongly negative such that only the contribution $-z$ in the bracketed part of eq. (6.2.13) remains. Thus the Mott–Schottky approximation is an approximation for fairly strong depletion layers at the surface of highly doped semiconductors. We write

$$\sigma_{\text{sc}} = \sqrt{\left(\frac{\varepsilon}{2\pi} N_D kT\right)} |z|^{\frac{1}{2}} \qquad (\varphi \gg kT; \; -z \gg 0). \tag{6.2.15}$$

For p-type material, where we must replace N_D by $N_A \; (= n_i \, e^{-u})$, a Mott–Schottky approximation is obtained when u is strongly negative and z strongly positive:

$$\sigma_{\text{sc}} = -\sqrt{\left(\frac{\varepsilon}{2\pi} N_A kT\right)} z^{\frac{1}{2}}, \qquad (-\varphi \gg kT; z \gg 0). \tag{6.2.16}$$

6.2.2. *The capacity*

We shall now distinguish between the charge σ_{sc} per unit area in the space charge region, and the charge σ_{ss} per unit area in the surface states. The case that $\sigma = \sigma_{\text{sc}}; \sigma = \sigma_{\text{ss}}$ can only be true for one particular value of ψ_0.

Consider a plate of semiconducting material, and a metal plate in parallel position at a distance d. Denoting the surface charge on the metal plate by σ, then the charge induced on the semiconductor is

$$-\sigma = \sigma_{\text{ss}} + \sigma_{\text{sc}}. \tag{6.2.17}$$

We denote the potential on the metal surface as ψ_m and that on the semiconductor surface as ψ_0.

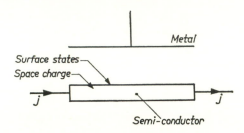

FIG. 6.6. Condenser system consisting of a metal plate and a semiconductor plate.

The reciprocal capacity of the condenser system is

$$C^{-1} = \frac{\partial \psi_m}{\partial \sigma} = \frac{\partial (\psi_m - \psi_0)}{\partial \sigma} + \frac{\partial \psi_0}{\partial \sigma}, \tag{6.2.18}$$

$$= C_0^{-1} + (C_{\text{ss}} + C_{\text{sc}})^{-1},$$

where

$$C_0 = (4\pi d)^{-1} \tag{6.2.19}$$

is a flat plate capacity and

$$C_{\text{ss}} = -\frac{\partial \sigma_{\text{ss}}}{\partial \psi_0} \quad \text{and} \quad C_{\text{sc}} = -\frac{\partial \sigma_{\text{sc}}}{\partial \psi_0}. \tag{6.2.20}$$

For C_{sc} one has

$$C_{\text{sc}} = -\frac{e}{2kT} \sigma_{\text{sc}} \frac{\sinh (u+z) - \sinh u}{\cosh (u+z) - \cosh u - z \sinh u}. \tag{6.2.21}$$

A plot of C_{sc} versus z or ψ_0 gives a minimum, depending somewhat on the value of u (see Fig. 6.7).

We readily obtain the small $|u|$, small $|z|$ approximation

$$C_{\text{sc}} = \frac{\varepsilon}{4\pi L}, \quad |u| \ll 1; \, |z| \ll 1. \tag{6.2.22}$$

This is the minimum value of the capacity at $u = 0$.

It is a "flat plate" capacity, the "gap width" between the "plates" being given by the Debye length.

The small $|u|$, large $|z|$ approximation is

$$C_{\mathrm{sc}} = \frac{e}{2kT}\,|\sigma_{\mathrm{sc}}| \qquad (|u| \ll 1;\, |z| \gg 1). \tag{6.2.23}$$

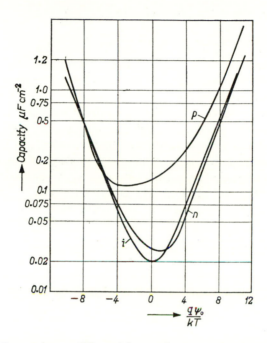

FIG. 6.7. Space charge differential capacity as a function of ψ_0 for p-type, n-type, and pure (intrinsic) Ge samples. (Bohnenkamp and Engell.[146])

The Mott–Schottky approximation gives

$$C_{\mathrm{M-S}} = \sqrt{\left(\frac{\varepsilon N_D\, e^2}{8\pi kT}\right)}|z|^{-\frac{1}{2}} \quad \text{or} \quad C_{\mathrm{M-S}} = \sqrt{\left(\frac{\varepsilon N_A\, e^2}{8\pi kT}\right)}z^{-\frac{1}{2}}. \tag{6.2.24}$$

Concerning C_{ss}, let us assume for simplicity that we have one set of donor surface states with density N_d per unit area, and one set of acceptor surface states with a density of N_a per unit area. In that case the charge in the surface states (per unit area) is

$$\sigma_{\mathrm{ss}} = +e[(1-f_d)N_d - f_a N_a]. \tag{6.2.25}$$

The energy E_a of an occupied acceptor state contains a contribution $-e\psi_0$. That means that in the Fermi distribution function f_a we write now, instead of $E_F - E_a$ in the exponential, $E_F - E_{a0} + e\psi_0$, where E_{a0} is the energy at $\psi_0 = 0$. A similar reasoning can be given for the donor states. Then the expression for C_{ss} becomes

$$C_{ss} = \frac{e}{kT} \left[f_a(1-f_a)N_a + f_d(1-f_d)N_d \right]. \tag{6.2.26}$$

The quantities $f_a(1-f_a)$ and $f_d(1-f_d)$ have maximum values of $0\cdot25$. Thus, C_{ss} tends to decrease at extreme ψ_0-values. This behaviour should be contrasted to that of the space charge capacity C_{sc} which *increases* at extreme ψ_0-values.

Large surface state densities therefore lead to large capacity values C_{ss}, and the electron energies in the surface states have rather *constant* values with respect to surface charge variations. This "pinning" effect is also manifest when the dope concentration of the semiconductor is varied. In that case the parameter φ (or u) is varied and with it (we now assume the metal plate at a large distance) the potential ψ_0 (or z). Since now for any φ-value $\sigma_{ss} + \sigma_{sc} = 0$ we have

$$d\sigma_{ss} = \left(\frac{\partial \sigma_{ss}}{\partial \psi_0}\right)_\varphi d\psi_0 + \left(\frac{\partial \sigma_{ss}}{\partial \varphi}\right)_{\psi_0} d\varphi \tag{6.2.27}$$

$$= -d\sigma_{sc} = -\left(\frac{\partial \sigma_{sc}}{\partial \psi_0}\right)_\varphi d\psi_0 - \left(\frac{\partial \sigma_{sc}}{\partial \varphi}\right)_{\psi_0} d\varphi$$

and one obtains

$$\frac{dz}{du} = e\frac{d\psi_0}{d\varphi} = -\frac{C_{ss}{}^u + C_{sc}{}^u}{C_{ss} + C_{sc}}, \tag{6.2.28}$$

where

$$C_{ss}{}^u = -e\left(\frac{\partial \sigma_{ss}}{\partial \varphi}\right)_{\psi_0} \quad \text{and} \quad C_{sc}{}^u = -e\left(\frac{\partial \sigma_{sc}}{\partial \varphi}\right)_{\psi_0}.$$

For large N_a and N_d we see that $dz \cong -du$. This means that any increase of u is accompanied by an equal *decrease* of z. Therefore the value of the electron energy in a surface state remains practically constant. For small N_a and N_d both C_{ss} and $C_{ss}{}^u$ are small. We find that also $C_{sc}{}^u$ becomes small because with few surface states z can not be large. However, C_{sc} remains finite. Thus $dz/du \to 0$, i.e. for small N_a and N_d the value of z, which under these circumstances is small, is not subject to important variations.

6.2.3. *Thermodynamics*

As derived in Chapter 2, the free energy of a double layer is

$$\Delta F_{el} = -A \int_0^{\psi_0} \sigma \, d\psi_0. \tag{6.2.29}$$

Assuming here that the double layer consists of a space charge σ_{sc} which compensates the charge $\sigma_{ss}(=\sigma)$ in the surface states, then we may use eq. (6.2.6) for the integration in eq. (6.2.29). The result can only be obtained numerically. Figure 6.8 gives curves for ΔF_{el} versus ψ_0 for various values of φ.

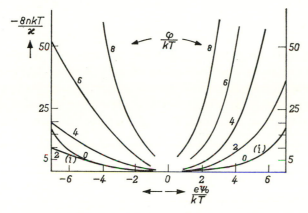

FIG. 6.8. ΔF_{el} as a function of ψ_0 for various values of φ.

We can rapidly obtain expressions for our three approximations, but we give only the Mott–Schottky approximation. The reason is two-fold. First, the small $|u|$ approximations do not offer important new insights and, second, for small $|u|$ the value of ΔF_{el} is small (of the order of $10^{12} \, kT$ per cm^2 or so for germanium in which $L = 10^{-4}$ cm and $n_i = 3 \times 10^{13}$ cm^{-3}).

The Mott–Schottky approximation gives

$$\Delta F_{el} = -\frac{1}{6N_D^2 L_D^2}\left[\left(\frac{\sigma_1}{e}\right)^3 - \left(\frac{\sigma_0}{e}\right)^3\right]kT, \tag{6.2.30}$$

where σ_0 is the initial value of the surface charge and σ_1 is the final value. We shall assume that $\sigma_0 < \sigma_1$. Furthermore,

$$L_D = \sqrt{\left(\frac{\varepsilon kT}{8\pi N_D e^2}\right)} \tag{6.2.31}$$

is defined as the Debye length for highly doped n-type material. We shall neglect the $(\sigma_0/e)^3$ term with respect to the $(\sigma_1/e)^3$ term. Nevertheless, in writing down eq. (6.2.30) we have assumed that the Mott–Schottky approximation is valid not only for σ_1 but also for σ_0. We drop the subscript "1" and note that ΔF_{el} is proportional to the *third* power of (σ/e). In Chapter 2 it was found that the small $|\psi_0|$ approximation (i.e. the Debye–Hückel approximation) led to an expression for ΔF_{el} which was proportional to the *second* power of (σ/e). The proportionality constant in that case played the role of a second virial coefficient and was denoted as B_2 (see Fig. 6.9). The proportionality constant in the Mott–Schottky case will be called a third virial coefficient and denoted as B_3:

$$B_3 = \frac{1}{6N_D^2 L_D^2}. \qquad (6.2.32)$$

With $N_D \cong 10^{17}$ cm^{-3}; $L_D \cong 10^{-5}$ cm, B_3 becomes of the order of 10^{-25} cm^{+4} and, with $(\sigma/e) = 10^{13}$ cm^{-2}, we see that $-\Delta F_{el}$ can easily become $10^{15}\, kT$ per cm^2.

Let us investigate the effect of the double layer, in particular of the Mott–Schottky case, on the adsorption and subsequent ionization of gas molecules on a semiconductor surface.

We assume the adsorption of O_2

$$O_2 \text{ (gas)} \leftrightharpoons O_2 \text{ (ads)}. \qquad (6.2.33)$$

The equilibrium condition is

$$\mu^g = \mu^{\text{ads}}. \qquad (6.2.34)$$

We assume the following ionization reaction:

$$O_2 \text{ (ads)} \leftrightharpoons O_2^- \text{ (ads)} + h^+. \qquad (6.2.35)$$

A possible mechanism of this ionization is one via the surface states. The adsorbing O_2 molecules may enhance the acceptor character of the states. However, the mechanism is irrelevant here. We are only interested in the equilibrium condition of the ionization reaction. This is

$$\mu^{\text{ads}} = \tilde{\mu}^{\text{ads}} - E_F. \qquad (6.2.36)$$

We write for μ^g, μ^{ads} and $\tilde{\mu}^{\text{ads}}$

$$\mu^g = \mu^{g0} + kT \ln p, \qquad (6.2.37)$$

$$\mu^{\text{ads}} = \mu^{\text{ads}0} + kT \ln \Gamma_1, \qquad (6.2.38)$$

$$\tilde{\mu}^{\text{ads}} = \tilde{\mu}^{\text{ads}0} \quad kT \ln \Gamma_2 - e\psi_0, \qquad (6.2.39)$$

where Γ_1 is the surface concentration of adsorbed O_2 molecules and Γ_2

that of ionized O_2 molecules. We write $\Gamma_1 = (1-\alpha)\Gamma$ and $\Gamma_2 = \alpha\Gamma$ where Γ is the total amount adsorbed.

The adsorption isotherm relates the amount adsorbed, Γ, to the pressure p. For the relation between p and Γ_1 we simply have

$$kT \ln p = (\mu^{\text{ads}0} - \mu^{g0}) + kT \ln \Gamma_1. \qquad (6.2.40)$$

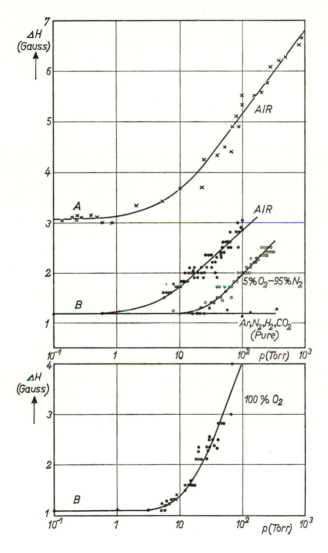

FIG. 6.9. Line width ΔH as a function of pressure. The line width is roughly proportional to the coverage of the oxidized silicon surface with paramagnetic oxygen. (From Müller et al.[96])

Concerning the relation between p and Γ_2 we take for $e\psi_0$ the Mott–Schottky approximation (we write $\Gamma_2{}^2$ for $(\sigma/e)^2$)

$$-e\psi_0 = \tfrac{3}{2}B_3\Gamma_2{}^2\,kT.\qquad(6.2.41)$$

Then we find for the relation between p and Γ_2

$$kT\ln p = (\tilde{\mu}^{\text{ads}0} - \mu^{g0} - E_F) + kT\ln \Gamma_2 + \tfrac{3}{2}B_1\Gamma_2{}^2kT.\qquad(6.2.42)$$

With $B_3 \cong 10^{-25}$ cm^4 the virial term in this equation becomes of the order of kT when Γ_2 is of the order of 10^{12} cm^{-2} or more. Now α must decrease when Γ is increased. From eq. (6.2.36) in which we insert eqs. (6.2.38) and (6.2.39) we find after differentiation

$$\frac{d\ln\alpha}{d\ln\Gamma} = -\frac{\beta}{1+\beta};\ \ \beta = 3\,(1-\alpha)\,B_3\,(\alpha\Gamma)^2\qquad(6.2.43)$$

For $\alpha \cong 1$ the effect of an increase of Γ upon the value of α is small and, for (almost) complete ionization, we expect a rapid saturation of the adsorption. However, although small, there is a decrease of α with rising Γ: when σ, and therefore ψ_0, increases, it becomes more and more difficult for newly arriving O_2 molecules to ionize. Once α is markedly smaller than 1, the quantity β is no longer negligibly small and, in particular for a high B_3-value, $d\ln\alpha/d\ln\Gamma \to -1$. In that case the quantity $\Gamma_2 = \alpha\Gamma$ and consequently the quantity $\tfrac{3}{2}kTB_3\Gamma_2{}^2$ tends to be constant when Γ is further increased.

According to eqs. (6.2.40) and (6.2.42) the adsorption has a Henry's law character, p being proportional to Γ.

There is one important flaw in this reasoning. When α starts to fall, the barrier potential ψ_0 can be so high (in the negative direction) that it is questionable whether the Mott–Schottky approximation is valid. Thus, starting with, say, a highly n-type semiconductor, the adsorbed $O_2{}^-$ ions create a positive space charge which may be largely formed by holes. The Mott–Schottky approximation is valid when $(u+z) \cong 1$. When, with $u \gg 0$, there is a *strongly* positive space charge such that even $(2u+z) \ll 0$, then, from eq. (6.2.6),

$$\sigma = \frac{\varepsilon kT}{4\pi\,eL_D}\,e^{-u}\,e^{\frac{1}{2}z};\qquad(6.2.44)$$

or, writing

$$\sigma = -e\Gamma_2 \quad\text{and}\quad \frac{\varepsilon kT}{4\pi\,eL_D} = -2N_DL_D\,e:$$

$$-e\psi_0 = 2\varphi + 2kT\ln\frac{\Gamma_2}{2N_DL_D}\,.\qquad(6.2.45)$$

We see that this approximation is valid when $\Gamma_2 \gg 2N_D L_D$. When we insert this form of $-e\psi_0$ into eq. (6.2.29) the saturation is less drastic than that provided by the virial term containing B_3. This means that both for a rapidly decreasing α and for an α-value remaining close to 1, the saturation region is restricted to a limited range of Γ-values.

Only when u can be varied over a wide range of values can one expect effects of the types described here. This is the case for semi-conductors with a large forbidden gap. Such a semiconductor is ZnO.

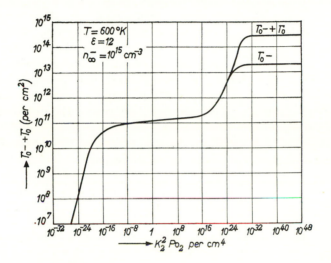

FIG. 6.10. Calculated adsorption isotherm for oxygen atoms (Γ_0) and oxygen ions (Γ_0^-) on ZnO (after Krusemeyer and Thomas[37]). The following equations were used:

$$\Gamma_0 = K_2 p_{O_2}^{\frac{1}{2}} \quad \text{and} \quad \Gamma_0^- = \Gamma_0 \exp(E_F - E_{v0})/kT \cdot \exp e\psi_0/kT$$

($\psi_0 = 0$ when $\Gamma_0^- = 0$); K_2 is the equilibrium constant for the equilibrium reaction $O_2(\text{ads}) \rightleftharpoons 2O(\text{ads})$.

Here the gap width is 3·3 eV. Krusemeyer and Thomas[37], following the work of Hauffe and Engell[38] and of Aigrain and Dugas[39], calculated the adsorption characteristics of one particular system: the O_2-adsorption on a ZnO surface. Their results are given in Fig. 6.10. Their approach was somewhat different from ours, but the interpretation of the curve can be given along the lines indicated above. Thus, the initial adsorption (very low pressure, very low coverage) follows Henry's law. The second, flat, portion of the curve corresponds to the saturation which we have ascribed to the B_3-term in eq. (6.2.42). The third, steeper, portion corresponds to the formation of a strongly p-type space charge, followed by a lowering of α. Finally, the fourth, flat, portion corresponds to a saturation of the adsorption of O_2 molecules. Since we assumed throughout a Henry-type adsorption of the molecules (see eq. (6.2.40)) this was not accounted for in our simple approach. However, when Γ becomes of the order of 10^{14} cm^{-2}, the simple assumption of an ideal behaviour of the adsorbed molecules is no longer correct and we should have assumed a Langmuir adsorption. This would have led to a saturation near monolayer coverage.

Returning to eq. (6.2.29), we have seen in Chapter 2 that ΔF_{el} is equivalent to a negative contribution (times A) $\Delta\gamma$ to the surface tension γ. This fact may have a bearing on the phenomenon of a spontaneous bending of thin (111)-cut InSb wafers, described by Hanneman et al.[40]. If the surface potentials on the 111 and the $\overline{1}\overline{1}\overline{1}$ faces are different, then also the contributions $\Delta\gamma$ on these faces are different. This may lead to a spontaneous bending. A difference of 50–200 mV for the surface potentials may explain the amount of bending. Of course, a purely "chemical" origin for the difference of the surface tensions can not be ruled out.

The integral $-\int_0^{\psi_0} \sigma \, d\psi_0$ can also be expressed in terms of surface state characteristics. From eq. (6.2.26) we derive

$$-\int_0^{\psi_0} \sigma \, d\psi_0 = -e\psi_0(N_d - N_a) - kT\left[N_a \ln \frac{f_{a0}}{f_{az}} + N_d \ln \frac{1-f_{d0}}{1-f_{dz}} \right] \quad (6.2.46)$$

where f_{a0} is the value of f_a at $\psi_0 = 0$. The other symbols have a similar meaning.

Equation (6.2.46) is pertinent to the case described in the previous subsection: a flat condenser system consisting of a semiconductor plate and a metal plate. In particular for large N_d or N_a (of course, combined with a strong electric field between the plates) the contribution $\Delta\gamma$ is almost entirely given by eq. (6.2.46).

It is worth pointing out that our considerations have only a bearing on equilibrium properties. They are only indirectly related to catalytic phenomena. In this connection it is useful to note that the surface charge on a semiconductor surface is only for strong band bending larger than $N_D L_D$(negative) or $N_A L_A$(positive) elementary charges per cm^2. The values of these products can be made of the order of 10^{12}–10^{13} per cm^2. Since the total number of surface atoms per cm^2 of the solid is about 10^{15}, chemical reactions depending only on the nature and (or) the geometric configurations of the surface atoms are practically independent of the semiconductor characteristics, but *electronic* processes (i.e. processes which need conduction electrons or holes) can be affected by varying these characteristics as explained by Wagner and Hauffe. For further discussions of this problem see ref. 41a, b.

6.2.4. *Degenerate surfaces*

When the edges of the valence band or of the conduction band are close to the Fermi level, i.e. when

$$|E_F - E_{a0} - e\psi| \lesssim kT \quad \text{or} \quad |E_F - E_{d0} - e\psi| \lesssim kT, \quad (6.2.47)$$

it is no longer permitted to use the Boltzmann approximation of the Fermi distribution function. Instead of a Poisson–Boltzmann equation a Poisson–Fermi equation should be solved. Seiwatz and Green[42] have attacked this problem and have compared their results with those of the Poisson–Boltzmann approach.

Another approximation may become dangerous for degenerate surfaces: the spacing between the energy levels *in* the conduction band or *in* the valence band may no longer be ignored. In fact a Schrödinger equation for electrons which are not only subject to the influence of the lattice, but also to that of a non-zero surface potential should be considered. Although this problem can only be solved when simplifying assumptions are introduced, its overall result must be that the space charge extends over a wider distance than in the classical case. It is no longer possible to accommodate the conduction electrons or holes in the same way as in the classical case, because fewer states are available. Quantization effects may become appreciable when the field strength $d\psi/dx$ becomes of the order of 10^5 V/cm or more.[43] In that case the density $(n+p)$ of the mobile charge carriers in the space charge region at a position where the potential is ψ $(=(kT/e)y)$, is becoming rather high. Thus, taking $(n+p) \simeq 2n_i \cosh(y+u)$, and writing

$$\frac{d\psi}{dx} = \frac{kT}{eL} \left[2(\cosh(y+u) - \cosh u - y \sinh u) \right]^{\frac{1}{2}},$$

we see that with $L = 10^{-4}$ cm, $u = 0$ (a pure crystal), $kT/e = 25$ mV, a field strength of 10^5 V/cm can be attained when $|\psi| = 280$ mV. Thus even pure Ge with strongly curved bands at the surface might show quantization effects.

An experimental method, applied by Fowler *et al.*[44a] and by Zemel and Kaplit[44b], which in principle must lead to more insight into the nature of a degenerate surface, is the one based on the concept of Landau levels. When free electrons are subjected to the influence of a strong magnetic field in, say, the z-direction, the electrons come into circular orbits in the x–y-plane and rotate with a frequency $\omega_H = eH/m^*c$. This is the cyclotron frequency. An effective mass m^*, different from the one used so far (see textbooks such as Ziman[1b]), is introduced for electrons moving in these orbits. (We note incidentally that the field strength H used in the Hall method is so weak that the dimensions of the crystal are prohibitive for the creation of orbits.) Theoretical considerations have led to the result, already envisaged by Landau, that the frequency ω_H can be considered as the frequency of a linear harmonic oscillator with energy $(n + \frac{1}{2})\hbar\omega_H$. The energy quantum $\hbar\omega_H$ is of the order of kT when $H = \frac{1}{2}T \times 10^4$ gauss and $m^* = 0.6 \times 10^{-27}$ gr. Thus, for $T = 2°\mathrm{K}$, $\hbar\omega_H = kT$ when $H = 10^4$ gauss. Fields of this magnitude can be provided by superconducting magnets.

Assuming that the electrons discussed here are conduction electrons in a semiconductor, we write, instead of

$$E = E_c + \frac{\hbar^2 k_z^2}{2m_e^*} + \frac{\hbar^2}{2m_e^*}\,(k_x^2 + k_y^2), \qquad (6.2.48)$$

which equation we assume valid in the absence of magnetic fields, now

$$E = E_c + \frac{\hbar^2 k_z^2}{2m_e^*} + (n + \tfrac{1}{2})\hbar\omega_H. \qquad (6.2.49)$$

We have seen that k_x, k_y and k_z were quantized in units $2\pi/L_x$, $2\pi/L_y$ and $2\pi/L_z$, leading to small energy quanta. However, the magnetic field imposes new, usually larger, quanta $\hbar\omega_H$ upon the system. It follows that there is a redistribution of states: the magnetic levels are degenerate, i.e. they can accommodate more than one electron. To calculate this degeneracy we consider, first in the absence of a magnetic field, and then in its presence, the number of levels, α, between the energies $E_2 = (h^2/2m_e^*)k_2^2$ and $E_1 = (h^2/2m_e^*)k_1^2$, where $k_1^2 = k_{x1}^2 + k_{y1}^2$ and $k_2^2 = k_{x2}^2 + k_{y2}^2$. It is seen that k_1 and k_2 are the radii of two circles in the $k_x - k_y$-plane. The area between these concentric circles is approximately equal to $2\pi k(k_2 - k_1)$ where k is half-way between k_2 and k_1. The number of levels corresponding to this area is $2\pi(L_x/2\pi)$ $(L_y/2\pi)k(k_2 - k_1)$. This is denoted as α.

Now apply a magnetic field such that

$$h\omega_H = E_2 - E_1 = \frac{\hbar^2}{2m_e^*}\,(k_2^2 - k_1^2). \qquad (6.2.50)$$

When k_2 and k_1 are not too far apart, it is easily found that

$$\alpha = \frac{m_e^*\omega_H}{2\pi\hbar}\,L_x L_y. \qquad (6.2.51)$$

Under experimental circumstances α is of the order of 10^{10}. All the levels previously present between E_2 and E_1 are now confined to a single level and the electrons in the conduction band are no longer quasi-continuously distributed. The population of the magnetic levels depends on their position with respect to E_F and an important point here is that this position can be varied by the experimentalist. This has interesting consequences when energy levels are present below and fairly near to E_F. Roughly, at the temperatures under investigation ($\sim 2°$K) all the levels lower than E_F are occupied. When H is increased in such a way that the position of a level is lifted up above E_F, it begins to empty and the electrons begin to occupy the nearest lower level. The free energy of the electrons decreases rather suddenly. Then it rises again with increasing H until the next level passes E_F. As clearly pointed out by Ziman this oscillatory behaviour is the basis of the interpretation of the De Haas–Van Alphen effect (oscillatory magnetic susceptibility) and the De Haas–Shubnikov effect (oscillatory electrical conductivity). This is the situation for free electrons or almost-free electrons such as are found in the conduction band and in non-degenerate space charges. In the case of bound electrons, and of electrons in degenerate surfaces, a Schrödinger equation must be solved, in which the potential energy includes a contribution arising from the interaction of the electron with the magnetic field. In a simple case, such as the one envisaged by Zemel and Kaplit, one can imagine that the solution of this equation consists of the energy values (E_i say) in the absence of a magnetic field plus contributions $(n + \frac{1}{2})\hbar\omega_H$. In that case the Fermi distribution function, here denoted as f_H, shows maxima when $E_F = E_i + (n + \frac{1}{2})\hbar\omega_H$. The pertinent capacity is proportional to $f_H(1 - f_H)$ and also shows maxima, in agreement with experimental findings.

6.3. Fast and slow surface states

Bardeen[45] was the first to realize the importance of surface states for semiconductors. He explained the fact[46] that for a metal/silicon contact the potential barrier in the silicon at the interface was practically independent of the work function of the metal by assuming that surface states at the interface acted as a "screen" against foreign influences upon the silicon. Surface states were also invoked by Shockley and Pearson[47] to explain the anomalously low field effect found for germanium. The field effect is the change of conductance of a sample caused by the application of a transverse field.

For a discussion of the field effect we can use the same model arrangement as the one discussed in connection with the capacity properties. The conductance of the semiconductor sample can be written as a sum of three contributions

$$\Lambda = \Lambda_b + \Lambda_{sc} + \Lambda_{ss} \qquad (6.3.1)$$

where Λ_b is the bulk contribution, Λ_{sc} is the contribution due to the conduction electrons and holes in the space charge layer—this contribution is zero when $\psi_0 = 0$—and Λ_{ss} is the contribution of the charges in the surface states. It is generally assumed, and this assumption is justified by experimental data, that these charges do not contribute to the conductance. Thus

$$\Lambda_{ss} = 0. \tag{6.3.2}$$

The effect of the field is measured by a change of the conductance in the space charge layer. The bulk properties are not affected. Thus

$$d\Lambda = d\Lambda_{sc} \quad \text{(field effect).} \tag{6.3.3}$$

As a general rule, when the field effect is small, the fraction of charge carriers immobilized in surface states must be large.

In the experiments by Shockley and Pearson, about 90% of the induced charges were immobilized in the surface states. Since a thin evaporated layer of germanium was used this figure does not allow a quantitative comparison with later experiments, which were all carried out with single crystals. The method of the field-effect measurement, initiated by Shockley and Pearson, has become one of the most important tools in semiconductor surface research.

Brattain and Shockley[48] noted that evidence for surface states could also be obtained by the measurement of contact potentials between n-type and p-type samples of the same semiconductor. In such measurements the Fermi levels in both samples are the same and the contact potential is determined by the electrostatic potential difference. Its value depends on the impurity content of the semiconductor and on the properties of the surface states. If the densities of those surface states which have energies not too far away from E_F is large there is a small change of the surface potential with a change of the value of $(E_C - E_F)$ and consequently there is a small contact potential change between an n-type and a p-type sample. This was actually the case in the experiments of Brattain and Shockley. For Si surfaces the surface state density was estimated at 10^{14} per cm^2.

After the invention of the transistor in 1948,[49] there was a rapid progress in semiconductor surface research. The reason was that it soon became clear that the characteristics of a transistor were highly dependent on the properties of the surface and on the surrounding atmosphere. Thus, it appeared that various gases and vapours and etchants had the effect of changing the conductivity type near the surface. For a review up to 1955 the reader is referred to Kingston.[50] Brattain and Bardeen[51] were the first to examine systematically the electrical consequences of the adsorption of various gaseous ambients, mainly ozone and water vapour, on etched germanium surfaces. (A well-known etchant is CP-4. It consists of 25 parts HNO_3 60%, 15 parts CH_3COOH 100%, 12 parts HF 38% and $\frac{1}{4}$ part Br_2 100% and gives the surface a brilliant appearance. It is noteworthy that it makes the surface hydrophobic, although this hydrophobicity is transient.) They proposed the model sketched in Fig. 6.11 which, with quantitative modifications, has appeared to be of quite general significance. For germanium Brattain and Bardeen found that the adsorption of water

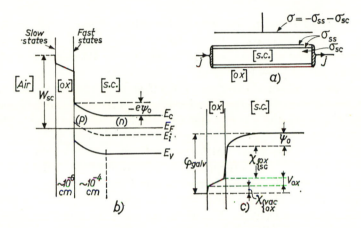

Fig. 6.11. (a) Brattain–Bardeen model.[51] (b) Band picture for this model. (c) Electrostatic potentials.

vapour led to a more n-type surface and the adsorption of dry ozone to a more p-type surface. This behaviour was reversible with respect to desorption and adsorption of these ambients and allowed "cycling" between these two surfaces (Brattain–Bardeen cycle).

During the cycle the contact potential varied over 0·3 V. Variations were measured by observing the change of the contact potential between the semiconductor sample and a metal electrode placed at some distance in the ambient atmosphere. However, the metal electrode was often unstable. A better method, applied by Morrison,[60] was to measure the variation of the conductance. These variations were, just as in the field effect measurements, ascribed to variations of the conductance due to mobile charge carriers in the space charge, i.e. to variations of the surface concentrations of the conduction electrons, Γ_e, and of the holes, Γ_h. The values of Γ_e and of Γ_h depend on the value of ψ_0. The relation between Γ_e, Γ_h and ψ_0 can be found in a way analogous to that followed in Chapter 3 for the case of excess ions present in the Gouy layer (see below). To determine whether the surface showed p-type or n-type conductivity, Brattain and Bardeen applied methods sensitive to the minority carrier density. These methods were the measurement of the surface recombination velocity, to be discussed below, and of the change of the contact potential under the influence of light. Creation of hole–electron pairs by light has the effect of flattening off the band curvature. This means that p-type space charges become less pronounced p-type as long as the light shines on the surface, leading to a less negative ψ_0-value and to an increase of the surface potential.

Variations of the contact potential resulting from variations of the ambient atmosphere were found to be almost independent of the conductivity type and this pointed, according to Bardeen, to a surface state density of at least 10^{13} per cm^2.

Time effects, observed during field effect measurements, led to the distinction between fast states and slow states. During the first seconds after the application of a transverse field the field effect is large, although not as large as could be expected on the basis of a complete absence of surface states. After these initial seconds the effect might die away almost completely. The interpretation is that first the induced charge resides largely in the space charge, a small fraction remaining in the fast states, and that afterwards the slow states immobilize almost all the induced charge carriers. The slow state density must be more than about 10^{13} per cm^2. Their influence can be eliminated by applying a.c. fields. Their frequency may not be chosen higher than about 10^5 Hz, relaxation times for fast states being of the order of 10^{-5} seconds. Brown[52] initiated a.c. field effect measurements and varied the gas atmosphere according to the Brattain–Bardeen cycle. In this way ψ_0 could be varied in two independent ways. Figure 6.12 gives a curve of the measured field effect as a function of the induced charge σ and also a theoretical curve of the surface conductivity as a function of the charge in the space charge region. Such a curve can be obtained as follows: The relation between Λ_{sc}, the surface conductance, and λ_{sc}, the surface conductivity, is

$$\Lambda_{\mathrm{sc}} = \frac{A}{l^2}\,\lambda_{\mathrm{sc}} \tag{6.3.4}$$

where A is the surface area and l is the length (i.e. the dimension in the flow direction) of the semiconductor sample. For λ_{sc} one has

$$\lambda_{\mathrm{sc}} = u_{\mathrm{sc}}(\Gamma_e + b_{\mathrm{sc}}\Gamma_h) \tag{6.3.5}$$

where u_{sc} is the mobility of the conduction electrons in the space charge region and u_{sc} b_{sc} that of the holes. These mobilities have probably, to within 20%, the same values as those in the bulk.[53] According to Schrieffer[53] they depend on the way the mobile charge carriers are reflected by the surface. When a diffuse scattering is assumed then the mobility in the surface region is somewhat reduced, when there is specular reflection, then the mobilities in the bulk and near the surface are the same. For diffuse scattering the prediction was made[54] that the surface mobility must decrease sharply near $\psi_0 = 0$. One crucial condition for such a behaviour was a large ratio of the mean free path (λ) of the charge carriers, and the Debye length L. Under these circumstances (i.e. $\lambda/L \gg 1$ and $|\psi_0| \ll (kT)/e$) a relatively large number of

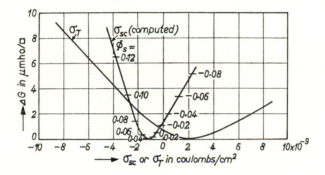

Fig. 6.12. Brown's method[52] for determining the flat band potential and the distribution of the density of states; $\sigma_T = \sigma_{ss} + \sigma_{sc}$.

charge carriers in the bulk can, per unit time, reach the surface and can be scattered there. For this explanation to apply it is necessary[54a] that also charge carriers incident at grazing angles are really scattered. This seems to be rarely the case and we therefore consider the original estimate (a surface mobility, which may be 0–20% lower than the bulk mobility) as valid.

For a low impurity content

$$\Gamma_e = \frac{\varepsilon kT}{4\pi\,e^2 L}\,(e^{\frac{1}{2}z}-1);\quad \Gamma_h = \frac{\varepsilon kT}{4\pi\,e^2 L}\,(e^{-\frac{1}{2}z}-1). \tag{6.3.6}$$

Assuming that u_{sc} and b_{sc} have their bulk values u and b, it is seen that the $\lambda_{sc} - \psi_0$ curve has a minimum which is given by

$$\lambda_{sc}(\min) = -u\,\frac{\varepsilon kT}{4\pi\,e^2 L}\,(1-\sqrt{b})^2;\quad \psi_0(\min) = \frac{kT}{e}\ln b, \tag{6.3.7}$$

and it rises almost exponentially on either side of the minimum. For higher impurity contents see Kingston and Neustadter.[34] Given the values of A, 1, u and b, values of the surface conductance lead to values of Γ_e and Γ_h and therefore, since the charge in the space charge region, σ_{sc}, is given by

$$\sigma_{sc} = e(\Gamma_h - \Gamma_e) \tag{6.3.8}$$

that part of the charge which is induced in the space charge region is known. Moreover, the measurement leads to knowledge of ψ_0. Subtraction of σ_{sc} from the induced charge σ leads to the charge, σ_{ss}, in the

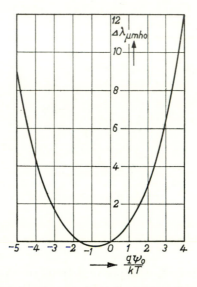

FIG. 6.13. Calculated surface conductivity $\Delta\lambda$ in μmho of intrinsic Ge at 25°C as a function of ψ_0, the barrier potential.

(fast) surface states, provided the applied a.c. field has suitable frequencies. The minimum of the measured curve of Fig. 6.13 was obtained in dry oxygen combined with a field of such value that $2\cdot5$ C cm^{-2} was induced in the fast surface states. The density of the fast states was about 10^{11} per cm^2 and their energies were distributed roughly half-way up the forbidden gap. There is only a slight effect of the gaseous ambient on the characteristics of the fast states. This adds to the evidence that fast states are to be found at the semiconductor/oxide interface. Slow states are found at the outer oxide surface. Capacity measurements led qualitatively to the same results.

Similar, and additional, information was obtained from measurements of the surface recombination velocity. The recombination, which is the transition of a conduction electron from the conduction band to the valence band, is facilitated when additional energy levels in the forbidden gap are available. In fact, the recombination velocity is almost entirely determined by the properties of the centres providing these levels. Impurity atoms and dislocations usually offer such levels and act as recombination centres. They are present in the bulk and at the surface. Shockley and Read[55] gave a theory for bulk recombination, which was worked out for surfaces by Stevenson and Keyes[56] and more completely by Many et al.[57] Detailed accounts of this theory and of non-equilibrium phenomena in general can be found in the books by Many et al.[10] and by Frankl.[11] Here we give only one representative result. Assume that a (neutral) centre is able to capture an electron, the capture probability being K_n. The capture probability for a hole, which actually is the probability that a captured electron vanishes into the valence band, is denoted by K_p. The surface recombination velocity s, which is the ratio of the rate of electron (or hole) flow into a unit surface area, to the excess carrier density in the bulk just beneath the surface (ref. 10, p. 197) becomes for one set of centres, density N_t, energy E_t:

$$s = \frac{N_t\sqrt{(K_nK_p)}\,(n+p)}{2n_i\left[\cosh\dfrac{E_t-E_i-e\varphi_0}{kT}+\cosh\dfrac{E_F-E_i+e(\psi_0-\varphi_0)}{kT}\right]} \tag{6.3.9}$$

where $e\varphi_0 = kT\ln\sqrt{(K_p/R_n)}$.

A plot of s against ψ_0 shows a maximum. The origin of this maximum can easily be visualized. The recombination depends on the availability of both electrons and holes. When the surface concentration of either holes or electrons is made very low, the rate of recombination must be low. This is the case for very large positive, or very large negative values of ψ_0. The conditions for surface recombination are optimal when $e\varphi_0 = e\psi_0 + (E_F - E_i)$ because here the effect of unequal capture probabilities is compensated by unequal surface concentrations. Thus, the effect of a low electron capture cross section ($\varphi_0 > 0$) must be compensated for by a high electron concentration near the surface, such that $E_F > E_i - e\psi_0$ and vice versa.

Parameters which are related to K_p and K_n are the capture cross sections A_p and A_n, where $K_p = v_pA_p$ and $K_n = n_nA_n$, v_p and v_n being the thermal velocities of a hole and a conduction electron. Typical values for etched Ge surfaces of A_p and A_n are of the order of 10^{-15} cm^2. Usually $A_p > A_n$.

The field effect and the surface recombination velocity can conveniently be measured by means of a double-bridge method ("Many-bridge") in which the semiconductor sample is part of both a wheatstone bridge and of a bridge containing a resistor (R) and a capacitor (C) as shown in Fig. 6.14. The field effect ΔR_f (R_f is the resistance of the sample) and the decay time $\tau = RC$ can be measured simultaneously with a null-indicator. The relation between τ and s is (viz. Subsection 6.5.4.1)

$$\frac{1}{\tau} = \frac{1}{\tau_b} + \frac{2s}{d} \qquad (6.3.10)$$

where τ_b is the bulk lifetime and d the thickness of the sample. To make $\tau \ll \tau_b$, d should be made 0·05 cm or less. It appeared that under those

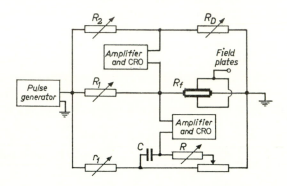

FIG. 6.14. "Many-bridge." Simplest arrangement, used by Many et al., for the measurement of the surface recombination velocity.[57] R_D = decade resistance; CRO = cathode ray oscillograph.

circumstances τ is of the order of 10^{-4} sec and therefore that s is typically of the order of 100–1000 cm/sec. As shown in Fig. 6.15 the predicted shape of the $s - \psi_0$ was obtained for germanium filaments.

The sign of $(E_t - E_i - e\varphi_0)$ which is not provided by eq. (6.3.9) can be found by carrying out the experiments at more than one temperature. The experiments of Many et al.[57] indicated that four sets of states, all with a density of about 10^{11} per cm^2, were present. Two of the sets had energies E_t close to the middle of the forbidden gap, the third set had an energy close to E_c and the fourth set had an energy close to E_v. A number of authors came, almost simultaneously, to this conclusion by using methods related to those mentioned here. Thus, photo effects were studied by Wang and Wallis,[58] "channels" by Statz et al.,[59] (if a single crystal contains two n-type regions, separated by a p-type region, an n-type "channel" along the surface of the p-type region may be made, for example by varying the gas atmosphere; also p-type channels may exist) and field effect techniques by Bardeen and Morrison.[60] Refinements were made by extending the temperature variation down to $100°K$.[61] However, in that case the condition $\tau \ll \tau_b$ is no longer fulfilled and another method, the pulsed field method, has to be applied: ΔR_f is measured as a function of the time elapsed after a d.c. field is switched on. The result was that E_t increased when the temperature was decreased and that A_n decreased according to an exponential law ($A_n \sim \exp - E_r/kT$, where $E_r \sim 0.2\,\mathrm{eV}$).

Measurements of the transverse magneto-resistance (the change of the resistance of a filament after application of a transverse magnetic

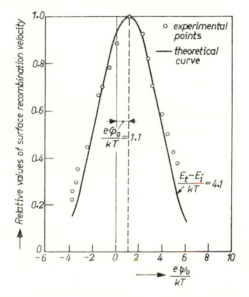

FIG. 6.15. Surface recombination velocity as a function of ψ_0 for CP-4 etched Ge sample (Many *et al.*[57]).

field) justified the given scheme.[62] The role of the light holes in silicon, with an effective mass of 0.04 times the electronic mass which is of importance in these measurements, was as predicted theoretically.

Most of the work reported here has been carried out during the first decade after the appearance of the transistor. The results are now widely used in industrial research and serve as background knowledge for the development of (usually) new Si-devices. As will be pointed out in the next sections, this does not mean that all the problems around fast and slow surface states are solved (Schlegel[63] has reviewed the situation up to 1968). Numerous results have been obtained in industrial research that may lead to a better appreciation of the nature of surface states. We mention here only the investigations concerning the role of

L

Na$^+$ ions regarding the charge in the oxide layer on Si. Sodium impurities are almost omnipresent and are, at least in part, responsible for the existence of positive charges in the oxide layer, the amount being 10^{11}–10^{12} per cm^2. These charges tend to create an n-type space charge (or to reduce a p-type space charge) in the semiconductor. Measurements of the capacity C as a function of the applied voltage V have led to this conclusion. The basic equation used was:[64]

$$\frac{1}{C} = \frac{1}{C_{\text{ox}}} + \frac{1}{C_{\text{Si}}} \qquad (6.3.11)$$

where $C_{\text{ox}} = \varepsilon_{\text{ox}}/4\pi\, d_{\text{ox}}$ is the capacity of the oxide layer. This layer has a dielectric constant ε_{ox} and a thickness d_{ox}. Furthermore, C_{Si} is the space charge capacity of the silicon. The capacity C is measured in a MOS structure. This is a three-layered system: a metal (M) layer, an oxide (O) layer of thickness d_{ox} ($d_{\text{ox}} \cong 1\ \mu\text{m}$), and a semiconductor (S) layer which is silicon in our case. The thickness of the silicon layer is large compared to a Debye length, e.g. 300 μm. There is a metal back contact for completing the electric circuitry. The positive charge in the oxide layer is denoted as σ_{ox} and it is assumed to be located at distance x from the metal/oxide interface. There are charged states at the oxide/Si interface, leading to an interface charge σ_{ss}, and there is a space charge σ_{sc} inside the silicon. The charge $(\sigma_{\text{ox}} + \sigma_{\text{ss}} + \sigma_{\text{sc}})$ is compensated by a charge σ_{m} in the metal or rather near the metal/oxide interface. The capacity at a given d.c. potential V is measured by superimposing an a.c. potential upon this d.c. voltage. The a.c. potential has a small amplitude but it must have a frequency which is of the order of 10 MHz or more. This value suffices to eliminate disturbing time effects arising from electronic processes in the fast and slow states. In this connection it should be said that the space charge capacity C_{Si} is frequency dependent. To see this, consider a silicon sample of an extrinsic conductivity type. The charge $d\sigma_{\text{sc}}$, which is necessary to follow an imposed potential variation $d\psi_{\text{sc}}$ (ψ_{sc} being the potential at the oxide/silicon interface) so that a capacity $d\sigma_{\text{sc}}/d\psi_{\text{sc}}$ can be measured, is provided by majority and minority charge carriers. The semiconductor can, as a rule, easily supply sufficient majority charge carriers whatever the time interval used for changing the potential over an interval $d\psi_{\text{sc}}$. This is not the case for minority carriers. Here, when the time interval is short, minority carriers are readily exhausted. Their supply depends on the diffusion from the bulk of the semiconductor and on the generation of new hole–electron pairs. For sufficiently high frequencies, in the MHz range, the diffusion and generation processes can be considered as infinitely slow. In that case the space charge capacity is determined by the majority charge carriers and its calculation is relatively easy. The complicated dispersion problem, i.e. the problem of C_{Si} as a function of the frequency, is considered in detail in the book by Myamlin and Pleskov[65] (cf. Fig. 6.16). Curves for the high-frequency capacities for accumulation-, depletion-, and inversion layers are given by Whelan.[64] The problem was first considered by Lindner.[66] At a given d.c. voltage V the measurement of C and the calculation of C_{Si} lead to knowledge of C_{ox}. Under flat band conditions ($\psi_{sc} = 0$) the voltage V can be identified without

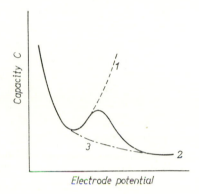

FIG. 6.16. Frequency dependence of space charge capacity at the surface of an extrinsic semiconductor. Curve 1: equilibrium (zero frequency ω). Curve 3: high-frequency limit ($\omega = \infty$). Curve 2: intermediate frequencies. The values of ω must be such that in the time ω^{-1} minority charge carriers can travel over a distance which is of the order of a Debye length. For strong inversion curves 2 and 3 coincide, as expected. (Myamlin and Pleskov.[65])

serious error with the voltage drop across the oxide layer, because V is found to be of the order of 10 V whereas χ-potentials are of the order of 0·1 V. We write

$$\sigma_{ss} + \frac{x}{d_{ox}}\sigma_{ox} = -C_{ox}V. \tag{6.3.12}$$

Typical values are $V = -10$ V and $C_{ox} = 3\frac{1}{2} \times 10^{-9}$ F. This leads to $(\sigma_{ss} + x/d_{ox}\sigma_{ox}) = 3\frac{1}{2} \times 10^{-8}$ C or about 2×10^{11} unit charges per cm². Without knowledge of x it is impossible to separate the contributions σ_{ss} and σ_{ox}.

Returning to the sodium problem, concentration profiles across the oxide layer, measured by radio active tracer techniques,[67,68] show a rather unequal distribution of sodium. There are high concentrations near the outer surface ($\sim 10^{18}$ Na atoms per cm³) and near the Si/SiO₂ interface ($\sim 10^{17}$ Na atoms per cm³, these atoms being present at both sides of the interface) whereas in the bulk of the oxide the Na-concentration is definitely lower. This distribution leads one to assume that sodium (ions or atoms, this is not revealed by the methods of investigation) shows a preference for both the outer oxide surface and for the Si/SiO₂ interface. Based on this assumption it may be suggested that sodium plays a role in the mechanism of the oxidation which is comparable to that of hydrogen. It is well known that water vapour has a positive effect upon the speed of oxide formation.[63,69] The hypothesis here is, that oxygen ions are accompanied by H⁺ or Na⁺ ions on their way across the oxide layer. The oxidation reaction takes place at the Si/SiO₂ interface. Most of the H⁺ and Na⁺ ions are neutralized and diffuse back to the outer oxide surface where complexes are formed with newly arrived oxygen. Thus the cycle is completed. Some of the cations remain at the Si/SiO₂ interface and form silicate complexes. A relatively small number of cations will reside in the bulk of the oxide. When all the reactions described here lead to an equilibrium situation, an "ionization-equilibrium" $(Na)_{ox} \leftrightarrows (Na^+)_{ox} + (e^-)_{Si}$ or $(H)_{ox} \leftrightarrows (H^+)_{ox} + (e^-)_{Si}$ may apply, but the experimental evidence so far obtained does not warrant the usefulness of a discussion of such ionization reactions.

Na⁺ ions inside the oxide seem to be more mobile than H⁺ ions and the instability which is frequently found in the systems discussed here, is generally ascribed to mobile Na⁺ ions. The instability can be reduced by working in a sodium-free atmosphere or by applying a sodium-"gettering" method. The formation of a phosphate glass by diffusion of P_2O_5 into the silicon oxide at 1000°C for a period of half an hour has turned out to be effective in reducing the instability.[67] In terms of eq. (6.3.12), the (partly ionized) sodium atoms are found at positions that vary in time. That means that x varies in time. These time variations have less effect when σ_{ox} is reduced, which is realized when sodium is removed. Concerning σ_{ss}, which arises from occupied acceptor states and empty donor states, this charge can be reduced by working in a wet atmosphere. Thus, thermally grown oxides in dry oxygen show up a state density of about 10^{12} per cm², whereas application of steam reduces this density to about 10^{11} per cm². The simplest explanation for this effect is in terms of a change of the physicochemical nature of the states brought about by hydrogen atoms. At the Si/SiO₂ interface not all the Si atoms are chemically saturated by Si—O bonds. Owing to surface heterogeneities on an atomic scale some of the Si atoms may have unpaired electrons and these Si atoms lead to the formation of states which can have both a donor, and an acceptor character. Hydrogen atoms can eliminate these states by forming Si—H bonds. Alternatively, the energies of the states, which were originally close to the conduction band and close to the valence band, are moved away from the forbidden gap into the valence band and into the conduction band.

The method of thermal growing, as developed by Atalla *et al.*[67] requires rather high temperatures. Unwanted diffusion of impurity atoms may take place at these temperatures. Oxide layers can also be prepared at lower temperatures, namely by thermal decomposition of gaseous silicon compounds, if necessary in combination with other gases. Examples are silane, silicon chloride, ethyltriethoxysilane (Schlegel's review papers[63] give detailed information).

Finally, a new method for obtaining samples with given impurity concentrations, and concentration distributions, may be mentioned, namely the method of ion implantation. In this method, ions of 50 keV or more are bombarding a surface in a direction given by a crystal axis (to avoid damage of the crystal) at room temperature. The ions penetrate the crystal to a depth depending on such parameters as the ion mass and its speed. The penetration depth is often of the order of micrometre. This is a good example of a method derived from well-known techniques which finds an application in an entirely different field.

6.4. Clean surfaces

6.4.1. *Cleaning methods. Methods of investigation*

The study of clean surfaces, that is of surfaces without foreign atoms, meets considerable problems. Thus, it is difficult to remove all the foreign atoms and to keep them removed during the experiments. Furthermore, at a supposedly clean surface it is difficult to demonstrate the complete absence of foreign atoms. Finally it is often difficult to find the cause of variations of surface properties, the detected effects being ascribable to foreign atoms but occasionally also to the nature of the surface itself.

Often used methods to obtain clean, or supposedly clean surfaces are:

(a) The ion bombardment and annealing technique, introduced by Farnsworth *et al.*[70] Ions, usually argon, accelerated by an electrostatic field directed normal to the surface, were used. Impurities were destroyed and the debris removed. According to Wolsky,[71] who determined weight losses (precision 10^{-8} g) of Ge- and Si-samples due to such bombardments, about one atom was removed per impinging argon ion, depending on the energy of the incident ions. Energies higher than 100 eV were required. Below 46 eV no weight loss was detected. Surface damage was, in this technique, repaired by annealing in high vacuum. Thus, for Ge surfaces the argon pressure should be about 10^{-3} torr; the argon current should be about 100 μA and be continued for 5 minutes at an energy of, say, 400 eV, and the annealing temperature should be about 500°C at a pressure of 10^{-9} torr. Annealing requires 15 minutes. This process should be repeated 5 to 10 times. For each material (semiconductor or metal) a somewhat different recipe was given or (as the claim was) could be given. In the light of recent developments of ion beam technology this cleaning method is a rather crude one. Energies and directions of ion beams can now be controlled and be measured with great precision and there is now a beginning of an understanding of the processes taking place after the impact of ions on solid surfaces. For low-energy ($E \lesssim 3$ keV) ion beams back scattering of the primary ions is preponderant. Knowledge of the energy (E_1) as a function of the scattering angle leads to information concerning the mass of the atoms present on the surface. It has turned out that this collision process can be reduced to a classical "billiard-ball" problem. The conservation of momentum and energy of a system, consisting of an impinging atom (or ion) with mass m_1 and an atom (mass m_t) on the surface leads to the relation $E_1 = E(m_t - m_1)/(m_t + m_1)$ (where $m_t > m_1$) when the scattering angle is 90°. Smith has reviewed[72] this field.† Mainly noble gas ions with current densities of less than 100 μA cm^{-2} are applied. When the energy of the impinging ions is about 3 keV or higher, secondary ions are formed, the mass of which can be detected. This is the method of secondary ion mass spectrometry (SIMS). Benninghoven[72a] applied 3-keV argon ions with a current density of 10^{-9} A cm^{-2} and detected, in this almost non-destructive method, the presence of foreign elements on the surface with a lower limit of, in favourable cases, about 1 ppm. An important parameter here is the efficiency with which the process of formation of secondary ions takes place. Yields of this process were measured by Werner[72b] and by Jurela[72c] and appeared to depend on such quantities as the ionization energies (secundary cations), electron affinities (secundary anions) and the work function of the surface under investigation.

(b) Cleavage of a single crystal. If the pertinent volume of the equipment is 1 litre, then, at a pressure of 10^{-9} torr, the number of gas atoms is about $3 \cdot 5 \times 10^{10}$, which is not enough to contaminate the surface (surface area is of the order of 1 cm^2) seriously. Amongst surfaces obtained in this way are the Si (111) plane, the InSb (111)

---

† By changing the angle of incidence of the ion beam, shadow effects can be measured. Brongersma and Mul[72d] found that Br-atoms adsorbed on Si(111) surfaces had a much larger scattering effect upon Ne$^+$ ions arriving at the surface at grazing incidence than at angles larger than 10°. This shows that the Br-atoms reside on top of the surface rather than between the Si surface atoms.

planes and so on. Cleavage is here relatively easy. The surfaces thus obtained may have a number of steps (height about 100 Å) as revealed by electron microscopy.[73] Allen[73] invented a technique to obtain cleaved surfaces with a high degree of smoothness.

(c) Gas-etching. Examples are Ge and Si. If oxygen is allowed to react with the surface volatile GeO or SiO may be formed when the temperature is sufficiently high, 600–700°C for Ge and 1100–1200°C for Si. Impurities at and underneath the surface are automatically removed. Cooling should take place in high vacuum. This method is particularly suitable for Si, whereas Ge surfaces are often still contaminated with carbon.

In all these methods thorough outgassing of equipment and crystals is essential. For example, in method (b) the unavoidable frictional movements in the equipment, which are necessary to cleave a crystal, may liberate gaseous impurities when outgassing was incomplete. Outgassing of the crystals is facilitated if at least one of the dimensions is small. If not, dissolved gases can not escape in time. However, a long-lasting outgassing procedure may be dangerous because unknown impurities may diffuse in. For the methods (a) and (c) it is important that the gases used are extremely pure. Thus, all the methods have their advantages and their drawbacks.[74]

We divide surface investigations into two groups: those leading to direct information of the electrical double layer and those aiming at a better knowledge of surface characteristics, such as the atomic structure of the surface, these characteristics providing only an indirect knowledge of the double layer. We are dealing here with this group of investigations and postpone double-layer effects to the next subsection. Much information is found in *Modern Methods of Surface Analysis*.[72] Of particular interest to our purpose are the contributions of Estrup and McRae on electron diffraction, Chang on Auger processes, and Smith.

An important method of investigating the surface is that of the *low energy electron diffraction* (LEED). In this method a beam of electrons with an energy of 150 eV or less is, in high vacuum, directed toward a solid surface and the angle of diffraction of these electrons is measured. Historically the method derives from the classical work of Davisson and Germer.[75] It has been worked out by Farnsworth and his school. Here a rotatable Faraday cage electron collector was used to determine the direction of the diffracted beams. In later years Germer and co-workers used a Willemite screen instead of a Faraday cage and measured the location of the light spots which were caused by the impact of the reflected electron beam. Farnsworth *et al.* were the first to find evidence for a surface structure which was different from the one to be expected on the basis of the bulk structure.

The energy of 150 eV is sufficiently low to prevent serious penetration of the electrons into the solid. Originally, the interpretation of the diffraction data made exclusive use of the wave character of the electron beam. Writing for the kinetic energy E of the electrons in vacuum

$$E = \frac{p^2}{2m} \qquad (6.4.1)$$

and using the Broglie equation in the form

$$p = h\lambda^{-1} \qquad (6.4.2)$$

the wavelength λ can be written as

$$\lambda = \frac{h}{\sqrt{(2mE)}} . \qquad (6.4.3)$$

If $E = 150$ eV, then $\lambda = 1$ Å. This is of the same magnitude as the cell dimensions of a crystal. It is also of the same magnitude as the wavelength of X-rays used for the structure analysis of crystals. For a long time LEED was considered as the two-dimensional counterpart of the (three-dimensional) X-ray analysis. The interpretation of LEED data based along these lines is called the "kinematical" interpretation. Whereas the X-ray diffraction by triperiodic structures is governed by three Laue conditions, electron diffraction was assumed, in the kinematical theory, to be determined by two. For normal incidence these are

$$m_x\lambda = a_s \cos \psi_x; \qquad m_y\lambda = b_s \cos \psi_y \qquad (6.4.4)$$

where m_x and m_y are integers and a_s and b_s are repeating distances in the x- and y-directions at the surface; ψ_x and ψ_y are the angles between the direction of the diffracted beam and the x- or y-directions. The observed diffracted beams coincide with the lines of interaction of the cones represented in the Laue conditions (6.4.4).

Wood[76] has compared surface crystallography with bulk crystallography, and has developed a coherent terminology which is now generally accepted. Thus, the "lattice" in triperiodic structures corresponds to the "net" in diperiodic structures; the "unit cell" corresponds to the "unit mesh". Let us assume that crystal surfaces are formed by an ideal cleavage experiment, i.e. by just breaking bonds in an ideal plane. Then, as a general rule, the atoms in the surface layers undergo displacements with respect to deeper-lying bulk atoms. These surface layers form the "selvedge". Chemisorption usually leads to a rearrangement of atoms in the selvedge. In the idealized case, where the surface atoms do not undergo displacements after cleavage, the net so obtained is the substrate net. This is used as the reference net. The substrate net of the (111) surfaces of the Ge and Si face-centred cubic crystals is hexagonal.

Whereas the orientation of a plane in a triperiodic structure is characterized by the three Miller indices h, k and l, the direction of rows on a diperiodic structure needs only two, h and k, indices. For hexagonal nets the two-index notation is maintained to avoid confusion. The third index, i, can be derived from the relationship $i = -(h+k)$. The mesh of the surface structure is defined with reference to the substrate mesh. When the lengths of the substrate mesh are given as a and b (unit mesh vectors $\mathbf{a}$ and $\mathbf{b}$) then those of the surface mesh, as deduced, for example, from LEED, are designated as a_s, b_s, $\mathbf{a}_s$ and $\mathbf{b}_s$. The notation is illustrated by a few examples. A (111) Si surface, obtained by cleavage in vacuum at room temperature[77] has $\mathbf{a}_s = 2\mathbf{a}$ and $\mathbf{b}_s = \mathbf{b}$. It is said that this surface has a 2×1 structure. This is written as Si (111) 2×1. The annealed (111) Si surface has a 7×7 structure, or briefly, a 7 structure: Si (111) 7×7. When foreign atoms are essential for obtaining a certain surface structure, its symbol is added. Thus, the presence of phosphorus atoms leads to a $6\sqrt{3}$ structure on the Si (111) surface: Si (111) $6\sqrt{3} - P$. Sometimes the directions of $\mathbf{a}_s$ and $\mathbf{a}$ and of $\mathbf{b}_s$ and $\mathbf{b}$ do not coincide. In that case we write $\mathbf{a}_s = \alpha\mathbf{a} + \beta\mathbf{b}$ and $\mathbf{b}_s = \gamma\mathbf{a} + \delta\mathbf{b}$ where the Greek symbols represent integers.

The kinematical interpretation has been questioned by a number of authors.[78] In a kinematical theory the potential in the crystal as "seen" by the incident electrons is considered as uniform. (This "inner potential" can be estimated at about 14–18 V.) Already Bethe[79] pointed out that this is an oversimplification. As a consequence, dynamical theories have been presented in which this potential was not uniform and in which the scattering properties of the individual atoms in the crystal lattice were considered in detail. These properties could provide for multiple scattering and for inelastic scattering. The latter phenomenon, to be precise, must probably be attributed to the interaction of the incident electrons with valence and conduction electrons. The scattering properties of surface atoms may be different from those of bulk atoms, but recent theoretical developments have indicated that this is unlikely.

For 100 eV electrons the scattering cross-section is of the order of an atomic area.[78a] For lower energies the cross-section tends to be smaller because impinging very low-energy electrons cannot penetrate the electron shells, i.e. they cannot come close to the positively charged ion core. Pendry[78] emphasized that the ion cores must be considered as anisotropic scatterers, which are mainly responsible for the back scattering of the incident electrons.

Inelastic scattering, which prevents the electrons from penetrating deeper than the first two atomic layers or so (this fact was tacitly incorporated in the kinematical theories), probably plays an important role. This was stressed by Duke and Tucker[78] and by Jones and Strozier.[78] It is largely responsible for the low reflectivity values, which were, at certain electron energies, a hundred times smaller than estimated when inelastic scattering was ignored. Also the broad peak width, found when plotting the electron energy against the reflectivity, must be ascribed to inelastic scattering.

Multiple scattering should lead to electron beams at scattering angles different from those of Bragg reflections. Since the intensities of Bragg peaks are often larger, it is believed that multiple scattering is relatively unimportant. Jones and Strozier pointed out that a modified kinematical theory must be considered as a good starting-point and that future developments must be regarded with optimism. A major step, of an experimental nature, must be mentioned: Webb *et al.*[78b] varied the angle of incidence of the electron beam and the azimuth of the substrate crystal (Ag, Ni {111} faces). Measurements leading to LEED patterns were repeated at 10° increments of azimuth for a range of 60° and at 2° increments in angle of incidence for a 12° range. An averaging procedure showed that the Bragg peaks were the only ones remaining, other reflections being reduced to an almost constant background level. In this way the kinematical interpretation has regained much of its utility, but the averaging procedure erases useful information.

In *all* the theories the existence of periodic surface structures is a prerequisite for the explanation of discrete diffracted beams. It is therefore useful to present some LEED data of semiconductor surfaces. In view of the large amount of available data we consider only the (111) surfaces of Ge and Si. Ge and Si crystals have the face-centred cubic (or diamond) structure. Each (tetravalent) Ge or Si atom is the centre of a tetrahedron, the nearest-neighbour distances being 2·44 Å for Ge and 2·35 Å for Si. The lattice constants are 5·65 Å (Ge) and 5·43 Å (Si). When a Ge or a Si crystal is finely divided by crushing, the proportions of the (111), the (110) and the (100) surfaces are about 70:5:25,[80,81] roughly corresponding to inverse ratios of the atom densities $\sqrt{2}:(4/3)\sqrt{3}:2$ at these surfaces.† Note, furthermore, that the formation of (111) planes by cutting a crystal requires the breaking of one bond per atom, and that for the formation of a (100) surface two bonds per atom must be broken.

Table 6.1 reviews the LEED data for the (111) Ge and Si surfaces.

---

† The number of atoms per cm² at the (111) Ge and the (111) Si surfaces are respectively $7·22 \times 10^{14}$ and $7·04 \times 10^{14}$.

TABLE 6.1†

Ge (111)	Refs.
Cleaved 2×1 (sometimes referred to as $\sqrt{3} \times 1$‡)	
2 structure	82
2×1 structure, 3 overlapping orientations	77
2×1 structure, after cleavage at $-195°C$	83
Annealed 2×8	
2 structure	84
8 structure	77
The following irreversible changes were observed:	
G (111)–2 (cleaved) $\xrightarrow{300°C}$ Ge (111)–12 → Ge (111)–8	
2×8 structure	85
8 structure, but no 12-structure	86
Annealed 1×1	
The following equilibrium was observed:	83
Ge (111) $2 \times 8 \xrightarrow{200–400°C}$ Ge (111) 1×1	
Si (111)	
Cleaved 2×1 (sometimes referred to as $\sqrt{3} \times 1$)	77
The following transitions were observed:	
Si (111) $3 \times 1 \xrightarrow{700°C}$ Si (111) 7 $\xrightarrow[\text{long annealed}]{600°C}$ Si (111) 5	
Annealed 7×7 (ion bombardment technique)	84
(cleavage technique)	77
7 structure, but no 5 structure	86
role of Fe in these structures	87, 88, 89
Annealed $\sqrt{19}$	82
structure induced by traces of Ni	90
discussion concerning role of Ni	88, 91, 89

† Thanks are due to Dr. F. Meyer, who has largely made this table.
‡ Depending on the orientation chosen for the substrate net.

Even from the viewpoint of the kinematical theory the interpretation of these, and of course other, LEED data offers considerable problems and at least three possible interpretations have been suggested for the Ge(111) and Si(111) data, the 2-structures. Those of Lander and Morrison[92] and of Seiwatz[93] require a top layer of substrate atoms whose two-dimensional density is less than that of a parallel layer in the bulk.

This seems improbable, although a large surface self-diffusion, leading to the required low surface density, can not be ruled out. The third possibility, offered by Haneman,[94] is attractive because it requires only minor displacements of the surface atoms after the formation of the surface. In Haneman's model three-fourths of the surface atoms are depressed by a small fraction of an angstrom (0·13 Å for Ge) and one-fourth are raised (by 0·17 Å in this model for Ge). There are minor displacements of the atoms directly underneath the top-layer. This situation is advantageous from an energetic point of view. The tetra-valent atoms in the bulk of the crystal have sp^3 hybridization. For atoms of the top layer one neighbouring atom out of four is missing and this may lead to "dangling bonds". When some of the top atoms are raised their hybridization may be somewhat lowered and the lone valence electron may obtain the characteristics of an s-electron. On the other hand, the valence electrons of the depressed atoms may have sp^2 hybridization, the remaining valence electron in this case becoming a "pure" p electron. The model was favoured by the fact that small internal splits made in Ge and Si crystals showed a "healing" effect, possibly on an atomic basis, after recontacting.[95] Haneman's model implies the existence of two sorts of surface states for the structure and he suggested that after admittance of oxygen to the surface, the first molecules attacked only the depressed atoms, this effect resulting in the well-known effect upon surface conductance and work function. Some experimental evidence to this selection was presented as an interpretation of electron spin resonance (e.s.r.) measurements, but here the situation is not very clear. E.s.r. measurements by Müller et al.[96] indicated that only very few centres were present at Si surfaces and a similar statement was made by Higinbotham and Haneman[96] for Ge surfaces. However, according to Haneman[96] about one centre per 10 Si atoms is present at the Si{111}–2 surface, this ratio being decreased when the 7 structure was formed. At an oxygen pressure less than 10^{-6} torr hardly any influence upon the e.s.r. spectrum could be detected. The interpretation by Haneman was that the top atoms remained free when oxygen molecules were present, but a fairly obvious alternative interpretation is according to a model in which the centres are in fact situated at some distance beneath the surface.

A number of thermal transitions have been observed (Table 6.1). These may be simply due to a change of the configurations of the raised and depressed atoms. Small amounts of impurity atoms, such as Ni, may also affect these configurations, even at a very low degree of coverage. In that case the presence of impurity atoms manifests itself only directly. A direct manifestation of impurity atoms occurs when they form an ordered adsorbate,[96] or when they obscure the original

LEED pattern. This effect begins to be noticeable at a degree of coverage of about 10^{-2}. We expect that the surface atom configuration of the pure or of the contaminated surface is connected with the surface state density. Possible connections are discussed in the next subsection.

An interesting corollary of the work with low-energy electrons is the following: Ibach,[97] using 1–100 eV electrons with 20 meV resolution, found that specularly reflected electrons suffered energy losses of (multiples of) 69 meV (ZnO surfaces), of 55 meV (clean Si(111)) and of 48 meV, 90 meV and 125 meV (oxidized Si(111)). These results were interpreted as caused by surface phonons with wavelengths of the order of 100 Å. In the Si case the atoms have a charge of about 0·1 e. Surface contaminations caused by other atoms than oxygen led to the same losses. This non-specificity is also found in surface conductance and ellipsometry measurements.

Whereas the diffraction of electron beams of given energy has received a great deal of attention in the past, the physics of the processes which can take place upon the impact of an electron with an atom at the surface is now gaining a wide interest. Apart from the diffraction there are processes leading to the emission of electrons with very low energies (the "true" secondary electrons), and to the emission of Auger[98] electrons, whose energies are determined by the nature of the target atoms. The true "secondary" electrons are not discussed. The Auger emission becomes more preponderant when the atoms involved have a lower atom number. For atomic numbers lower than 20 (calcium) the Auger electron emission is calculated to be more than 90% of the total emission.[99] This makes Auger emission a suitable tool for the study of light elements. The book by K. Siegbahn et al.[100] gives a clear review of the field. In an Auger process one has the following sequence. First an atom is brought to a state of high energy through the interaction with a photon or an electron. The scheme pictured in Fig. 6.17 gives an illustration. Here the state of high energy is brought about by the removal of an electron from a deep lying K-shell. Second, the atom attempts to reach a lower state of energy. This de-excitation can be effected by a process in which an electron is brought from the nearby upper level (L_I) down to the K-shell. In this process sufficient energy may be released for the emission of another electron (L_{III}), thus restoring the energy balance. This is the Auger transition KL_IL_{III}. Alternatively a photon may be emitted, where the frequency ν is an X-ray frequency. The distribution between Auger emission and X-ray quanta is governed by the atomic number, see above.

Lander[101] studied in 1953 the secondary emission current $I(E)$ as a function of the energy E of incident electrons. As substrates he used carbon, a number of metals and a number of metal oxides. The energy

E was varied from a hundred to a few thousand electron-volts. In order to obtain Auger electrons with maximum efficiency (which is about 10^{-5}) the energy V of the incident electron beam is, as a rule of thumb, chosen at about three times the ionization energy of the atoms. A typical value of the incident electron current is 10 μA, an energy precision of $0\cdot3\%$ of V can be obtained. The spectra obtained were

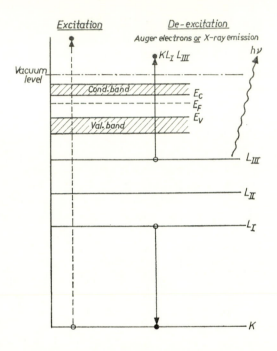

Fig. 6.17. Schematic energy diagram illustrating excitation and de-excitation process. $\bigcirc$: empty electron states; $\bullet$: occupied electron states.

characteristic of the nature of the surface atoms. Harris[102] succeeded in obtaining directly the derivative of the electron distribution with respect to energy. In this modification the Auger electrons appeared as peaks in the secondary-electron distribution. This meant an important increase of the sensitivity in the study of surfaces, in particular for the detection of light elements. Harris' work has smoothed the way for surface studies by means of the secondary electron emission. We mention one example: Palmberg and Rhodin[103] combined LEED with Auger spectroscopy. In one of their experiments they evaporated silver onto the (100) surface of gold, which had a 1×5 structure.† This

---

† This structure has been questioned by Fedek and Gjostein.[104]

structure changed into an Au (100) 1 × 1–Ag structure after a certain time of deposition, which was considered to correspond to completion of monolayer coverage. Secondary electron emission was characteristic for silver when four or more monolayers of silver were deposited. Consequently the escape depth was 8 Å. The escape depth varied with varying energy of the impinging electrons. The 362 eV silver peak corresponded to four Ag layers, the 72 eV silver peak could already clearly be observed after the deposition of two Ag layers.

A technique, closely related to that described here, is the one called[100] ESCA (*e*lectron *s*pectroscopy for *c*hemical *a*nalysis). In the ESCA technique a mono-energetic X-ray is applied for the excitation of a target atom. This excitation leads to the emission of electrons from the target atoms, their kinetic energy being determined by the wavelength of the X-ray and by their original energy level in the target atom. Also Auger electrons are emitted, but their kinetic energy is solely determined by the energy levels in the target atom.

A further method which leaves the surface unaffected is that of ellipsometry. This method uses the fact that the reflection of a (monochromatic) light beam of known ellipticity against a surface (with or without an adsorbate), leads to a change of the ellipticity.[105] The measurement of this effect leads to information concerning the optical constants of the substrate material at the surface, and, if an adsorbed layer is present, information is obtained concerning a combination of these constants and of those of the layer together with its thickness. The method is often used in a differential fashion: the data before and after an adsorption process has taken place are compared.

The elliptic light beam can be resolved into two components (s) and (p), respectively normal to, and in the plane of incidence. The ellipticity is characterized by an amplitude ratio A_p/A_s and a phase difference $(\delta_p - \delta_s)$. Both A_p/A_s and $(\delta_p - \delta_s)$ change upon the reflection. These changes can be measured in a variety of ways.[96] Figure 6.18 gives an illustration. A plane polarized monochromatic light beam obtained by a polarizer, mounted on a goniometer, passes through a quarter wave plate, which has a fixed position. The ellipticity is chosen in such a way (by rotating the polarizer) as to give a plane-polarized reflected beam. This light can be extinguished by a second polarizer, called the analyser, which again is mounted on a goniometer. The angle where extinction takes place can be determined with a precision of about 0·01°. As was first pointed out by Archer,[107] this limit is set by the mechanical stability of the equipment. A discussion of possible errors was given by Lukes (see ref. 106). After some trivial calculations the data obtained by the polarizer and the analyser readings can be transformed into values of amplitude ratio and phase difference. Alternatively, the same information concerning the change of the ellipticity as a result of the reflection can be obtained by using the phenomenon of Faraday rotation instead of using the classical optical devices. Faraday rotation is the rotation of the plane of polarization produced when plane polarized light is passed through a substance in a magnetic field, the light travelling in a direction parallel to the lines of force. The rotation depends on

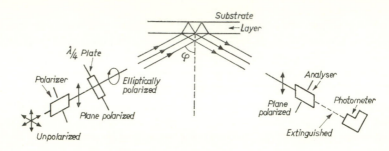

Fig. 6.18. Arrangement for ellipsometric measurements.[110]

the nature of the substance, the path length of the light in the substance and the magnetic field strength. In either case the change Δ of the phase difference and the change $\tan \psi$† of the amplitude ratio can be determined

$$\Delta = (\delta_p{}^r - \delta_s{}^r) - (\delta_p{}^t - \delta_s{}^i), \qquad (6.4.5)$$

$$\tan \psi = \frac{A_p{}^r / A_s{}^r}{A_p{}^i / A_s{}^i} \qquad (6.4.6)$$

where the superscripts r and i indicate the reflected and the incident beams.

The determination of Δ and of ψ leads to knowledge of the optical properties of the substrate material provided the surface is clean and the atoms in the selvedge are on the same lattice positions as those in the bulk. Since *these* optical properties can be obtained by other methods, we pay here only attention to the effect of adsorbed layers on Δ and ψ. Drude[105] gave a macroscopic theory for this effect. When the thickness d of the adsorbed layer was small compared to the wavelength λ (yet large compared to atomic dimensions) he obtained

$$\bar{\Delta} - \Delta = \delta\Delta = \zeta d, \qquad (6.4.7)$$

$$\bar{\psi} - \psi = \delta\psi = -\eta d \qquad (6.4.8)$$

---

† There should be no confusion between the use of ψ here to indicate the amplitude ratio and its use elsewhere for potential. We have adapted the usual notation.

where ζ and η are constants, ζ being of the order of $0 \cdot 3$ Å^{-1} and η being of the order of $0 \cdot 03$ Å^{-1}. Thus the shift of the phase difference is of more importance than that of the amplitude ratio. Writing $\tilde{n}_1 = n_1 - ik_1$ and $n_2 = n_2 - ik_2$ for the complex refractive indices of layer and substrate material, ζ and η are found to be functions of $\sin^2 \varphi$ or $\cos^2 \varphi$, where φ is the angle of incidence, of the wavelength λ, and of $n_1{}^2$, $n_2{}^2$, $k_1{}^2$ and $k_2{}^2$, all taken at the same λ.

The macroscopic theory cannot be expected to be valid when the layer thickness becomes of the order of atomic dimensions. Strachan[108] and Sivukhin[109] have gone into this problem and found that, to a fair approximation, the linear $\delta\Delta$ vs. d and $\delta\Delta$ vs. d relationships could be extended to formal "thicknesses" smaller than 1 angstrom. Atomic properties such as the atomic polarizability of adsorbed atoms must in that case replace the macroscopic refractive index and it becomes appropriate to use a degree of coverage, θ, instead of a "thickness" d. Only when the distance between the adsorbed atoms becomes of the order of λ, is a non-linear relationship to be expected. For physisorption on Ge and Si surfaces Bootsma and Meyer[110] confirmed the discussed linear relationships. They carried out simultaneous measurements of the change of ellipticity and (on crushed samples) of θ. A fairly good fit was obtained, when the atomic polarizability of the gaseous adsorbate was inserted. Noble gases and some other non-reacting gaseous compounds were used as the adsorbates. Although these results were interesting, they had no bearing on double-layer characteristics, for one thing because the surface conductance was left unaffected by the physisorption. However, when the adsorbate reacted chemically with the semiconductor substrate surface, things were different. In that case $\delta\psi$ turned out to be positive, and its absolute value was an order of magnitude higher. It reached a limit at $\delta\psi \cong 0 \cdot 2°$ for a number of different chemisorbing vapours. Figure 6.19, taken from the work of Bootsma and Meyer, gives some examples. The fact that the same drastic change takes place upon the chemisorption of various ambients was explained by Bootsma and Meyer by assuming that the selvedge of the clean surface had an abnormally high k-value. When the selvedge thickness was assumed to be 10 Å, then $k = 0 \cdot 4$. After chemisorption the atoms in the selvedge were then rearranged in such a way that the k-value was lower and tended to the bulk value of $0 \cdot 028$. Such a rearrangement, although not observed by LEED, was plausible, because chemisorption at the Ge and Si surfaces involves bond saturation. As we shall see, chemisorption usually leads to a sharp decrease of the fast surface state density. Therefore, a high k-value may be connected with a high surface state density. Further experiments, at wavelengths between $0 \cdot 35$ μ and $1 \cdot 8$ μ, led Meyer to compare the optical constants of the selvedge of the clean (111), (110) and (100) Si and Ge surfaces with those of amorphous Si and Ge and a rather striking similarity was found. There was only one important deviation: at the (100) Si surfaces the k-value at $\lambda = 1\mu \pm 0\cdot 3\mu$ was found higher than in amorphous Si. This was ascribed to a surface state effect. On the whole the similarity between the amorphous state and the state of

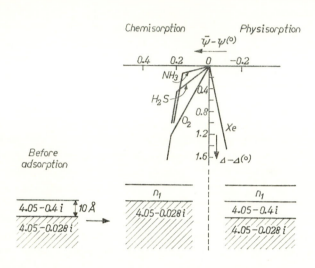

Fig. 6.19. Change of amplitude ratio $\bar{\psi} - \psi$ and change of phase difference $\bar{\Delta} - \Delta$ (see eqs. (6.4.5)–(6.4.8)) for a wavelength $\lambda = 5461$ Å, which accompany the adsorption of Xe, H_2S, NH_3 and O_2 on a clean (111) Si surface.[110] Below: a possible[110] interpretation of the results.

the crystal in the selvedge indicates that the density in the selvedge may be somewhat less than in the bulk crystal, or alternatively the average interatomic distance near the surface must be somewhat larger than in the bulk of the crystal.

It is a well-known fact that the value of the dielectric constant of a substance changes when an electric field is applied. Since the refractive index is closely related to the dielectric constant measured at light frequencies, it may be expected that the ellipticity is a function of an electric field applied to the surface under investigation. This has turned out to be the case. Buckman[106] has observed this effect for gold electrodes in a 1 M KCl aqueous solution. A potential difference was applied between the gold electrode and an auxiliary Pt foil electrode immersed in the same liquid. When the incident light is non-polarized then, for reasons similar to those valid for the effect on the ellipticity, the reflectance R is modified as a result of the application of a field. This effect, the electroreflectance effect, has been studied by Seraphin,[111] who used semiconductor samples. The field penetration is in this case larger than for metals although in both cases a precise knowledge of a potential–distance curve across the interface is lacking and, if known, its relation to optical constants is difficult to establish. Only for band structure analysis are fairly good estimates available.[112] Therefore the electroreflectance effect is only of secondary importance for double-layer studies, although it may be useful for the detection of electrostriction in surface layers.[113] Measured electroreflectance effects have a relative magnitude $\Delta R/R$ of, usually, 10^{-5} or less. The electroreflectance effect is closely related to the Franz–Keldysh effect, which is the modulation of the transmission as a result of the application of an electrical potential difference. Another optical method is that of multiple internal reflections, worked out by Harrick.[114] This is a useful method, when large and uniform surfaces are available.

Lastly we mention the measurement of gas-adsorption isotherms. The analysis of isotherms according to the BET[115a] method or to some other method leads to an estimate of the surface area. It is also possible to obtain such information as the difference between the behaviour of adsorbed argon atoms on clean and on oxidized Ge surfaces.[115b] Here a precise knowledge of the surface area is not necessary. Adsorbed argon atoms are more mobile on clean than on oxidized Ge surfaces. However, as already mentioned, physical adsorption methods have little to say about double-layer characteristics. Therefore they are typically auxiliary methods.

6.4.2. *Double-layer effects*

The physical quantities of a cleaned semiconductor surface, which we wish to discuss, are schematically given in Fig. 6.20 which shows a metal plate and, electrically connected to it, a semiconductor plate. There is band bending in the semiconductor, corresponding to a barrier potential ψ_0, which we ascribe to the existence of surface states. In principle space charge and surface states at cleaned surface can be investigated in the same way as those at etched surfaces, although the experiments at cleaned surfaces must be carried out in an ultra high vacuum. As already mentioned, and in anticipation of what will be said later on, cleaned Ge and Si surfaces have a very high surface state density, of the order of 10^{14}–10^{15} per cm^2. A high surface state density corresponds to small field effect and such a small field effect cannot reveal much about the energies of the surface states as may be seen from the following analysis.

Assume the presence of N_{Ai} acceptor states per cm² with energy E_{Ai} and assume that the total number of acceptor states per cm² is $\sum N_{Ai}$ ($i = 1 \ldots n$). Likewise we have $\sum N_{Dj}$ ($j = 1 \ldots m$) donor states per cm² with energies E_{Dj}. The surface charge σ_{ss} is

$$\sigma_{ss} = e\left[-\sum N_{Ai}f_{Ai} + \sum N_{Dj}(1-f_{Dj})\right] \qquad (6.4.9)$$

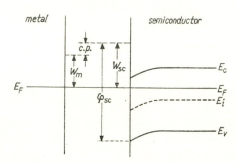

FIG. 6.20. Physical quantities related to semiconductor surface, the sample being in electronic equilibrium with a metal (Fermi energy E_F uniform). W_{sc}: work function semiconductor; W_m: work function metal; c.p.: contact potential; ϕ_{sc}: photoelectric threshold.

where

$$f_{Ai}^{-1} = 1 + \exp\frac{E_{Ai}-E_F}{kT} \quad \text{and} \quad f_{Dj}^{-1} = 1 + \exp\frac{E_{Dj}-E_F}{kT}. \qquad (6.4.10)$$

One contribution to E_{Ai} and E_{Dj} is the electrostatic energy per electron $-e\psi_0$. Therefore

$$\frac{\partial E_{Ai}}{\partial \psi_0} = -e = \frac{\partial E_{Dj}}{\partial \psi_0}.$$

The application of a potential difference between the sample under investigation and an auxiliary electrode leads to a change $\Delta\psi_0$ of the surface potential. For cleaned Ge and Si surfaces it is impossible to make $|\Delta\psi_0|$ larger than a small fraction of kT/e. Therefore the change $\Delta\sigma_{ss}$ can be written as (compare with eq. (6.2.26)):

$$\Delta\sigma_{ss} = \frac{e\Delta\psi_0}{kT}\left[\sum N_{Ai}f_{Ai}(1-f_{Ai}) + \sum N_{Dj}f_{Dj}(1-f_{Dj})\right] \qquad (6.4.11)$$

whereas for $\Delta\sigma_{sc}$ and the measured quantity $\Delta\lambda$ we can write

$$\Delta\sigma_{sc} \cong -\frac{e\Delta\psi_0}{2kT}\,\sigma_{sc} \qquad (6.4.12)$$

and

$$\Delta\lambda \cong -\frac{e\Delta\psi_0}{2kT}\,\lambda\,. \qquad (6.4.13)$$

usually $\Delta\lambda$ is proportional to the applied voltage. This means that the bracketed part in eq. (6.4.11) is a constant and it is difficult to attach numerical values to n, m, f_{Ai} and f_{Dj}. All we can say is that $(\sum N_{Ai} + \sum N_{Dj})$ is of the order of $\Delta\sigma_{ss}/\Delta\psi_0 . (4kT/e)$ because $f(1-f) \lesssim \frac{1}{4}$. Only when there are state energies fairly close to E_F will $f(1-f)$ have an appreciable value. That means that, in order to manifest themselves, the energies of the states must lie around the middle of the forbidden gap. Chemical interaction may lead to a shifting of these energies, i.e. to a decrease of $f(1-f)$, or to a change of the state densities. It is often difficult to distinguish between these alternatives.

Furthermore, the semiconductor surface is characterized by a work function W and a photoelectric threshold φ. The work function is counted from the Fermi level, the photoelectric threshold, being the energy $(h\nu_0)$ which is necessary to liberate a valence electron, is counted from the top of the valence band near the surface.

The contact potential, $\Delta c.p.$ is the difference between two work functions. In Fig. 6.20 it is the difference between the work function (W_{sc}) of the semiconductor and the work function (W_m) of the metal

$$\Delta c.p. = W_m - W. \qquad (6.4.14)$$

Both $\Delta c.p.$ and W_m can be measured and consequently W is an experimentally accessible quantity. The contact potential $\Delta c.p.$ can be measured by means of the Kelvin vibrating electrode method, while the work function W_m can be measured by using the photoelectric effect. We note that for metals W_m is equal to the photoelectric threshold φ_m. We write

$$W_m = \varphi_m = h\nu_{0m}. \qquad (6.4.15)$$

Fowler[116] derived an equation which relates the photocurrent, I, at temperature T, to the energy $h\nu$ of incident photons. At $T \to 0$ no electrons would be emitted at frequencies ν lower than the threshold frequency ν_{0m}. This is no longer the case at temperatures other than zero. Fowler obtained, using the free electron model of metals,

$$\ln\frac{I}{T^2} = \ln M + f(x); \quad x = \frac{h(\nu - \nu_{0m})}{kT} \qquad (6.4.16)$$

where M is the emission constant, which may be several electrons per photon $\times$ (degree)^2, depending on the nature of the metal and its surface, and where the function $f(x)$, derived by Fowler, contains ν_{0m} as an adjustable parameter. For $x \gg 1$, $f(x)$ reduces to a simple form and instead of eq. (6.4.16) we write

$$I = \frac{M}{2k^2} h^2(\nu - \nu_{0m})^2; \quad \nu - \nu_{0m} \gg kT/h. \tag{6.4.17}$$

Alternative methods of obtaining ν_{0m} are the field-emission method which is due to Fowler and Nordheim[117] and the thermal emission or Richardson[118]–Dushman[119] method. This is a less accurate method.

The threshold frequency ν_0 ($= \varphi/h$) for semiconductors can be obtained by photoemission methods, but the Fowler theory is not valid for semiconductors, free electrons being relatively scarce. Additional information can be obtained when not only the total electron yield is measured, but also the velocity distribution of the electrons emitted upon the incidence of photons with energy $h\nu$.[120]

Specifying the case of cleaned Ge and Si surfaces of pure crystals, a large number of workers have shown that these surfaces have a p-type space charge. This naturally corresponds to a negatively charged surface, which therefore should have an acceptor character. The first of these workers was Handler.[121] He measured the Hall effect and found that mainly positive charge carriers were mobile. Other early indications were those by Dillon and Farnsworth[122] who found for Ge(110) faces that the work function W and the photoelectric threshold φ were about equal. This pointed to a degenerate p-type space charge. Later workers found on the whole a less pronounced p-type space charge. Thus, Palmer et al.[123] reported a surface conductivity which was an order of magnitude smaller than that observed by Handler et al.[121] From field effect measurements they concluded that the space charge was p-type. The difference between these results may be due to different sample treatments. Palmer et al.[123] cleaved a Ge crystal roughly along a (111) plane in a high $\sim (10^{-9}$ torr) vacuum, whereas Dillon and Farnsworth, and Handler used argon bombardment techniques to clean the surface.

A "gas-etching" technique showed that a heat treatment alone of non-etched Ge surfaces led to qualitatively the same results[124] as those of Palmer et al. Oxygen was admitted to the Ge surface and use was made of the reaction $2Ge + O_2 \rightarrow 2GeO \uparrow$. The oxidation reaction was carried out at 600°C where the GeO vapour pressure was 10^{-3} torr.[125] In this way Ge atoms could be removed and it appeared possible to decrease the thickness of, for instance, Ge cylinders (Fig. 6.21), from an initial value of 1 mm to about 50 μ. In the actual experiments, described in ref. 124, cylindrical pure Ge single crystals were taken,

These were pulled in the (111) direction, their length being 25 cm. Only a central 5-cm length was heated, the oxygen pressure being 10^{-3} torr. The glass in these experiments did not contain boron. Boron is dangerous, because it may introduce acceptor states in the surface region of the crystal.[126]

The conductance of the cylinder could be measured at arbitrary values of the diameter, after interrupting the gas etching procedure by pumping off the oxygen, followed by cooling. Field effects could be measured by introducing a Pt cylinder enveloping, in vacuum, the Ge cylinder and using it as an electrode. It should be noted that in this way only the Ge (111)-8 structure is obtained. According to Henzler[127] the transition from the low-temperature 2×1 structure to an 8 structure is accompanied by a decrease of the surface conductance.

The omnipresent problem of the removal of foreign atoms from the surface makes itself noticeable when electrical measurements are carried out. Impurities may diffuse along the surface from semi-conductor/metal junctions which are necessarily present and from those parts of the surface which cannot be sufficiently cleaned. It is difficult to eliminate all sources of impurities from the semiconductor/metal junctions. Heat treatments are obviously very dangerous. In the experiments of ref. 124 the junction area was about 10 cm away from the area which was given the gas-etching treatment. Henzler[127c] checked on the absence of impurity effects in the following way. Like Palmer et al. he used one-half of a cleaved crystal as an auxiliary electrode for field effect measurements at the surface of the other half. But Henzler also varied the distance, d, between the two halves. The field strength should, in the ideal case of the flat condenser, be constant when the applied voltage is varied in proportion to d. The flat condenser geometry should be approached, when d is lowered. If not, the effect of the adjoining, uncleaned, surfaces was still present in some way or other, for instance as patches of diffusing impurity atoms over the cleavage planes.

Just after cleavage the surface potential is of the order of -275 mV[127] and, after heating above 200°C followed by cooling, about -200 to -240 mV, the latter value being in agreement with results by Boonstra,[128] who followed the same experimental procedures as indicated in ref. 124 and Fig. 6.21. That means that in these cases the p-type space charge contained, depending somewhat on assumptions concerning surface roughness and hole mobility in the surface region, about $3-4 \times 10^{11}$ cm^{-2} mobile charge carriers in the case of a pure crystal. For silicon this is even less, about 10^{10} cm^{-2}, corresponding to a ψ_0-value of about -50 mV. The field effect was invariably found small. Thus, in the equipment sketched in Fig. 6.21, a voltage of 10 kV,

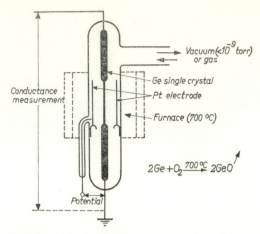

Fig. 6.21. Arrangement for measuring surface conductance and field effect.[124]

applied between the Pt cylinder (diameter 1 cm) and the Ge single crystal wire (diameter 2×10^{-2} cm) led to an induced charge density of 1.5×10^{11} elementary charges per cm², but the relative conductivity change $\Delta\lambda/\lambda$ was only 0.002, i.e. the change of the charge density in the space charge was only about 2×10^8 charges per cm². Thus, most of the induced charges resided in the surface states. As shown in Fig. 6.22a,

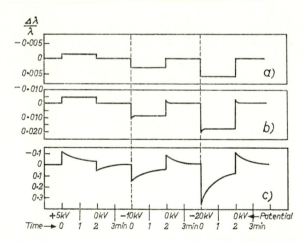

Fig. 6.22. Field effect measurements. Sample: intrinsic (111) Ge (cylindrical, see Fig. 6.21).[124] (a) Clean Ge surface. (b) After oxidation at 10^{-6} torr followed by evacuation. (c) After oxidation at 10^{-1} torr followed by evacuation. See also Fig. 6.24.

the effect was found for both a positive and a negative voltage. That means, at once, that the density of both acceptor states and donor states at the surface must be large. For instance, if only acceptor states were present, the induction of extra holes in the semiconductor by the application of a positive voltage should have led to a large change in the space charge density and to very few holes stored in the surface states. According to the analysis of eqs. (6.4.9)–(6.4.13),

$$\frac{e|\Delta\psi_0|}{2kT} = 0 \cdot 002$$

at the given conditions. Since at the same time $\Delta\sigma_{ss} = 1 \cdot 6 \times 10^{11}\ e$, we obtain for $(\sum N_{Ai} + \sum N_{Dj})$ a value which is of the order of $3 \times 10^{14}\ cm^{-2}$. This is somewhat higher than the values found by Henzler, who obtained, just after cleavage, a value of about $10^{14}\ cm^{-2}$ with a tendency toward a decrease after a heat treatment (and the appearance of the Ge (111) 8 structure).

For silicon qualitatively the same results were obtained, the surface state density being found of the order of 10^{14}–$10^{15}\ cm^{-2}$.[129]

These high values of the surface state densities were in agreement with those obtained from measurements of the photoelectric threshold[130] and the work functions of cleaved Ge(111) and Si(111) surfaces. A large number of Ge and Si crystals, the conductivity varying from highly p-type to highly n-type, were used. The observed variation of the difference $(\varphi - W)$ which accompanied the varying bulk position of the Fermi level was small. In fact, Gobeli and Allen[131] found for freshly cleaved Ge(111) surfaces that $W = \varphi$ for all the samples, although few samples were available with low impurity content. This fact pointed to a degenerate p-type surface with 10^{14}–10^{15} surface states per cm^2, whereas after heating at 350°C the work function W dropped from 4·79 eV to 4·61 eV, φ remaining at the same value. The result was, that after heating $(\varphi - W) \sim 0 \cdot 13$ eV for almost all the samples.

A typical example of an electron yield (y) vs. energy (hv) curve is given in Fig. 6.23. The low energy "tail" emission was ascribed by Scheer and Van Laar[130] to emission from surface states because the threshold value was found to be stable, i.e. it was independent of the impurity bulk concentration, the stable value for Si being 4·85 eV. This stability corresponded to the stability of the Fermi level at the surface this being just 4·85 eV with respect to vacuum.[131] Gobeli and Allen did not interpret the tail emission in terms of surface state emission. In some of their experiments they used polarized u.v. light and observed the direction of the emitted electrons as a function of the direction of polarization. The emission could come from two sources:

the surface states (low $h\nu$) and the bulk (high $h\nu$). The escape depth for excited bulk electrons in Si to overcome the surface barrier is about 200 Å. If there were two sources one could expect this to be detectable as a difference of the polarization effect between high $h\nu$ and low $h\nu$ values. In the absence of such a difference Gobeli and Allen concluded that there was only one source of photoemission, namely the bulk solid.

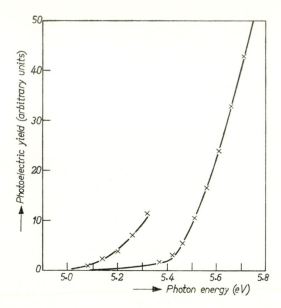

FIG. 6.23. Spectral yield of photoemission from a clean (111) Si surface, p-type, 10^{14} holes per cm^3. The tail is also shown on a $10 \times$ enlarged scale.[138]

Those authors who admitted oxygen at room temperature to the Ge or Si surface invariably found a transient and irreversible increase of the surface conductance, followed by a decrease to a value corresponding to almost the flat band condition. Thus, for the Ge surface used in the equipment of Fig. 6.21 a maximum surface conductance corresponding to 7×10^{11} mobile charge carriers (holes) per cm^2 was observed (Fig. 6.24). At the same time the field effect increased continuously and time effects were observed (Fig. 6.22). The field effect was measured after pumping off the oxygen, thus cutting off the chemical interaction at the surface. The increase of the field effect was interpreted as a decrease of the sum

$$\left(\sum N_{Ai} f_{Ai}(1 - f_{Ai}) + \sum N_{Dj} f_{Dj}(1 - f_{Dj}) \right)$$

which we denote as the effective state density.

At the conductance maximum the oxygen coverage was, by independent adsorption measurements, found to be about $0 \cdot 1$ of a mono-layer and from field effect measurements an effective state density of about 10^{13} cm^{-2} was found. For monolayer coverage the space charge contained $2 \cdot 5 \times 10^{11}$ charges per cm^2, whereas the effective state density (measured directly after the application of the voltage) was 5×10^{11} per cm^2. These are fast states with (probably) the same relaxation times of 10^{-7}–10^{-8} sec as the fast states at the semiconductor/oxide interface.

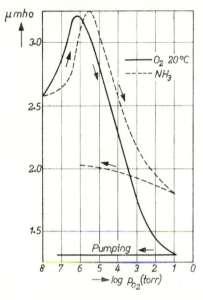

FIG. 6.24. Conductance against oxygen pressure (drawn curve) and against ammonia pressure (dotted curve) for a (111) Ge surface (intrinsic conductivity type). The arrows indicate that the pressure is first increased and then, at about $0 \cdot 1$ torr, decreased by pumping. For ammonia the adsorption is somewhat more reversible than for oxygen. Part (b) of Fig. 6.22 corresponds to the conductance maximum of the oxygen curve.[124]

Figure 6.22 shows that slow states also were formed as a consequence of the oxygen admittance, their effective density being about $2 \cdot 5 \times 10^{12}$ per cm^2. This was found for Ge. Qualitatively the same results were found for Si(111) surfaces.[128,129]

It is relevant to note that the chemical interaction between oxygen, and the surface atoms of Ge and Si samples, which leads to the anni-hilation of surface states (or/and to a strong shift of the state energies) and to the formation of slow states, is accompanied by a large heat of adsorption. Brennan et al.[132] found integral heats of adsorption of 510 kJ mol^{-1} (= 122 kcal mol^{-1}) (O$_2$/Ge) and 913 kJ mol^{-1} (= 218 kcal mol^{-1}) (O$_2$/Si). The sticking probability of oxygen on Ge and Si surfaces was surprisingly low. Schlier and Farnsworth[70] reported a value of 10^{-3} at 300°K on Ge(111). More detailed data were given in later years.[134] The sticking probability will decrease at lower temperatures, and at 77°K reversible physisorption can be observed.

Although these data will be useful for a full understanding of the electrical effects due to oxygen adsorption, the observed transient increase and the subsequent decrease of the surface conductance is primarily an effect of the substrate surface, because other chemisorbing gases and vapours, for example NH_3, show up similar effects, as shown in Fig. 6.24. For NH_3 also a transient increase of the conductance, followed by a (slightly more reversible) decrease, was observed. The similarity of the behaviour of various vapours with respect to the surface conductance of the substrate is in agreement with the ellipsometric observation that all the investigated chemisorbing vapours show up the same large increase of the angle ψ, the optical parameters of the substrate selvedge changing from those of the "amorphous" state into those of the normal crystalline state.

Although in this way some insight is gained into the process of chemisorption, a number of questions remain. Thus, we have seen that the surface state density is already drastically affected, when the oxygen coverage is only about 10% of a monolayer. This phenomenon, the induction of an important effect by relatively few adsorbate molecules, points to a co-operation between the substrate atoms. At the same time it indicates that the nature of a surface state cannot be fully understood on the basis of the often used theoretical model of a single row of atoms terminating at the surface. To date it seems impossible to present a full explanation of all the phenomena involved.† Therefore we restrict ourselves to some examples which are pertinent here.

Combined[128] conduction and adsorption[135] measurements showed that the coverage of only 10^{-3}–10^{-4} of a monolayer of CO molecules on the Ge(111) surface led to a 30% increase of the surface conductance. This adsorption was largely reversible. A direct ionization of adsorbed CO molecules into negative CO^- ions and holes which are sent into the space charge seems improbable because sufficient donor surface states are available for hole consumption. An alternative explanation of the conductance effects is again in terms of an electrically important rearrangement of surface atoms in such a way that the acceptor character of the surface states is enhanced. Such a rearrangement would also explain the rapid saturation of the CO adsorption: the adsorbing CO molecules not only enlarge the surface charge, but they also must

† The book by V. F. Kiselev, *Surface Phenomena in Semiconductors and Dielectrics* (Moscow, 1970 (Russ.)), gives much general information.

provide for the surface rearrangement.[136] Another example of the induction of an important effect by few foreign atoms is the Si(111) $\sqrt{19}$-Ni structure (see Table 6.1) induced by 10^{10} Ni atoms per cm^2 or even less.

These phenomena are not restricted to Ge or Si surfaces. For CdS Mark[17] found, through conductance measurements, that incident light with a photon energy larger than that of the bandgap (which is about 2·5 eV) had the effect of releasing 10^{14} electrons per cm^2. This was attributed to a large state density. After terminating the illumination the original situation was restored in 10^2–10^3 sec, depending on such parameters as the temperature and the amount of adsorbed oxygen. Only 10^8 electrons per cm^2 were released when oxygen was adsorbed. This is remarkable, because the pertinent oxygen adsorption was certainly less than 10^{14} molecules per cm^2.[137] Mark's experiments might have a bearing on those of Mead.[18] Mead, studying CdS and CdSe surfaces, prepared a number of crystals with different S/Se ratios. The surface barrier potential at a number of CdSe/metal interfaces was found constant (~ 300 mV) irrespective of the metal chosen, whereas large fluctuations were found at the CdS/metal interface. Thus, there was more surface stabilization in the CdSe case than in the CdS case. This is in agreement with Shockley's point of view, see Subsection 6.1.3. In fact, the surfaces studied by Mark and by Mead were not strictly comparable; therefore, and in view of what has been said about the large effects induced by small numbers of foreign atoms, the different statements as to the surface state densities may be reconciled in the future. Further experiments by Mark *et al.*[137a] showed a photodesorption of CO_2 after shining greater-than-bandgap light on CdS surfaces of powdered samples. Finally, it is interesting to note that contact potential measurements by Gatos *et al.*[137b] carried out after prolonged exposure of a $\{11\bar{2}0\}$ CdS surface to water vapour (24 hr, 2×10^{-8} torr) indicated that waterdipoles are adsorbed with the negative side to the surface.

Mead established a "rule" derived from numerous experimental findings: When stabilization of E_F with respect to the surface positions of E_c and E_v takes place, then $E_c - E_F \cong 2(E_F - E_v)$. An interesting case in which this "rule" appeared to be obeyed and which at the same time served as an example of a phenomenon of the induction of a large effect by few foreign atoms is that of the stabilization of GaAs(111) surfaces found by Van Laar and Scheer.[138] The surface state density of the freshly cleaved crystals was small as could be shown by studying the photo-yield vs. photon energy curves and (or) by comparing these curves obtained for crystals containing various dopes. Small amounts of foreign atoms, for instance Cs, in the ambient atmosphere led to a great increase of the surface state density, leading to a stabilization according to Mead's "rule".

This stabilization must be attributed to the adsorption of only few foreign atoms or molecules. It may be relevant to note here that small amounts of Hg atoms, evaporated on the Ge(111) surface, show much

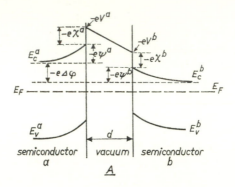

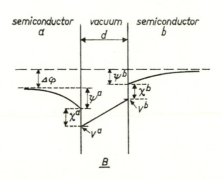

FIG. 6.25. (A) Band scheme for two semiconductor plates in electronic equilibrium (E_F uniform) at distance d. (B) Electrostatic potentials. $\Delta\phi$: Galvani potential difference; the barrier potentials ψ^a and ψ^b are denoted as ψ_0 elsewhere.

the same behaviour, i.e. an increase of the hole concentration near the surface, as adsorbed CO molecules. Larger quantities, about a mono-layer or more, of adsorbed atoms, in particular of metal ions of metals with a low ionization energy (Cs), have the effect of lowering the work function. As may be seen from Fig. 6.25, two contributions to W can be distinguished, one arising from the space charge barrier given by the potential ψ_0, and one arising from a surface dipole layer which we may identify with the χ-potential. When the ionization energy of the metal under consideration is low, it is probable that atoms of this metal, deposited on the surface, easily lose electrons and act as donor atoms.

The electrons donated by these atoms are distributed over the space charge, affecting ψ_0, and over the surface dipole layer, affecting χ. Also the photoelectric yield is affected by these foreign metals. Scheer and Van Laar[139] reasoned that the combination of the application of certain foreign metals (cf. Cs) and of a certain conductivity type (p-type) of the semiconductor may provide for a good photoemitter. If Cs is adsorbed on a p-type semiconductor, the bands bend sharply down near the surface. When light with a wavelength shorter than $hc/E_c - E_v)$ is shining on such a surface, it may liberate valence electrons from the bulk region beyond the space charge. This is the case for Si or GaAs where the penetration depth is about 1 micron but the Debye length for heavily doped material can be made less than one-hundredth of this value. Therefore, excited electrons are found in the conduction band which in this case is situated well above E_F. Indeed, the bottom of the conduction band and the vacuum level can almost be made to coincide. Quantitatively this means for caesiated GaAs surfaces that light with a wavelength of 900 nm or less liberates electrons which are then capable of escaping the solid, the work function being 1·4 eV. It has turned out that the quantum yield can be made close to 1 for Si and GaAs surfaces, in particular when certain Cs-compounds are taken instead of pure Cs.[139a]

Cho and Arthur[140] found a very interesting low-temperature photo effect. They used light with relatively low frequencies ($v < 1/h(E_c - E_v)$) and found, by shining this light at 77°K on $(\overline{111})$ GaP surfaces, a complete discharge of the surface states. According to Cho and Arthur, the light was capable of constantly keeping the electrons out of the acceptor states and the holes out of the donor states.

We conclude this section with a typical double-layer problem, that of the law of attraction between two electrically connected, parallel flat semiconductor plates a and b. The situation is sketched in Fig. 6.25.[141] The force f, per unit area, is

$$f = -\frac{(V^a - V^b)^2}{8\pi d^2}.\qquad (6.4.18)$$

For two metal plates this is a simple d^{-2} law, because the potential difference $V^a - V^b$ is constant with respect to variations of the distance d. This is no longer the case when two semiconductor plates are taken. Here $V^a - V^b$ diminishes when d is lowered as shown in Fig. 6.26. The smaller the capacities $C_{ss}{}^{a,b}$ and $C_{sc}{}^{a,b}$, the more marked is the diminution of the potential difference with decreasing d. Therefore, f increases less rapidly with decreasing d than in the case of metals. When the flat plate capacity $C_0 = (4\pi d)^{-1}$ is of the same order of magnitude of $C_{ss}{}^{a,b}$ and of $C_{sc}{}^{a,b}$ or larger, then the effect is noticeable.

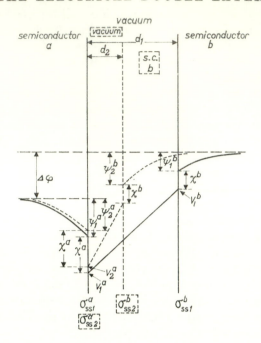

FIG. 6.26. Electrostatic potentials of two semiconductor plates at a distance d_1 (drawn curves) and at a distance d_2 (dotted curves).

When $d \sim 10$ Å this is the case for an effective state density of roughly 10^{10} per cm². Knowledge of this effect may lead to insight into the process of adhesion.[142]

6.5. The interface between a semiconductor and an aqueous electrolyte

6.5.1. *Introduction. Capacities*

In contrast to the AgI/aq. soln. and the Hg/aq. soln. systems in which thermodynamic equilibrium can be established, this is not so for semiconductor/aq. soln. systems. Among the best investigated of these semiconductor systems are the Ge/aq. soln. and Si/aq. soln. systems. It is, therefore, appropriate to include briefly a discussion of some important non-equilibrium properties, the more so as these non-equilibrium properties are of themselves extremely interesting. The foundation of this branch of electrochemistry was laid by Brattain and Garrett[143] in their study of the anodic dissolution of Ge. This was stated by Myamlin and Pleskov in their excellent book on the electrochemistry of semiconductors.[65]

Double-layer systems at the semiconductor/aq. soln. interface consist of two space charges, one at the semiconductor side (σ_{sc}) and one at the solution side (σ_{Gouy}). Furthermore, there are charges located in surface states (σ_{ss}), which we distinguish from charges in ionized groups at the interface (σ_{ion}). The latter charges may be due to slow states. In fact, a physical description of the origin of σ_{ion} may be one in terms of "slow states" whereas a chemical description would make use of the concept of ionized groups at the interface, not unlike, for instance, Payens had done in his treatment of weakly ionized adsorbed fatty acids. Finally there are charges (ions) in the Helmholtz layer (σ_H). The whole system is reminiscent of the charged interface between two adjoining liquids considered by Verwey.[144]

In discussing double-layer systems of this kind, the notation must be distinguished from that used in the simpler cases so far considered. The total potential difference between the bulk of the solution and the bulk of the semiconductor is denoted as ψ

$$\psi = \psi_s + \chi + \psi_{ox} + \psi_{soln} \tag{6.5.1}$$

where ψ_s is the potential difference across the semiconductor part, ψ_{soln} that across the solution part of the double layer. When an oxide layer is present, there is a contribution ψ_{ox}. Finally, there will be rather complicated dipole layers present between the semiconductor and the oxide, and between the oxide and the solution. In eq. (6.5.1) χ collects the contributions arising from these dipole layers.

We assume that the surface charge σ_{soln} exactly compensates the surface charge σ_s at the semiconductor side:

$$\sigma_{soln} = \sigma_H + \sigma_{Gouy} \quad \text{and} \quad \sigma_s = \sigma_{sc} + \sigma_{ss} + \sigma_{ion}$$

and the neutrality condition becomes

$$\sigma = \sigma_{soln} = -\sigma_s. \tag{6.5.2}$$

The expression for the differential capacity becomes more complicated than the one derived in Subsection 6.2.2. Following a procedure analogous to that used by Dewald[145] for the ZnO/aq. soln. interface, we write

$$C = \frac{\partial \sigma}{\partial \psi} = -\frac{\partial(\sigma_{ss} + \sigma_{sc})}{\partial \psi_s} \cdot \frac{\partial \psi_s}{\partial \psi} - \frac{\partial \sigma_{ion}}{\partial \psi_{soln}} \cdot \frac{\partial \psi_{soln}}{\partial \psi}. \tag{6.5.3}$$

Assuming $d\chi = 0$, and introducing $C_{ion} = \partial\sigma_{ion}/\partial\psi_{soln}$ and $C_s = \partial(\sigma_{ss} + \sigma_{sc})/\partial\psi_s$, this becomes

$$C = C_s\left(1 - \frac{\partial\psi_{soln}}{\partial\psi} - \frac{\partial\psi_{ox}}{\partial\psi}\right) - C_{ion}\frac{\partial\psi_{soln}}{\partial\psi}. \tag{6.5.4}$$

M

Introducing $C_{\text{ox}} = \partial\sigma/\partial\psi_{\text{ox}}$ and $C_{\text{soln}} = \partial\sigma_{\text{ion}}/\partial\psi_{\text{soln}}$ we obtain

$$\frac{1}{C} = \frac{1}{C_s}\left(1 + \frac{C_{\text{ion}}}{C_{\text{soln}}}\right) + \frac{1}{C_{\text{ox}}} + \frac{1}{C_{\text{soln}}}. \tag{6.5.5}$$

This equation, derived from a crude model, should be used only to estimate orders of magnitude. Thus, as argued in Chapters 3 and 4, C_{soln} is of the order of 10 μF/cm^2 unless the solution is very dilute. This is a much higher value than that of C_{sc} (see Subsection 6.2.2) which often is of the order of 0·1 μF/cm^2. Therefore

$$C_{\text{sc}} \ll C_{\text{soln}} \tag{6.5.6}$$

and this inequality holds true for widely varying applied potentials. In

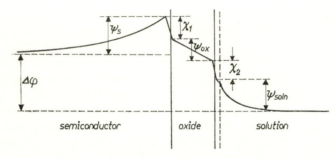

FIG. 6.27. Potential distribution across semiconductor/oxide/solution system. Vertical dotted line: Stern–Gouy interface.

this connection it should be noted that C_{soln}, which is closely connected to C_{Gouy}, has a minimum, which occurs at a zero value of ψ_{soln} and that C_{sc} has its minimum at $\psi_s = 0$ (flat band condition). The minimum value of C_{soln} is usually much larger than any C_{sc} value occurring in the experiments. It is usually also larger under most experimental circumstances than C_s ($= C_{\text{ss}} + C_{\text{sc}}$, see Subsection 6.2.2), unless the surface state density is of the order of 10^{12} cm^{-2}.

Concerning C_{ion}, if the ionized groups at the interface follow the same laws as those derived for weakly ionized adsorbed fatty acids, then

$$C_{\text{ion}} = (kT/e)\sigma_{\text{ion}}, \tag{6.5.7}$$

which is of the same order of magnitude as C_{soln}, if $\sigma_{\text{ion}} = 10^{11}$–$10^{12}$ e per cm^2.

Concerning C_{ox}, if the thickness of the oxide layer is greater than about 100 Å, then, at least in the case of Ge and Si, changes of ψ are largely taken over by ψ_{ox} and C is almost entirely determined by C_{ox}.

In a number of cases this is already the case when oxide layers, thinner than 100 Å, were present,[136] showing that the simple equation $C_{ox} = \varepsilon_{ox}/4\pi \delta_{ox}$ (where δ is the thickness and, for Si, $\varepsilon_{ox} \cong 4$) is not valid. In those cases charges were present inside the oxide layer, just as in the etched surface layers, considered in Section 6.3.

According to eq. (6.5.5), the measured capacity C can never be made smaller than C_{sc}. Other models will give a similar result. Surface states only make the capacity larger. Since a good theory exists for the $C_{sc} - \psi_s$ relationship, it can be checked whether the measurements provide C_{sc} alone or in combination with other capacities.

6.5.2. Fast states

Measurements of the capacity were first carried out by Bohnenkamp and Engell.[146] They used a three-electrode system, namely a Ge-electrode in combination with a platinized Pt-electrode, and a Ag/AgCl electrode, all in contact with the same solution (1 M KOH, or $\frac{1}{2}$ M H_2SO_4, or 0·25 M Na_2SO_3 or 1 M KCl). A d.c. potential was applied between Ge and Pt and its effect on the Ge-electrode was measured with the auxiliary Ag/AgCl electrode, which was provided with a Haber–Luggin capillary. At each applied potential the capacity of the cell Ge/aq. soln./Pt was measured by a bridge method, the measurements being carried out with the aid of superimposed, small a.c. voltage in the frequency range of 10^3–10^6 Hz. The recorded capacity was interpreted as that of the Ge/aq. soln. interface because the surface area of the platinized Pt electrode was comparatively large. Reproducible values were only obtained after removal of surface layers and impurities and this removal was carried out by an anodic polarization at a potential against saturated calomel (E_E) of about 0·4 V for a few minutes. This voltage amounted to passing an anodic current of 0·1–1 mA cm^{-2} across the sample. Water-soluble oxides were formed and removed from the surface by this treatment. Hoffmann-Perez and Gerischer[147] carried out similar experiments, although their electronic circuit was somewhat different (it included a potentiostat). Measurements were also carried out by Brattain and Boddy,[148] who applied a pulse technique. A current pulse of known value and duration was applied and the decay followed on an oscilloscope. They purified their solution by using Ge powder as a substrate for the adsorption of impurities and arrived at a capacity minimum, which was only 1·3 times that of the calculated value of the space charge capacity. The factor 1·3 was ascribed to surface roughness. All the workers arrived at the conclusion that the anodically cleaned Ge surface contains very few fast states.

That impurities could be dangerous indeed was shown by Boddy and Brattain, who measured the effect of the presence of Cu^{++} ions down to concentrations of 10^{-7} M. They inferred from measurements of the capacity, the surface conductance and the surface recombination velocity that two charge traps may be present with energies half-way between E_c and E_v and at densities of 10^{10}–10^{11} per cm^{-2}. One of them acted as a recombination centre.[150] In a purified solution the surface

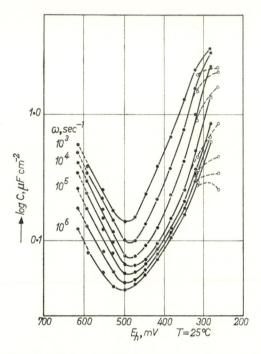

FIG. 6.28. Capacity against applied potential E_h (versus saturated H_2 electrode) at seven different frequencies for a slightly anodized Ge electrode. (From Bohnenkamp and Engell.[146])

state density could be made less than 10^9 cm^{-2} by anodic polarization: such a surface was called by Brattain and Boddy a "perfect" surface.

When such a surface was *cathodically* polarized, at a potential between -0.4 V and -0.6 V with respect to the saturated calomel electrode (note that H_2 is formed at -1.2 V) for less than about 2 minutes, 10^{10}–10^{11} surface states per cm^2 were formed. This was shown by Memming and Neumann[151] by means of measurements of the surface recombination velocity s. When the cathodic polarization was maintained for 5 minutes or more, the states almost disappeared, as was inferred from a ten- to twenty-fold decrease of s. There appeared to be only little effect of the surface states upon the shape of the capacity–voltage curves, these being in all cases close to the predicted $C_{sc} - \psi_s$ curves. However, the measurements of the capacity showed, in all those cases where both anodic and cathodic polarization experiments were carried out, that the flat band potential E_{E0} was strongly dependent on the state of polarization. Memming and Neumann found for anodized Ge-electrodes (having a "perfect" surface) at a pH of 2, that $E_{E0} \sim 50$ mV, and for the same Ge-electrode, but now cathodically polarized for more than 5 minutes, that $E_{E0} \sim -500$ mV. This difference can be understood, when it is assumed that the potential drop at the liquid side of the interface (χ-potential plus potential drop across Helmholtz layer) is different in the two cases, the difference being about 550 mV.

The fact that a difference is found, and that it decreases when the (pre-anodized) electrode is cathodically polarized for less than about 2 minutes, in short that electrochemical reactions are constantly taking place, has a bearing on the validity of capacity measurements. When the capacity is measured as a function of widely varying potentials, the potential sweep must be sufficiently fast to avoid unwanted changes in the χ-potential and/or the potential drop across the Helmholtz layer. Memming and Neumann used a sweeping speed of 0·35 Vs^{-1} and obtained capacity values, which were about equal to calculated C_{sc} values over a range of about 0·5 V.

When Si is taken instead of Ge, the measurements are more difficult to carry out, because an insoluble oxide film forms rapidly. In strongly alkaline solutions and after cathodic polarization, or, in particular, after etching with HF, an interface may be obtained which is somewhat comparable to the Ge-electrodes just described and which may contain only few states.[152] Memming and Schwandt[153] studied Si-electrodes and found that the capacity was determined by C_{sc}, at least for frequencies higher than 140 Hz, when the electrode was placed in a highly concentrated HF solution, for instance 10 M. For frequencies higher than 140 Hz, a simple equivalent circuit, consisting of a capacitor (C_{sc}) and a resistor, placed in series, could be found. Surface recombination velocity measurements showed that surface states were present. Although this conclusion is justified it is insufficient to explain all the observations, because the effect of electrochemical processes, which naturally involve the annihilation or creation of electron–hole pairs (Subsection 6.5.4) cannot be eliminated, and because the interaction between the surface states and the conduction and valence bands is rather complicated. Quite often[153-4,165] Si- and Ge-electrodes showed low capacity values when measured at high frequencies and in those potential regions in which an inversion layer could be expected. We encountered this effect already in Subsection 6.2.2. The explanation must again be sought in a lack of supply of the necessary charge carriers. Whereas previously this lack was entirely determined by a slow diffusion, we have in the case of electrodes also the possibility of the consumption of these charge carriers in the course of an electrochemical reaction.

An interface, for which good investigations turned out to be possible, is that between single crystals of ZnO and a borate buffered (pH ≈ 8.5) 1 M KCl solution. Dewald[145] measured the capacity of this interface for a large number of samples, containing various amounts of impurities and for a wide range (about 2 V) of applied potentials. Prior to the measurements the ZnO crystals were etched in a H_3PO_4 solution or in a KOH solution. The result was that only few states were present, because the space charge alone could well account for the results. Thus, the Mott–Schottky approximation applied when a depletion layer was present and it did so over a range of about 2 V (the bandgap of ZnO is 3.3 V) as shown in Fig. 6.29. When an accumulation layer was present

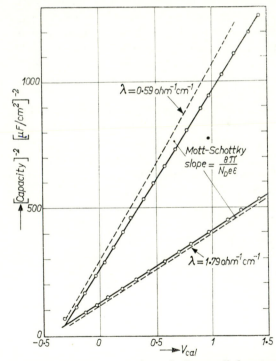

FIG. 6.29. (Capacity)$^{-2}$ as a function of the applied potential V_{cal} (versus normal calomel) for vapour-phase grown ZnO crystals. Dotted lines: theoretical Mott–Schottky slopes. The intercepts of the linear plots afford a quantitative measure of the electrode potential at the flat-band condition. (From Dewald.[145])

and more particularly in the case of degeneracy, the measured capacity coincided with that calculated on the basis of a Poisson–Boltzmann and a Poisson–Fermi equation. After etching in KOH the flat band potential was 130 mV more negative than after etching in H_3PO_4. As explained by Dewald, this was caused by a change of the potential drop across the Helmholtz layer: in acid solutions the ZnO surface will contain excess positive charges compensated by anions in the Helmholtz layer and in the solution, and after etching with an alkaline solution the surface will contain excess negative charges (ZnO_2^{--}) compensated by cations.

6.5.3. *Role of pH. Role of Cu^{++} ions. Slow states. Contact angles*

Bohnenkamp and Engell made the observation that the flat band potential of the pre-anodized Ge-electrode was, in a 1 M KOH solution, 750–800 mV more negative than in a $\frac{1}{2}$ M H_2SO_4 solution. This observation, amounting to a displacement of the flat band potential of about 60 mV per pH unit, was confirmed by other workers.[147,154] The cathodic and anodic overpotentials of the Ge-electrode were studied by Harvey.[155]

Electromotive force measurements of the cell Ge/aq. soln./sat. calomel electrode showed that the Ge-electrode acted as a hydrogen electrode,[156] although, in particular at low pH values, as a rather bad one. The slope of the e.m.f.–pH curve was 50–60 mV/pH, but below pH $\approx$ 2 the slope was less and the data were less reproducible. For an ideal hydrogen electrode at room temperature one could expect a slope of 59 mV/pH according to the reversible reaction

$$\{H\} \leftrightarrows H^+ + e^- \quad \text{or} \quad p^+ + \{H\} \leftrightarrows H^+ \tag{6.5.8}$$

where the symbol $\{H\}$ denotes, in the case of a classical hydrogen electrode, hydrogen atoms in equilibrium with hydrogen molecules, but which denotes in the case of a Ge-electrode a molecular complex formed through a reaction between Ge, H_2O and O_2. This complex should have a chemical potential which is independent of the pH and of the Fermi energy, and the reaction should be completely reversible. Only when these three conditions are obeyed can a slope of 59 mV/pH be expected. The fulfilment of these conditions was ensured when H_2O_2 was added to the solution. In that case a slope of exactly 59 mV/pH was found.[157]

Also, since the Ge–H_2O–O_2 complex is independent of the impurities in the Ge-crystal, a dependence of the e.m.f. upon φ ($= E_F - E_i$) need not be expected. This is different for the capacity versus φ data as shown by Brattain and Boddy.

Information concerning the chemical nature of the component $\{H\}$

can be obtained from measurements of the zeta potential as a function of the pH. The zeta potential of the Ge/H₂O interface turned out to be negative for each pH-value.[158]

According to electro-osmosis experiments, carried out with a plug of n-type (not p-type) Ge crystallites the zeta potential was -40 mV at pH $= 3.90$ and -10 mV at pH $= 3$, the slope of the potential vs. pH curve being about 30 mV/pH (Fig. 6.30). At pH values higher than 4 no liquid could be forced through the plug, probably because gel-like

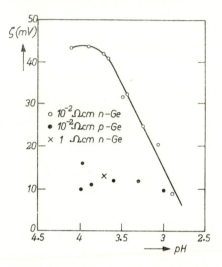

FIG. 6.30. Zeta potential as a function of pH (no salt added) for n-type Ge and p-type Ge. (From Sparnaay.[158])

germanates were formed. At pH values lower than 2 the ionic strength was too high and the effect of the displacement of the liquid was undetectable. The effect of the H⁺ ions upon the zeta potential was much larger than that of other cations or anions. Thus, the addition of KNO₃ to a HNO₃ solution of pH $= 3.9$ until a K⁺-ion concentration of 3×10^{-3} M was reached left the zeta potential unaffected. After the addition of more KNO₃, to a K⁺-ion concentration of 10^{-2} M, the zeta potential became -30 mV.

The liquid displacements measured for plugs formed by p-type material were always much less than for n-type material. The reason was that in the case of p-type plugs most of the current transport was taking place through the crystallites and not through the liquid, the electrochemical reactions at the Ge/H₂O interface involving a higher resistance in the n-type case than in the p-type case. As discussed in the next subsection, current–voltage characteristics have shown that this is indeed the case. (Note that the specific resistance of the Ge samples was many orders of magnitude lower than that of the liquid.)

The zeta potential, obtained for n-type Ge, can be used to obtain an approximate value of the charge σ_{sp} at the solid side of the slipping plane. According to Gouy's theory one has, with the Debye approximation $\sinh x = x$, and with a concentration c in mol cm³,

$$\sigma_{sp} = 1 \cdot 2 \times 10^{12} \, e \sqrt{c} \, \frac{e\zeta}{kT}. \tag{6.5.9}$$

Since $\zeta < 0$, also $\sigma_{sp} < 0$. In purely acid solutions, where c was determined by the HNO_3 concentration, an increase of c was accompanied by a decrease of $|\zeta|$ and therefore σ_{sp} tended to a constant value with respect to pH variations. Thus, $\sigma_{sp} \cong -0 \cdot 7 \times 10^{12}$ electrons per cm² at pH $= 3$ and $\sigma_{sp} \cong -0 \cdot 8 \times 10^{12}$ electrons per cm² at pH $= 3 \cdot 9$. The situation was entirely different, when KNO_3 was added. In a 10^{-2} M KNO_3 solution at pH $= 3 \cdot 9$, the value of σ_{sp} was as high as $-4 \cdot 5 \times 10^{12}$ electrons per cm².

The result can be explained by assuming the presence of acidic groups attached to the Ge surface, in agreement with a rather common concept in colloid chemistry. In this concept $\sigma_{sp} \cong -N e\alpha$, where N is the total number of acidic groups per cm², and α is the degree of ionization.

Acidic groups at the surface are formed by a surface oxidation, followed by a reaction with water and an ionization reaction, which is written in the following scheme:

$$2Ge_s + xO_2 \rightarrow 2(GeO_x)_s \tag{6.5.10}$$

$$2(GeO_x)_s + xH_2O \rightarrow 2(GeO_xH_x)_s \tag{6.5.11}$$

$$(GeO_xH_x)_s \leftrightarrows (GeO_xH_{x-1})_s^- + H^+ \tag{6.5.12}$$

where x is the degree of oxidation ($1 \leqslant x \leqslant 2$) and where the subscript s indicates an atom or a complex, attached to the surface. The groups denoted as $(GeO_xH_{x-1})_s^-$ provide the negative surface charge. The oxidation reaction (6.5.10) is assumed to be irreversible and so is, in an aqueous environment, the reaction with water. The ionization reaction (6.5.12) is assumed to be an equilibrium reaction. It can be handled in the same way as the ionization equilibrium in the case of weakly ionized long-chain fatty acids, adsorbed at the interface between oil and an aqueous solution with variable pH. Just as in that case (Payens, *loc. cit.*, chap. 5) there was a pH region in which α was a constant, and there was an increase of α upon the addition of neutral electrolyte.

Surface oxidation can also take place by means of anodic reaction.

Then, instead of eqs. (6.5.10) and (6.5.11) we have a reaction mechanism, which we may write in two alternative ways:

$$Ge_s + xH_2O + xh^+ \rightarrow (GeO_xH_x)_s + xH^+$$

$$Ge_s + xH_2O \rightarrow (GeO_xH_x)_s + xH^+ + x\ e^- \qquad (6.5.13)$$

Reactions of this type will be discussed in the next subsection.

A third way of oxidizing the surface is through the addition of Cu^{++} ions (or cations of more noble metals) to the adjoining solution. The ionization reaction for Cu^{++} ions is

$$Cu \leftrightarrows Cu^{++} + 2e^- \qquad E_M = 0.34 \text{ V} \qquad (6.5.14)$$

where E_M is the standard potential.[159]†

Combination of eq. (6.5.13) and eq. (6.5.14) gives

$$2Ge_s + 2xH_2O + xCu^{++} \leftrightarrows (GeO_xH_x)_s + 2xH^+ + xCu \qquad (6.5.15)$$

This equation indicates that, at sufficiently high H^+ ion concentrations, a reduction of the surface must take place, observable through the formation of Cu^{++} ions, and vice versa: the addition of Cu^{++} ions leads to a lowering of the pH. This has been shown to be the case.[160] At a pH value of 1 the equilibrium concentration of Cu^+ ions was of the order of 10^{-3} eq./l. The Cu atoms, deposited on the surface led, according to Brattain and Boddy, to the formation of surface states. Since usually more than 10^{15} atoms were deposited per cm^2, and the density of the states was of the order of 10^{11} per cm^2, only those atoms which were deposited on irregularities formed the states. Memming's[151] capacity measurements showed that the states only became detectable at pH values higher than 5. We have here a case of a rather direct correlation between the chemical and the physical properties of an interface system. The oxidation brought about by the addition of Cu^{++} ions can be considered as the result of a competition for electrons, expressed by eqs. (6.5.13) and (6.5.14). The standard potential for eq. (6.5.13) should be close to that of (6.5.14).

From the σ_{sp}-data it is possible to estimate an ionization constant of the reaction (6.5.12). We assume that σ_{sp} gives the surface concentration of ionized $(GeO_xH_{x-1})_s^-$ groups, i.e. we ignore the contribution of the space charge inside the crystal. This contribution is probably negligibly small, σ_{sp} being of the order of 10^{12} electrons per cm^2 whereas as we have seen, the space charge σ_{sc} is usually of the order of

---

† The positive sign is used for electrode reactions more noble than that of the reactions $H_2 \leftrightarrows 2H^+ + 2e^-$.

10^{10} electrons per cm^2. We also ignore ionic charges adsorbed at the solid side of the slipping plane. This neglect may aggravate the merit of the estimate of the ionization constant, for one thing because the adsorption may be a function of the ionic strength. Assuming that about 7×10^{14} $(GeO_xH_x)_s$ complexes per cm^2 are available, an ionization constant can be found, which, between pH $\cong$ 2 and pH $\cong$ 4, is of the order of 10^{-6} eq/l. This is rather (a hundred times) higher than the ionization constant of dissolved H_2GeO_3.[161]

Equation (6.5.8), which characterized the behaviour of a hydrogen electrode, can now be specified in the following way:

$$(GeO_xH_x)_s \leftrightarrows (GeO_xH_{x-1})_s + H^+ + e^-. \qquad (6.5.16)$$

This means that

$$\Delta\tilde{\mu}_{\{H\}} = \tilde{\mu}_{H^+} + E_F \qquad (6.5.17)$$

where $\Delta\mu_{\{H\}}$ is the difference between the chemical potentials of the components $(GeO_xH_x)_s$ and $(GeO_xH_{x-1})_s$. Since it was assumed throughout that conduction electrons and holes are subject to the equilibrium condition $h^+ + e^- \leftrightarrows 0$, eq. (6.5.16) could equally well have been written with holes instead of electrons.

For the Ge electrode to operate as a good hydrogen electrode, we require that the reaction is reversible and that, taking now $x = 1$ for simplicity, the difference of the chemical potentials of $(Ge—OH)_s$ and $(Ge—O)_s$ is constant. This means that their degree of ionization, being already small, must vary little when the pH or the E_F is varied. As mentioned, H_2O_2, added to the solution, helps to fulfil these requirements. Analogous requirements for Si electrodes could not be met in a simple way.

Much work has been done to elucidate the nature of the oxidation products on the Ge surface. Plane specificity with regard to the effects of etchants has been investigated. Worthy of mention is the work of Gatos et al.,[162] who, leaving aside oxidation problems in an arrow sense, obtained interesting results with 3–5 compounds. Soluble oxides, as formed by mild anodization, seem to be related to the hexagonal form of GeO_2. Insoluble oxides, formed by strong anodization, or by a number of etching procedures followed by exposure to air, seem to be related to tetragonal GeO_2. A number of problems remain to be solved.[164] However, it is certain that at the oxide/aq. soln. interface both ionized and non-ionized GeO_xH_x groups are present, as shown by e.m.f. vs. pH measurements. The dry oxide surface is the seat of slow states and it is tempting to correlate these with $(Ge—O)_s$ or, more generally, with $(GeO_x)_s$ groups. If wetted, slow states may be correlated with $(GeO_xH_x)_s$, $(GeO_xH_{x-1})_s$ and with $(GeO_x)_s$ groups. That indeed the oxide/aq. soln. interface can be slowly filled with electrons or holes, indicating the presence of slow acceptor and slow donor states at the interface, was shown by placing a droplet of an aqueous electrolyte on a Ge surface. Freshly etched Ge surfaces are hydrophobic,[163] the contact angle being close to 180°. The wettability increases spontaneously during the first 15 minutes or so, leading to a contact angle of about 90°.

The wettability could be greatly enhanced when a potential difference of 5 V or higher was applied between the droplet and the Ge substrate. In that case, whatever the sign of the potential difference, the contact angle could be made zero in the course of a few minutes.[164] In this stage, the solid surface was covered by a liquid surface. Part of the potential difference was taken over by the oxide layer, and part of it was used for the formation of a double layer at the interface. The increased wettability pointed to a decreased value of γ_{SL}, the interfacial tension between solid and liquid, which could amount to the value of the surface tension of the aqueous electrolyte, which is 70×10^{-13} N m^{-1} ($= 70$ erg cm^{-1}). Ascribing this decrease to the "adsorption" of electrons in slow acceptor states in the case of a negative Ge-substrate, and to "adsorption" of holes in slow donor states when the substrate is positive, one arrives at an acceptor and a donor state density of about 10^{14} per cm^2 and a relaxation time of the order of minutes. Similar results were obtained when the substrate was an etched Si or metal surface. Thus, the measurement of the contact angle as a function of a potential difference reveals the properties of slow states. We note that this sort of measurement leads to essentially the same information as the measurement of the electrocapillary curve. Here also, a decrease of the interfacial tension (of the Hg/aq. soln. interface) is observed as the result of the application of a potential difference which is positive or negative.

6.5.4. *Electrode reactions*[166,12]

Brattain and Garrett discovered that the anodic current of an n-type Ge-electrode showed a sharp saturation: at potentials more anodic than a certain value E_{lim}, the current did not rise above a certain limiting value i_{lim}. For p-type Ge anodes there was no such limiting current. Light, shining on the surface of the electrode and creating electron–hole pairs, led to an important increase of i_{lim} for an n-type electrode; for p-type electrodes there was only a small effect. These phenomena showed that the rate-determining step of the anodic reaction was provided by the minority charge carriers, the holes. The limiting current can, in this view, be considered as that current at which the depletion of the holes at the interface is complete, every hole formed by the incident light or by the anodic process itself, being consumed. For the interpretation of their results Brattain and Garrett applied kinetic concepts which were used in transistor technology and which were, at

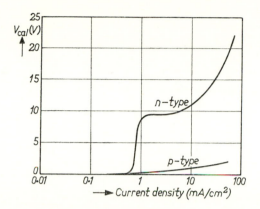

FIG. 6.31. Anodic current against applied potential V_{cal}. (From Turner.[167])

the time of the publication of their results, also the basis of semi-conductor surface research. Although we have applied these concepts implicitly previously, notably in the discussion of surface recombination data and of capacity of inversion layers, we give here a brief outline of some relevant aspects of these concepts.

6.5.4.1. KINETIC CONCEPTS; CURRENT EQUATION AND CONTINUITY EQUATION

The concepts are based on two equations. The first equation is the flow or current equation. In fact, this is a Nernst–Planck equation which we have met previously (eq. (5.5.21)). The flow $v_i n_i$ of a component i is

$$v_i n_i = -\frac{u_i}{e} n_i (\text{grad } \tilde{\mu}_i - f_i) \quad (T \text{ uniform}) \tag{6.5.18}$$

where u_i is the mobility of component i (u_e for conduction electrons, and u_h for holes) which is usually expressed in cm^2 V^{-1} s^{-1}; f_i is an external force acting on a particle of component i, v_i is the average velocity, n_i is the average number of particles per unit volume and $\tilde{\mu}_i$ is the electrochemical potential.

The equation is valid for uniform temperature. It can be derived by using the principles of irreversible thermodynamics. The current J_i is related to the flow as follows:

$$J_i = z_i \, e v_i n_i. \tag{6.5.19}$$

The second equation (eq. (6.5.20)) is the continuity equation. For semi-conductors in which recombination takes place, this is written in a schematized form[10,11] and disregarding surface effects for the moment, as follows:

$$\text{div}(v_i n_i) + \frac{\delta n_i}{\tau_b} + \frac{\partial n_i}{\partial t} = 0. \tag{6.5.20}$$

The contribution $(\delta n_i / \tau_b)$, which is absent in the more common forms of the continuity equation, is the net rate of (bulk) recombination, i.e. the number of particles of component i, disappearing through a recombination process per unit time and unit volume; τ_b is the average time between the formation and the recombination of holes and conduction electrons in the bulk, or the bulk lifetime, and δn_i expresses the deviation from equilibrium

$$\delta n_i = n_i - n_{i0}. \tag{6.5.21}$$

For $\delta n_i \ll n_{i0}$ the linear form of the recombination term as given in eq. (6.5.20) is valid. This has been pointed out in refs. 10 and 11.†
Note that $\partial n_i / \partial t$ in eq. (6.5.20) can be replaced by $\partial \delta n_i / \partial t$.

---

† While we are only communicating consequences of the existence of a linear re-combination term $(\delta n_i / \tau_b)$ the *origin* of a recombination term may be discussed by suggesting that recombination can be described by a sudden change of the momentum p_{ki} of the kth particle of component i to almost zero. In this view the particle does not vanish, but is merely immobilized. We then state that $(dn_i/dt) = 0$ and use a reasoning followed by Landau and Lifshitz (ref. 28, p. 9) in their discussion of the continuity equation and Liouville's theorem:

The motion of the kth particle is described by the time t, the coordinate q_{ki} and the momentum p_{ki}. For each particle the coordinate and the momentum has three components in space, so the system has $6n_i + 1$ variables. By partial differentiation we find:

$$\frac{dn_i}{dt} = \sum v_{ki} \frac{\partial n_i}{\partial q_{ki}} + \sum f_{ki} \frac{\partial n_i}{\partial p_{ki}} + \frac{\partial n_i}{\partial t} = 0 \tag{6.5.20a}$$

where $v_{ki} = \dot{q}_{ki} = (\partial q_{ki}/\partial t)$ is the velocity of the kth particle and where $f_{ki} = \dot{p}_{ki}$ is the force upon the particle.

Equation (6.5.20a) is rewritten by considering the derivatives of $v_{ki}n_i$ and of $f_{ki}n_i$:

$$\frac{\partial(v_{ki}n_i)}{\partial q_{ki}} = v_{ki} \frac{\partial n_i}{\partial q_{ki}} + n_i \frac{\partial v_{ki}}{\partial q_{ki}} \; ; \quad \frac{\partial(f_{ki}n_i)}{\partial p_{ki}} = f_{ki} \frac{\partial n_i}{\partial p_{ki}} + n_i \frac{\partial f_{ki}}{\partial q_{ki}}.$$

In view of the Hamilton equations

$$\dot{q}_{ki} = \frac{\partial H}{\partial p_{ki}} \quad \text{and} \quad \dot{p}_{ki} = -\frac{\partial H}{\partial q_{ki}}$$

At equilibrium the rate of recombination equals the rate of formation of hole–electron pairs. At equilibrium also $\partial \delta n_i = 0$, and $v_i n_i = 0$. The condition for the stationary (not necessarily equilibrium) case is

$$\text{div}(v_i n_i) = 0. \tag{6.5.22}$$

In that case $v_i n_i$ is a non-zero constant, and

$$\delta n_i = \delta n_{i(t=0)} \, e^{-t/\tau_b}. \tag{6.5.23}$$

For the gradient of the electrochemical potential we write

$$\text{grad } \tilde{\mu}_i = kT \text{ grad ln } n_i + z_i \, e \text{ grad } \psi' \tag{6.5.24}$$

where ψ' is the electrostatic potential in the non-equilibrium case. Now

$$v_i n_i = -D_i \text{ grad } n_i - z_i u_i n_i \left(\text{grad } \psi' - \frac{1}{e} f_i \right) \tag{6.5.25}$$

where, just as in Chapter 5, the diffusion coefficient D_i is

$$D_i = \frac{u_i}{e} kT \tag{5.5.6}$$

and u_i is the mobility which we count positive for any sign of charge. The divergence of $v_i n_i$ is

$$\text{div } (v_i n_i) = -D_i \nabla^2 n_i - z_i u_i \text{ div } \left\{ n_i \left(\text{grad } \psi' - \frac{1}{e} f_i \right) \right\}. \tag{6.5.26a}$$

At equilibrium $f_i = 0$, and ψ' has its equilibrium value ψ. Then

$$0 = D_i \nabla^2 n_{i0} + z_i u_i \text{ div } \{n_{i0} \text{ grad } \psi\}. \tag{6.5.26b}$$

We subtract eq. (6.5.26b) from eq. (6.5.26a) and ignore the contribution $z_i u_i \text{ div } \{n_i \text{ grad } \psi' - n_{i0} \text{ grad } \psi\}$. The result is inserted in eq. (6.5.20). This gives

$$D_i \nabla^2 \, \delta n_i - z_i \frac{u_i}{e} \text{ div } (f_i \, \delta n_i) - \frac{\delta n_i}{\tau_b} = +\frac{\partial \delta n_i}{\partial t}. \tag{6.5.26}$$

The right-hand side is often put equal to zero.

---

we see that $(\partial v_{ki}/\partial q_{ki}) + (\partial f_{ki}/\partial p_{ki}) = 0$. Therefore eq. (6.5.20a) becomes

$$\text{div}_q(v_i n_i) + \text{div}_p(f_i n_i) + \frac{\partial n_i}{\partial t} = 0 \tag{6.5.20b}$$

where

$$\text{div}_q(v_i n_i) = \sum \frac{\partial(v_{ki} n_i)}{\partial q_{ki}} \quad \text{and} \quad \text{div}_p(f_i n_i) = \sum \frac{\partial(f_{ki} n_i)}{\partial p_{ki}}.$$

We have identified $\text{div}_q(v_i n_i)$ with $\text{div}(v_i n_i)$ and $\text{div}_p(f_i n_i)$ with the recombination term.

Equation (6.5.26) has served as a basis for most considerations concerning non-equilibrium properties of semiconductors. We note that, dividing thee quation by D_i, an important parameter, $D_i \tau_b$, appears. We write

$$L_i = \sqrt{(D_i \tau_b)} \tag{6.5.27}$$

which is the definition of the diffusion length. For pure Ge τ_b can be, at room temperature, 10^{-3} sec or more,[1c] and so L_i is of the order of 10^{-2} cm or more.

Since numerous applications have been worked out in textbooks, we restrict ourselves to only a few simple cases. These pertain to minority charge carriers and this is quite natural because for minority charge carriers the ratio $|\delta n_i|/n_{i0}$ can be made relatively large. For this reason the detection and measurement of kinetic properties (for instance, of τ_b) is often only possible for minority charge carriers. A good example is the decisive role played by the holes in the kinetics of n-type Ge electrodes as observed by Brattain and Garrett.

(a) Equation (6.3.10) $(1/\tau = 1/\tau_b + 2s/d)$ is justified by considering a thin wafer whose thickness in the x-direction is d and whose dimensions in the z- and y-directions are large. We assume that there is only an electric field applied in the z-direction: $z_i u_i f_i = e u_i E_z$.

Surface recombination takes place at the surfaces $x = \frac{1}{2}d$ and $x = -\frac{1}{2}d$ and is ignored at other surfaces. The surfaces $x = \frac{1}{2}d$ and $x = -\frac{1}{2}d$ act as a "sink" for the minority charge carriers and we give the diffusion flow toward these surfaces simply as $s \, \delta n_i$ where s is the surface recombination velocity. This flow represents a current of the minority charge carriers, i, disappearing at these surfaces per unit time and per unit area, equal to the average number of excess particles i which is present between the planes $x = 0$ and $x = \frac{1}{2}d$ or between $x = 0$ and $x = -\frac{1}{2}d$, multiplied by s. This product, $s(2\delta n_i/d)$, must be equal to the negative divergence of these particles in the x-direction

$$D_i \frac{\partial^2 \, \delta n_i}{\partial x^2} = -s \frac{2\delta n_i}{d} . \tag{6.5.28}$$

Here it is assumed that s is sufficiently small so that δn_i can be taken as approximately constant through the sample. Introducing eq. (6.5.28) in eq. (6.5.26) gives

$$D_i \left(\frac{\partial^2 \, \delta n_i}{\partial y^2} + \frac{\partial^2 \, \delta n_i}{\partial z^2} \right) - u_i E_b \frac{\partial \, \delta n_i}{\partial z} - \left(\frac{2s}{d} + \frac{1}{\tau_b} \right) \delta n_i = \frac{\partial \, \delta n_i}{\partial t} , \tag{6.5.29}$$

where u_i is a bulk property. In eq. (6.3.10) the sum of $1/\tau_b$ and $2s/d$ is denoted as $1/\tau$ and this is the quantity accessible to measurements.

(b) Consider a wafer, thickness d $(0 \leqslant x \leqslant d)$ which is not necessarily small with respect to L_i and assume that there is an electric field in the x-direction. For convenience we now take the $x = 0$ plane as one surface and $x = d$ as the other. Minority charge carriers are formed or consumed by some electrochemical process at or near the plane $x = 0$. We are only interested in δn_i as a function of x. For the unidimensional case eq. (6.5.26) becomes, with the right-hand side put equal to zero and taking $\tau_b = \tau$:

$$D_i \frac{\partial^2 \, \delta n_i}{\partial x^2} - u_i E_x \frac{\partial \, \delta n_i}{\partial x} - \frac{\delta n_i}{\tau} = 0. \tag{6.5.30}$$

It proves useful to define not only a diffusion length $L_i = \sqrt{(D_i \tau)}$ but also a drift length $L_E = |u_i E_x| \tau$. The drift length is an average distance traversed by the charge carriers under the influence of an electric field before it recombines. Assuming first that $L_i \ll L_E$ the solution of eq. (6.5.30) is

$$\delta n_i = \alpha \, e^{-x/L_E} \tag{6.5.31}$$

where α is a constant.

Next, assuming that $L_E \ll L_i$, i.e. assuming that E_x is relatively small, the solution of eq. (6.5.30) is

$$\delta n_i = \alpha_1 \, e^{-x/L_i} + \alpha_2 \, e^{+x/L_i}, \tag{6.5.32}$$

where α_1 and α_2 are constants, the values of which are determined by the conditions of the problem under consideration.

When the condition is $\delta n_i = 0$ at $x = \infty$, then $\alpha_2 = 0$ and there is a simple exponential decay of δn_i when x is increased.

When $n_{i0} = 0$ at $x = 0$, i.e. when the component i is entirely consumed by the electrochemical process at the surface, then $\delta n_i = -n_{i0}$ at $x = 0$ and α_1 becomes $-n_{i0}$. The partial current J_i at $x = 0$ is in this case given by

$$J_i(x = 0) = z_i \, e D_i \frac{\partial \, \delta n_i}{\partial x} \bigg|_{x=0} = z_i \, e \frac{D_i}{L_i} n_{i0}, \tag{6.5.33}$$

and this contains only bulk parameters. This is a limiting case of a situation which occurs frequently under experimental conditions.

As a second case, assume that d is finite and that there is an ohmic contact at $x = d$ in such a way that $\delta n_i = 0$ at $x = d$.

$$\alpha_2 = -\alpha_1 \, e^{-2d/L_i} \quad \text{and} \quad \alpha_1 = \alpha_0 \frac{e^{d/L_i}}{2 \sinh d/L_i}, \tag{6.5.34}$$

where α_0 is the value of δn_i at $x = 0$.

Equation (6.5.34) leads to

$$\delta n_i = \alpha_0 \frac{\sinh (d-x)/L_i}{\sinh d/L_i}. \tag{6.5.35}$$

The partial current at $x = 0$ is

$$J_i(x = 0) = -z_i \, eD_i \frac{\alpha_0}{L_i} \coth \frac{d}{L_i}. \tag{6.5.36}$$

The partial currents of eq. (6.5.33) and of eq. (6.5.36) decrease when x is increased, because recombination with other charge carriers takes place. When no current flows in the y- or z-directions and when the flow of component i across the plane $x = 0$ is the only source of the current J_i, then the measured current at $x = d$ is equal to this current. This is also true for the previous case ($\alpha_2 = 0$) since without violation of the imposed conditions an ohmic contact can be made at the $x = d$ surface.

6.5.4.2. CURRENT MULTIPLICATION. ROLE OF VARIOUS SOLUTES

For the understanding of the mechanism of the anodic dissolution of Ge it was essential to control the hole concentration of an n-type sample near the interface. This was done by Brattain and Garrett not only by shining light on the Ge/aq. soln. interface, but also by using a thin-slice technique directly adopted from transistor technology. A

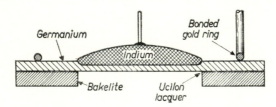

FIG. 6.32. Experimental arrangement used by Brattain and Garrett[143] for the study of Ge dissolution.

thin slice of n-type Ge, whose thickness was small (0·1 mm) compared to the diffusion length of the holes, was at one side in contact with the solution while the other side was covered with a rectifying junction made by the indium alloying technique. Extra holes were created in the n-type base material by passing a hole current from the rectifying junction in the forward direction. The non-equilibrium hole concentration, p, could be taken as uniform over the thin n-type sample. Measurements of the floating potential V_f, which is the potential difference between the alloyed junction and the n-type base at zero current (for non-zero current an ohmic $J \times R$ contribution comes in), led to relevant values of p, V_f and p being correlated according to $V_f = kT/e \ln p/p_0$. This technique has found a wide application and we mention two important modifications. Pleskov[65] replaced the p-type junction by a second electrolyte. Thus the thin slice used by Pleskov was really a

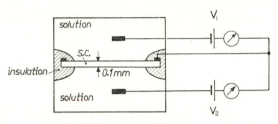

Fig. 6.33. Pleskov's thin slice arrangement.[65]

double-sided electrode. One of the sides was anodically polarized. It produced holes and in this way it resembled the p–n junction used by Brattain and Garrett. Pleskov's version is used in those cases in which the experimental circumstances are unfavourable for the introduction of an alloyed p–n junction. The second modification was introduced by Harten.[165] Holes, produced by light shining on the electrolyte side of a Si slice (and penetrating about 0.25μ), diffused to the back side where

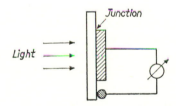

Fig. 6.34. Harten's thin slice arrangement.[165]

they modified the floating potential. The method was used for measurements of the surface recombination velocity s at the electrolyte side of the thin slice. The higher s, the lower the number of holes that reach the junction at the back side of the slice and so the measured signal was inversely proportional to s. These measurements were carried out at various values of the potential difference V_{pt} between the thin slice electrode and an auxiliary Pt electrode. Harten used Na_2SO_4 solutions and found on the whole results similar to those obtained for dry Si surfaces. Bell-shaped V_{pt-s} curves were obtained. Part of the applied voltage was taken over by an oxide layer between the Si sample and

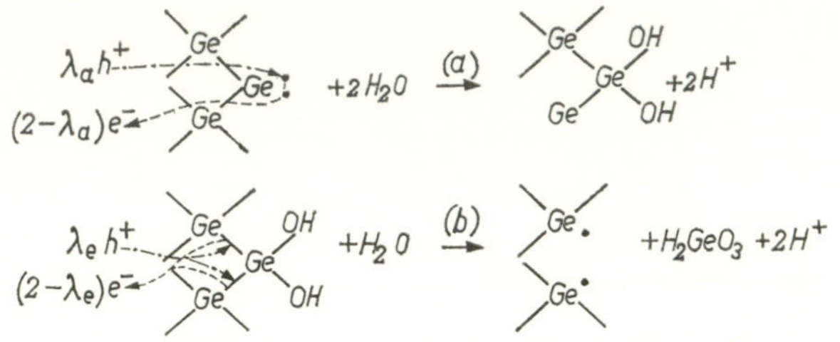

FIG. 6.35. Possible dissolution mechanism of Ge. See Gerischer[166] and Turner.[167]

the electrolyte. Returning to the experiments of Brattain and Garrett, these workers found that, when the rate of supply of the holes was increased by a given factor, the value of i_{lim} was increased by 1·4–1·8 times this factor. This is the phenomenon of current multiplication. In the case of the anodic dissolution of Ge, the interpretation of this phenomenon must be that not only holes, but also conduction electrons, participated in the reaction.

According to Gerischer,[166] who refined a model by Turner,[167] the dissolution mechanism may be as schematically given in Fig. 6.35. Holes are transported from the bulk to the interface and conduction electrons flow in the opposite direction. Physically this means that valence electrons are transported away from the interface, leaving behind a non-stable complex in which other valence electrons are activated and become conduction electrons. Each of the two reaction schemes of Fig. 6.35 may be decomposed, in this view, into a part, consuming holes, and a part, producing conduction electrons. When *reduction* takes place this reaction mechanism may be found in reversed order. Thus Memming,[168] studying the reduction of $S_2O_8^{--}$ ions at the GaP electrode, found that an electron brought to the conduction band by light excitation led to a reduction of the persulphate ion. However, the observed reduction current was twice as large as that measured during the hydrogen evolution in H_2SO_4. Therefore it was postulated that the reduction through the conduction electron initially led to the formation of a radical ion $\cdot SO_4^-$ which was further reduced, to SO_4^{--}, by consuming an electron from the *valence* band. For the reduction of

H_2O_2 at GaP electrodes Memming postulated likewise as a first step the reaction

$$H_2O_2 + e^- \rightarrow OH + OH^-, \qquad (6.5.37)$$

and as a second step

$$OH \rightarrow OH^- + h^+. \qquad (6.5.38)$$

The current multiplication factor in these cases was 2. For the overall anodic reaction at the Ge/soln. interface one has

$$Ge + (\lambda_a + \lambda_e)h^+ + 3H_2O \rightarrow H_2GeO_3 + 4H^+ + (4 - \lambda_a - \lambda_e)\,e^- \quad (6.5.39)$$

and the multiplication factor, f, is given by

$$f = 1 + \frac{4 - \lambda}{\lambda} \qquad \lambda = \lambda_a + \lambda_e. \qquad (6.5.40)$$

Experiments by Pleskov[65] and by Gerischer[166] showed that, on an increase of the hole injection, f decreased ultimately to values close to 1. From these and other measurements Pleskov inferred that the multiplication factor f can only be given a statistical meaning. For a detailed discussion, see the book by Myamlin and Pleskov.[65]

Beck and Gerischer[169,166] showed that the value of the saturation current was not only altered by the influence of light or the injection of holes, but also by the presence of redox solutions. For example, working with a Ce^{4+}/Ce^{3+} solution, the Ce^{4+} ions were reduced to Ce^{3+} at the Ge surface, thereby injecting a hole

$$Ce^{4+} \rightarrow Ce^{3+} + h^+. \qquad (6.5.41)$$

That holes were injected, was found by observing an increase of the limiting anodic current as a result of the addition of the redox ions. Other redox systems led to a decrease of i_{lim}. This was the case for the $Fe(CN)_6^{3-}/Fe(CN)_6^{4-}$ system in 0·5 M H_2SO_4 solutions. Obviously an oxidation instead of a reduction was taking place here.

No such hole transfer mechanism was observed by Pleskov and Kabanov[170] for V^{3+}/V^{2+} solutions. Here the (oxidation) mechanism uses conduction electrons

$$V^{3+} + e^- \rightarrow V^{2+}. \qquad (6.5.42)$$

Some insight into the problem whether a hole mechanism or a conduction electron mechanism prevails was given by Gerischer.[166] He considered the equilibrium reaction

$$Ox^{z+} + e^- \leftrightarrows Red^{(z-1)+}. \qquad (6.5.43)$$

In view of the equilibrium reaction in the semiconductor $h^+ + e^- \leftrightarrows 0$, eq. (6.5.43) could also have been written as

$$Ox^{z+} \rightleftharpoons Red^{(z-1)+} + h^+.\tag{6.5.44}$$

From either equation it follows that

$$E_F = E_F{}^0 + kT \ln \frac{c_{red}}{c_{ox}}\tag{6.5.45}$$

where $E_F{}^0 = (\mu_{red}{}^0 - \mu_{ox}{}^0)$ is the standard oxidation potential of the redox system.

In equilibrium the electrochemical potential of the electrons, E_F, so obtained, must have the same value as the Fermi energy in the semi-conductor. Drawing the parallelism further, the oxidized species Ox^{z+}, containing allowed but unoccupied electronic levels, must be compared with the conduction band, whereas the reduced species $Red^{(z-1)+}$ must be compared with the valence band. There is a forbidden energy gap between the levels of Ox^{z+} and $Red^{(z-1)+}$ just as in the semiconductor, although owing to the liquid structure, there will be no sharp transition between forbidden and allowed levels. With the aid of these concepts, Fig. 6.36 can be constructed. Electron transfer is possible when the empty states of the redox solution are at the same energy level as those of the conduction band, so that the electrons need not pass forbidden levels. Hole transfer is possible, when the occupied states are at the same level as the valence band.

The electron exchange current j_n can be represented as

$$j_n = a \times c_{ox} \times n\tag{6.5.46}$$

where a is a frequency factor.

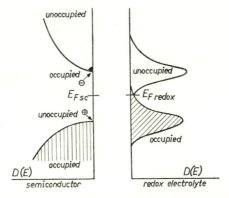

FIG. 6.36. Band models of semiconductor and redox solution compared. (Gerischer.[166]) $D(E)$ is the state density as a function of the energy E. When conduction electrons and (or) holes can reversibly be exchanged between the two phases, E_{Fsc} and E_F redox are at the same level.

The hole exchange current, j_p, is

$$j_p = a' \times c_{\text{red}} \times p \qquad (6.5.47)$$

where a' is a frequency factor. For a discussion of the frequency factors a and a', see Gerischer.[166]

One may rewrite these equations by using eq. (6.5.45) and by using the relations $n = n_i \exp (E_F - E_i)/kT$ and $p = n_i \exp (E_i - E_F/kT$. Before doing this it should be borne in mind that E_F^0 refers to the bulk of the electrolyte and E_i to the bulk of the semiconductor. Therefore, instead of E_i, a term $E_i - e\Delta\varphi$ should be considered where $\Delta\varphi$ is the Galvani potential difference between the two phases. After elimination of E_F one obtains

$$j_n = a \times c_{\text{red}} \times n_i \, e^{(E_F^0 - E_i)/kT} \qquad (6.5.48)$$

and

$$j_p = a' \times c_{\text{ox}} \times n_i \, e^{-(E_F^0 - E_i)/kT}. \qquad (6.5.49)$$

According to this theory the value of E_F^0 is an important parameter. Take $c_{\text{ox}} = c_{\text{red}}$ and assume that for a number of redox systems a/a' and $\Delta\varphi$ have the same values. Then for $E_F^0 \ll 0$ (a highly oxidizing system) the hole transfer mechanism is preferred, whereas for $E_F^0 \gg 0$ a charge transfer via the conduction band is more likely, the "turning point" being at $E_F^0 \sim 0$ V. A fairly large number of redox solutions obeyed these predictions, but there was at least one important exception, namely H_2O_2. The oxidation reaction here is

$$H_2O_2 + 2H^+ + 2e^- \leftrightarrows 2H_2O \qquad (6.5.50)$$

with $E_F^0 = +1\cdot77$ V. The Ge/soln. interface does not use the expected hole transfer mechanism in this case. This points to a more complicated reaction mechanism than assumed for the other redox reactions. To elucidate this point, Gerischer et al. returned to the anodic and cathodic reactions at the interface between Ge and an aqueous solution of given pH but without a redox system. Gerischer et al.[171] proposed the following reversible reactions:

$$\text{>Ge—OH} + e^- + H^+ \leftrightarrows \text{>Ge} \cdot + H_2O \qquad (6.5.51a)$$

and

$$\text{>Ge} \cdot + H^+ \leftrightarrows \text{>Ge—H} + p^+. \qquad (6.5.51b)$$

The oxidation complex >Ge—OH was formed after pre-anodization; the reduction complex >Ge—H was formed after pre-cathodization. The chemical representations given above were regarded as tentative.

The first reaction used conduction electrons because illumination of p-type pre-anodized Ge electrodes led to an increased cathodic current compared with the dark-current at the same potential. The second reaction used holes, because illumination of pre-cathodized n-type Ge-electrodes led to an increased anodic current. The scheme is the simplest one that could be imagined and there is no direct proof for the existence of the surface complexes $\rangle$Ge—OH, $\rangle$Ge—H, and of the radical $\rangle$Ge$\cdot$.

The addition of H_2O_2 to the solution led Gerischer and Mindt to suggesting the following scheme:[172]

$$\rangle Ge\text{—}OH + e^- + H^+ \leftrightarrows \rangle Ge\cdot + H_2O, \qquad (6.5.52a)$$

$$\rangle Ge\cdot + H_2O_2 + e^- \rightarrow \rangle Ge\text{—}OH + OH^-. \qquad (6.5.52b)$$

The reduction reaction (6.5.52b) was slower at p-type than at n-type Ge electrodes. Therefore a mechanism using conduction electrons was suggested. We see that there is again a transient formation of the radical $\rangle$Ge$\cdot$. Such a radical reminds one of the situation at the clean Ge surface exposed to vacuum. In both cases there must be a lone pair electron connected with the outer Ge-atom. Since at the Ge/vacuum surface numerous surfaces states were found, an argument in favour of the existence of radicals would be the finding of surface states. We have seen that Memming and Neumann did indeed find surface states so that these states must be ascribed to a physicochemical complex intermediate between the oxidized and the reduced surface complex, the latter two being state-free. The interaction of Ge with H_2O_2 involves the disintegration of the redox compound and is of an entirely different nature from the interaction with other redox systems. This explains why the behaviour of H_2O_2 does not fit predictions based on eq. (6.5.48) and eq. (6.5.49). Memming's findings, the two-step reduction of the $S_2O_8^{--}$ ion and of H_2O_2 at GaP electrodes are other examples of a strong interaction which cannot be incorporated in the "weak" interaction model, leading to the eqs. (6.5.48) and (6.5.49). In fact, Memming introduced an "oxidation potential" for each step, the macroscopic (and tabulated) oxidation potential being an average between those two.

Turning to another type of "weak" interaction between solute molecules and a semiconductor, very interesting phenomena have been observed in studies of the interaction with dissolved dye molecules. These molecules were excited by light with a photon energy which was less than the band gap. So assuming the absence of energy levels in the forbidden gap, no photo-excitation took place in the semiconductor. Thus, only wide-gap semiconductors can be used.

The situation, for adsorbed and excited dye molecules is as depicted in Fig. 6.37. The excited dye molecule can act as a donor (a) or as an acceptor (b), depending on the relative positions of the electron energies in the dye and in the semiconductor. Situation (a) is not unlike the one for a redox system with a high positive E_F^0 value, the electron energy of the excited electron in the dye molecule being at the same height as the energies in the conduction band. Situation (b) corresponds to that of a redox system with low E_F^0. Rhodamin B is a good example, the electron injection mechanism being found at the ZnO and CdS electrodes and the hole injection mechanism at the GaP and the Cu$_2$O electrodes.[173] In the case of a ZnO/1 M KCl soln. interface with adsorbed rhodamin B, the band-to-band transition in ZnO corresponds to a wavelength of 370 nm. When illuminated with light at wavelengths between 400 nm and 700 nm, the anodic current is greatly enhanced and

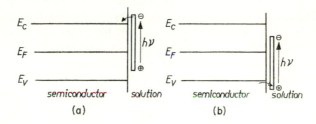

FIG. 6.37. An adsorbed dye molecule, excited by a photon hν ($< E_c - E_v$) may lead to the injection of a conduction electron (left) or of a hole (right).

shows peaks at 530 nm and 580 nm, corresponding to the adsorption spectrum of rhodamin B. The rhodamin B molecules are, by this injection, oxidized. The quantum efficiency of this reaction can be enhanced to values up to 1, by the addition of substances such as hydroquinone which reduce the dye. These substances are super-sensitizers. The mechanism by which they act is not yet known. Ideas can be found in a review article by Gerischer[174] on charge transfer processes at semiconductor/electrolyte interfaces.

It is quite probable that the field of overlap between electrochemistry and photochemistry will be profoundly studied in the future. We mention the work by Mulder[175] who studied the anthracene/aq. soln. interface. Light in the visible and u.v. region ($\lambda < 420$ nm) produces, in anthracene crystals, excitons. These are anthracene molecules in an excited state. The excitation energy can be transferred to a neighbouring molecule. In this way the excitons are mobile. They have a diffusion length of about 400 Å and produce light emission at $\lambda = 420$ nm and at

$\lambda = 450$ nm, when annihilated. Excitons in conventional inorganic semiconductors are mobile hole–electron pairs. Since anthracene crystals are held together by means of weak Van der Waals forces it is somewhat dangerous to identify excitons in anthracene with excitons in inorganic semiconductors and it is equally dangerous, in spite of the fact that mobile charge carriers in anthracene have been detected, to apply a band model to this crystal.

When rhodamine B was adsorbed at the interface between anthracene and an (alkaline) aqueous solution to an amount which was less than one molecule per 1000 Å^2, illumination ($\lambda < 420$ nm) of the anthracene crystal at the side of the adsorbed rhodamine B led to a light emission, which had the colour of the fluorescence of rhodamine B, the emission maximum being at $\lambda = 605$ nm. At the same time an enhanced photoconductance could be observed provided a positive potential was applied to the rhodamine covered side. This effect was ascribed to the formation of holes. Also the anthracene fluorescence was quenched as a consequence of the rhodamin adsorption. All these processes required the presence of O_2. Rhodamine molecules are too large to penetrate into the anthracene crystal and Mulder proposed the following surface reactions (A = anthracene, A* = excited molecule):

$$A^* + H^+ \rightarrow (AH^+)^*$$

$$(AH^+)^* + O_2 \rightarrow A^+ + HO_2$$

or, in one equation,

$$exciton + O_2 + H^+ \rightarrow hole + HO_2$$

where the radical HO_2 rapidly decomposes into an O_2 molecule and an H atom. This type of surface reactions, initiated by excited molecules of the adsorbent, has opened a new field of research where interesting phenomena may be revealed.

REFERENCES

1. (a) C. KITTEL, *Introduction to Solid State Physics*, 3rd ed., Wiley.
   (b) J. M. ZIMAN, *Principles of the Theory of Solids*, Cambridge U.P., 1964.
   (c) J. S. BLAKEMORE, *Semiconductor Statistics*, Pergamon Press, 1962.
   (d) W. SHOCKLEY, *Electrons and Holes in Semiconductors*, Van Nostrand, 1950.
   (e) N. B. HANNAY (ed.), *Semiconductors*, Reinhold, 1959.
   (f) J. C. SLATER, *Quantum Theory of Molecules and Solids*, Vol. 2, McGraw-Hill, 1965.
2. e.g. S. RAIMES, *The Wave Mechanics of Electrons in Metals*, Univ. of Pennsylvania Press, 1961.

3. J. BRADLEY and P. GREENE, *Trans. Faraday Soc.*, 1966, **62**, 2069; 1967, 63, 424, 2516.

4. J. H. KENNEDY and F. CHEN, *J. Electrochem. Soc.*, 1968, **116**, 207.

5. M. FARADAY, *Experimental Researches in Electricity*, I (1839), in Everyman's Library, 1922, p. 44.

6. G. L. PEARSON and W. H. BRATTAIN, History of semiconductor research. *Proc. I.R.E.*, 1955, **43**, 1794.

7. e.g. A. B. LIDIARD, Ionic conductivity, in *Encyclopedia of Physics*, 1956, **10**, 246 (ed. S. FLÜGGE, Berlin).
K. HAUFFE, *Reaktionen in und an festen Stoffen*, 2nd ed., Berlin, 1966.
N. F. MOTT and R. W. GURNEY, *Electronic Processes in Ionic Crystals*, 2nd ed., Oxford U.P., 1948.
F. A. KRÖGER, *The Chemistry of Imperfect Crystals*, North Holland Publ. Co., Amsterdam, 1964.
R. E. HOWARD and A. B. LIDIARD, Matter transport in solids. *Repts Progress in Physics*, 1964, **27**, 161.
W. JOST, *Diffusion in Solids, Liquids, Gases*, 3rd ed., Academic Press, New York, 1960.

8. C. WAGNER, *Z. Elektrochemie*, 1956, **60**, 4.

9. C. TUBANDT, *Z. Anorg. Allgem. Chem.*, 1926, **115**, 109, 113; *Handbuch der Experimentalphysik*, Leipzig, 1933 (eds. W. WIEN and F. HARMS), **12**, part 1, p. 412.

10. A. MANY, I. GOLDSTEIN and N. B. GROVER, *Semiconductor Surfaces*, North Holland, Amsterdam, 1965.

11. D. R. FRANKL, *Electrical Properties of Semiconductor Surfaces*, Pergamon Press, 1967.

12. M. GREEN, in *Modern Aspects of Electrochemistry*, Vol. II (ed. J. O'M. BOCKRIS), Butterworth, 1959, p. 343.

13. I. TAMM, *Phys. Z. Sowjetunion*, 1932, **1**, 733.

14. I. BARTOS, *Surface Sci.*, 1969, **15**, 94.

15. W. SHOCKLEY, *Phys. Rev.*, 1939, **56**, 317.

16. J. KOUTECKY, *J. Phys. Chem. Solids*, 1960, **14**, 233; J. KOUTECKY and M. TOMACEK, *J. Phys. Chem. Solids*, 1960, **14**, 241.

17. P. MARK, *R.C.A. Review*, 1965, **26**, 461.

18. C. A. MEAD, *Solid State Electronics*, 1966, **9**, 1023.

19. F. FORSTMANN, *2 Physik*, 1970, **235**, 69.

20. F. FORSTMANN and J. B. PENDRY, *2 Physik*, 1970, **235**, 74.

21. F. FORSTMANN and V. HEINE, *Phys. Rev. Letters*, 1970, **24**, 1419.

22. *Colloquium on the Optical Properties and Electronic Structure of Metals and Alloys*, Paris, 1965 (ed. F. A. ABELÈS), North Holland, Amsterdam, 1966, papers by W. E. SPICER and C. N. BERGLUND.

23. E. W. PLUMMER and J. W. GADSUK, *Phys. Rev. Letters*, 1970, **25**, 1493.

24. N. F. MOTT, *Proc. Phys. Soc.*, 1949, **62**, 28.

25. M. HENZLER, *Surface Sci.*, 1971, **25**, 650.

26. J. D. LEVINE and S. G. DAVISON, *Solid State Physics* (eds. EHRENREICH, SEIK and TURNBULL), Academic Press, 1970, **25**, 2.
P. MARK in *Clean Surfaces* (ed. G. GOLDFINGER), Marcel Dekker, 1970, p. 307; S. G. DAVISON and M. STESLICKA, *Intern. J. Quantum Chem.*, 1971, **4**, 455.

27. *Surface Sci.*, 1971, **25**, no. 1, *Proceedings of the Symposium on Modern Methods of Surface Analysis* (eds. MARK and LEVINE).

28. L. B. LANDAU and E. M. LIFSHITZ, *Statistical Physics* (Vol. 5 of the *Course of Theoretical Physics*), Pergamon Press, 1959, p. 152 (2nd edn., revised and enlarged, published 1968).

29. J. W. GIBBS, *Collected Works*, II, p. 192 (Yale U.P., 1948).

30. R. C. TOLMAN, *The Principles of Statistical Mechanics*, Oxford U.P., 1950.

31. T. L. HILL, *Statistical Mechanics*, McGraw-Hill, 1956.

32. P. T. LANDSBERG, *Proc. Phys. Soc.,* A 1952, **65**, 604; 1953, **66**, 662.

33. E. A. GUGGENHEIM, *Proc. Phys. Soc.*, A, 1953, **66**, 121.

34. E. H. KINGSTON and S. F. NEUSTADTER, *J. Appl. Phys.,* 1955, **26**, 718.

35. N. F. MOTT, *Proc. Roy. Soc. (London)* A, 1939, **171**, 27.

36. W. SCHOTTKY, *Z. Physik*, 1939, **113**, 367; 1942, **118**, 539; W. SCHOTTKY and R. SPENKE, *Wiss. Veroeffentl. Siemens*, 1939, **18**, 3.

37. H. J. KRUSEMEYER and D. G. THOMAS, *J. Phys. Chem. Solids*, 1958, **4**, 78.

38. H. J. ENGELL and K. HAUFFE, *Z. Elektrochemie*, 1952, **56**, 366; 1953, **57**, 762.

39. P. AIGRAIN and C. DUGAS, *Z. Elektrochemie*, 1952, **56**, 363.

40. R. E. HANNEMAN, M. C. FINN and H. C. GATOS, *J. Phys. Chem. Solids*, 1958, **4**, 78.

41. C. WAGNER and K. HAUFFE, *Z. Elektrochemie*, 1938, **44**, 172.

41a. C. G. B. GARRETT, *J. Chem. Phys.*, 1960, **28**, 966.

41b. *Discussions Faraday Soc.*, 1966, **41**.

42. R. SEIWATZ and M. GREEN, *J. Appl. Phys.,* 1958, **29**, 1034.

43. F. DEWALD, *Ann. N.Y. Acad. Sci.,* 1963, **101**, 872; N. ST. J. MURPHY, *Surface Sci.*, 1964, **2**, 86.

44a. A. B. FOWLER, F. F. FANG, W. E. HOWARD and P. J. STILES, Proc. Int. Conf. on the Phys. of Semiconductors, Kyoto, *Phys. Soc. Japan Suppl.*, 1966, **21**, 331; A. B. FOWLER, F. F. FANG and W. E. HOWARD, *Phys. Rev. Letters*, 1966, **16**, 901.

44b. M. KAPLIT and J. N. ZEMEL, *Phys. Rev. Letters,* 1968, **21**, 212; J. N. ZEMEL and M. KAPLIT, *Surface Sci.*, 1969, **13**, 17.

45. J. BARDEEN, *Phys. Rev.,* 1947, **71**, 717.

46. W. E. MEYERHOF, *Phys. Rev.,* 1947, **71**, 727.

47. W. SHOCKLEY and G. L. PEARSON, *Phys. Rev.,* 1948, **74**, 232.

48. W. H. BRATTAIN and W. SHOCKLEY, *Phys. Rev.,* 1948, **72**, 345.

49. J. BARDEEN and W. SHOCKLEY, *Phys. Rev.,* 1948, **74**, 230; 1949, **75**, 203, point contact transistor; W. SHOCKLEY, M. SPARKS and G. K. TEAL, *Phys. Rev.*, 1951, **83**, 151, *n–p–n* and *p–n–p* transistor. For an excellent early review see G. L. PEARSON and W. H. BRATTAIN, History of semiconductor research, *Proc. I.R.E.*, 1955, **43**, 1794.

50. R. H. KINGSTON, *J. Appl. Phys.,* 1956, **27**, 101.

51. W. H. BRATTAIN and J. BARDEEN, *Bell Syst. Techn. J.,* **32**, 1953, 1.

52. W. L. BROWN, *Phys. Rev.*, 1959, **100**, 590; W. L. BROWN, W. H. BRATTAIN, C. G. B. GARRETT and H. C. MONTGOMERY, *Semiconductor Surface Physics* (ed. R. H. KINGSTON), Philadelphia, 1957, p. 111.

53. J. R. SCHRIEFFER, *Phys. Rev.,* 1955, **97**, 641. T. B. WATKINS, review paper "The electrical properties of semiconductor surfaces" in: *Progress in Semiconductors*, 1960, **5**, Heywood, London.

54. R. F. GREENE, D. R. FRANKL and J. N. ZEMEL, *Phys. Rev.,* 1960, **118**, 967. R. F. GREENE, *Phys. Rev.*, 1963, **131**, 592.

54a. R. F. GREENE, *Phys. Rev.*, 1966, **141**, 690.

55. W. SHOCKLEY and W. T. READ, *Phys. Rev.,* 1952, **87**, 835.

56. D. T. STEVENSON and R. J. KEYES, *Physica*, 1954, **20**, 1041.

57. A. MANY, E. HARNIK and Y. MARGONINSKI, *Semiconductor Surface Physics* (ed. R. H. KINGSTON), Philadelphia, 1957, p. 85; A. MANY and D. GERLICH, *Phys. Rev.,* 1957, **107**, 404.
58. S. WANG and G. WALLIS, *Phys. Rev.,* 1957, **105**, 1459.
59. H. STATZ, G. A. DeMARS, L. DAVIES and A. ADAMS, *Phys. Rev.,* 1956, **101**), 1272; 1957, **106**, 455; *Semiconductor Surface Physics* (ed. R. H. KINGSTON), Philadelphia, 1957, p. 139.
60. J. BARDEEN and S. R. MORRISON, *Physica*, 1954, **20**, 873.
61. G. RUPPRECHT, *J. Phys. Chem. Solids*, 1960, **14**, 208.
62. J. N. ZEMEL and R. PETRITZ, *Phys. Rev.,* 1958, **110**, 1263; *J. Phys. Chem. Solids*, 1959, 8; J. N. ZEMEL, *Phys. Rev.,* 1958, **112**, 762; W. A. ALBERS and J. E. THOMAS, *J. Phys. Chem. Solids*, **14**, 1960, 181.
63. E. S. SCHLEGEL, *I.E.E.E. Trans. on Electron Devices* ED, 1967, **14**, 728; ED, 1968, **15**, 951.
64. M. V. WHELAN, *Philips Res. Repts.*, 1965, **20**, 620.
65. V. A. MYAMLIN and YU. V. PLESKOV, *Electrochemistry of Semiconductors*, Plenum Press, 1967 (Russian edition 1965).
66. R. LINDNER, *Bell Syst. Techn. J.,* 1962, **41**, 803.
67. M. M. ATALLA, E. TANNENBAUM and E. J. SCHEIBNER, *Bell Syst. Techn. J.,* 1959, **38**, 749.
68. E. KOOI, *Philips Res. Repts.*, 1965, **20**, 306, 578, 595.
69. F. M. FOWKES and T. E. BURGESS, in *Clean Surfaces* (ed. G. GOLDFINGER), Marcel Dekker, 1970, p. 351.
70. R. E. SCHLIER and H. E. FARNSWORTH, *J. Chem. Phys.*, 1959, **30**, 917; *Semiconductor Surface Physics* (ed. R. H. KINGSTON), Philadelphia, 1957, p. 3.
71. S. P. WOLSKY, *Phys. Rev.,* 1957, **108**, 1131; S. P. WOLSKY and E. J. ZDANUK (eds.) *Ultra Microweight Determination in Controlled Environments*, New York, Interscience, 1969.
72. D. P. SMITH, *Surface Sci.,* 1971, **25**, 171, a review paper appearing in *Modern Methods of Surface Analysis* (Proceedings of a Conference).
72a. A. BENNINGHOVEN, *Z. Physik*, 1970, **230**, 403; A. BENNINGHOVEN and S. STORP, *Z. Angew. Physik*, 1970, **31**, 31.
72b. H. W. WERNER, in *Developments in Applied Spectroscopy*, 1969, **7a**, 239.
72c. Z. JURELA, "Atomic Collision Phenomena in Solids", Proceedings of an International Conference, Brighton, 1969, North Holland Publ. Co., 1970, p. 339; and *Fizika*, to be published.
72d. H. H. BRONGERSMA and P. M. MUL, *Chem. Phys. Lett.*, 1972, **14**, 380.
73. J. A. DILLON, *Ann. N.Y. Acad. Sci.,* 1963, **101**, 634; G. W. GOBELI and F. G. ALLEN, *J. Phys. Chem. Solids*, 1960, **14**, 23.
74. R. W. ROBERTS, *Brit. J. Appl. Phys.,* 1963, **14**, 537.
75. C. J. DAVISSON and L. H. GERMER, *Phys. Rev.,* 1927, **30**, 705.
76. E. A. WOOD, *J. Appl. Phys.,* 1964, **35**, 1306.
77. J. J. LANDER, G. W. GOBELI and J. MORRISON, *J. Appl. Phys.,* 1963, **34**, 2298.
78. R. M. STERN, J. J. PERRY and D. S. BOUDREAUX, *Rev. Mod. Phys.,* 1969, **41**, 275; C. B. DUKE and C. W. TUCKER, *Surface Sci.,* 1969, **15**, 231; C. B. DUKE, J. R. ANDERSON and C. W. TUCKER, *ibid.,* 1970, **19**, 117; R. O. JONES and J. A. STROZIER, *Phys. Rev. Lett.,* 1969, **22**, 1186; J. A. STROZIER and R. O. JONES, *Phys. Rev. B*, 1971, **3**, 3228; J. B. PENDRY, *J. Phys. C. (Proc. Phys. Soc.)*, 1969, **2**, 2273, 2283; 1971, **4**, 2501, 3095; K. KAMBE, *Z. Naturf.,* 1967, **22a**, 322, 422; E. G. McRAE, *Surface Sci.*, 1968, **11**, 479, 492.

78a. J. J. LANDER, *Progr. Solid State Chem.*, 1956, **2**, 26.
78b. M. B. WEBB and M. G. LAGALLY, Lecture Notes *Fifth Low-Energy Electron Diffraction Seminar*, Washington D.C., 1971.
79. H. BETHE, *Ann. d. Physik*, IV, 1928, **87**, 55.
80. A. J. ROSENBERG, *J. Phys. Chem. Solids*, 1960, **14**, 175.
81. F. MEYER, *J. Phys. Chem.*, 1969, **73**, 3844.
82. H. E. FARNSWORTH, *Ann. N.Y. Acad. Sci.*, 1963, **101**, 658.
83. P. W. PALMBERG, *Surface Sci.*, 1968, **11**, 153.
84. R. E. SCHLIER and H. E. FARNSWORTH, *J. Chem. Phys.*, 1959, **30**, 917.
85. P. W. PALMBERG and W. T. PERIA, *Surface Sci.*, 1967, **6**, 57.
86. F. JONA, *IBM J.*, 1965, **9**, 375.
87. R. M. BROUDY and H. C. ABBINK, *Appl. Phys. Lett.*, 1968, **13**, 212.
88. F. BAUER, *Phys. Lett.*, 1968, **26A**, 530.
89. N. J. TAYLOR, *Surface Sci.*, 1969, **15**, 169.
90. A. J. VAN BOMMEL and F. MEYER, *Surface Sci*, 1967, **8**, 467.
91. H. L. LINTZ, *Surface Sci.*, 1968, **12**, 390.
92. J. J. LANDER and L. MORRISON, *J. Chem. Phys.*, 1962, **37**, 729; *J. Appl. Phys.*, 1963, **34**, 1463; also ref. 77.
93. R. SEIWATZ, *Surface Sci.*, 1964, **2**, 473.
94. D. HANEMAN, *Phys. Rev.*, 1961, **121**, 1093; N. HANSEN and D. HANEMAN, *Surface Sci.*, 1964, **2**, 566.
95. D. HANEMAN, W. D. ROOTS and J. T. P. GRANT, *J. Appl. Phys.*, 1967, **38**, 2203; D. HANEMAN, J. T. P. GRANT and R. O. KHOKHAR, *Surface Sci.*, 1969, **13**, 119.
96. **Si**: K. A. MÜLLER, P. CHAN, R. KLEINER, D. W. OVENALL and M. J. SPARNAAY, *J. Appl. Phys.*, 1964, **35**, 2254.
     Ge: D. HANEMAN, *Surface Sci.*, 1970, **19**, 39; J. HIGINBOTHAM and J. J. LANDER and J. MORRISON, *Surface Sci.*, 1964, **2**, 553, ordered structure.
     D. HANEMAN, *Phys. Rev.*, 1968, **170**, 705; M. F. CHUNG and D. HANEMAN, *J. Appl. Phys.*, 1966, **37**, 1879; A. J. VAN BOMMEL and F. MEYER, *Surface Sci.*, 1967, **8**, 381, P-structure.
97. H. IBACH, *Phys. Rev. Lett.*, 1970, **24**, 1416 (ZnO); 1971, **27**, 253 (Si).
98. P. AUGER, *J. Phys. Radium*, 1925, **6**, 205.
99. H. L. HAGEDOORN and A. H. WAPSTRA, *Nucl. Phys.*, 1960, **15**, 146; H. E. BISHOP and J. C. RIVIÈRE, *J. Appl. Phys.*, 1969, **40**, 1740.
100. K. SIEGBAHN *et al.*, *ESCA*, Uppsala, 1967.
101. J. LANDER, *Phys. Rev.*, 1953, **91**, 1382.
102. L. A. HARRIS, *J. Appl. Phys.*, 1968, **39**, 1419, 1428.
103. P. W. PALMBERG and T. N. RHODIN, *J. Appl. Phys.*, 1968, **39**, 2425.
104. D. G. FEDEK and N. A. GJOSTEIN, *Surface Sci.*, 1967, **8**, 77.
105. P. DRUDE, *Wied. Ann.*, 1889, **36**, 865.
106. Ellipsometry in the Measurement of Surfaces and Thin Films, Symposium Proceedings, Washington 1963 Nat. Bur. Standards, Miscell. Publ. 1964; *Surface Sci.*, 1969, **16**, Conference Proceedings.
107. R. J. ARCHER, *J. Opt. Soc. Amer.*, 1962, **52**, 970.
108. C. STRACHAN, *Proc. Cambridge Phil. Soc.*, 1933, **29**, 116.
109. D. V. SIVUKHIN, *Zh. Eksperim. Teor. Fiz. SSSR*, 1948, **18**, 976; 1951, **21**, 367; *Soviet Phys. J.E.T.P.*, 1956, **3**, 269.
110. G. A. BOOTSMA and F. MEYER, *Surface Sci.*, 1969, **14**, 52; F. MEYER, E. E. DE KLUIZENAAR and G. A. BOOTSMA, *ibid.* 1971, **27**, 88; F. MEYER, *ibid.*, 1971, **27**, 107.

111. B. O. SERAPHIN, *Surface Sci.*, 1969, **13**, 136.
112. B. O. SERAPHIN and N. BOTTKA, *Phys. Rev.*, 1966, **145**, 144.
113. B. J. HOLDEN and F. G. ULLMAN, *J. Electrochem. Soc.*, 1969, **116**, 280.
114. N. J. HARRICK, *Internal Reflection Spectroscopy*, Wiley, New York, 1967.
115a. S. BRUNAUER, P. H. EMMETT and E. TELLER, *J. Amer. Chem. Soc.*, 1938, **60**, 309.
115b. M. J. SPARNAAY, *Solid State Phys. Electronics Telecomm.* (Conf. Brussels 1958), 1959, **1**, 613. The degree of coverage, expressed on the vertical axis, should be reduced by a factor 2. This result does not affect the interpretation of the results.
116. R. H. FOWLER, *Phys. Rev.*, 1931, **38**, 45; L. A. DUBRIDGE, *ibid.*, 1933, **43**, 727. See also R. H. FOWLER and E. A. GUGGENHEIM, *Statistical Thermodynamics*, Cambridge U.P., 1949, p. 481.
117. R. H. FOWLER and L. NORDHEIM, *Proc. Roy. Soc. London* A, 1928, **119**, 173; L. NORDHEIM, *ibid.*, 1928, **121**, 626.
118. O. W. RICHARDSON, *The Emission of Electricity from Hot Bodies*, Longmans, 1921.
119. M. DUSHMAN, *Phys. Rev.*, 1923, **21**, 625.
120. T. E. FISCHER, *Surface Sci.*, 1969, **13**, 30.
121. P. HANDLER and W. M. PORTNOY, *Phys. Rev.*, 1959, **116**, 516; P. HANDLER, *Semiconductor Surface Physics* (ed. R. H. KINGSTON), Philadelphia, 1957.
122. J. A. DILLON and H. E. FARNSWORTH, *J. Appl. Phys.*, 1957, **28**, 174.
123. D. R. PALMER, S. R. MORRISON and C. E. DAUENBAUGH, *J. Phys. Chem. Solids*, 1960, **14**, 27.
124. A. H. BOONSTRA, J. VAN RULER and M. J. SPARNAAY, *Proc. Kon. Nederland Akad. Wetenschap.* B, 1963, **66**, 64, 70.
125. W. L. JOLLY and W. M. LATIMER, *J. Amer. Chem. Soc.*, 1952, **74**, 5757.
126. F. G. ALLEN, T. M. BUCK and J. T. LAW, *J. Appl. Phys.*, 1960, **31**, 979.
127. (a) M. HENZLER, *Surface Sci.*, 1968, **8**, 31; (b) *J. Appl. Phys.*, 1969, **40**, 3758; (c) *Phys. Stat. Sol.*, 1967, **19**, 833.
128. A. H. BOONSTRA, *Philips Res. Suppl.*, 1968, no. 3 (thesis 1967).
129. G. HEILAND and H. LAMATSCH, *Surface Sci.*, 1964, **2**, 18.
130. J. VAN LAAR and J. J. SCHEER, in *Proc. Intern. Conf. on Physics of Semiconductors*, Exeter, 1962 (The Institute of Physics and The Physical Society, London), p. 827.
131. G. W. GOBELI and F. G. ALLEN, *Surface Sci.*, 1964, **2**, 402.
132. D. BRENNAN, D. O. HAYWARD and B. M. W. TRAPNELL, *J. Phys. Chem. Solids*, 1960, **14**, 117.
133. C. A. CAROSELLA and J. COMAS, *Surface Sci.*, 1969, **15**, 303.
134. R. F. LEVER and H. R. WENDT, *Surface Sci.*, 1970, **19**, 430.
135. For adsorption measurements see I. A. KIROVSKAYA and L. G. MAIDONOVSKAYA, *Russ. J. Chem. Phys.*, 1967, **41**, 1194.
136. M. J. SPARNAAY, *Surface Sci.*, 1969, **13**, 99.
137a. G. A. BOOTSMA, *Surface Sci.*, 1968, **9**, 396 (CdS).
137a. S. BAIDYAROV, W. R. BOTTOMS and P. MARK, *Surface Sci.*, 1971, **28**, 517.
137b. C. L. BALESTRA and H. C. GATOS, *Surface Sci.*, 1971, **28**, 56.
138. J. J. SCHEER and J. VAN LAAR, *Solid State Comm.*, 1967, **5**, 303; J. VAN LAAR and J. J. SCHEER, *Surface Sci.*, 1967, **8**, 342.
139. J. J. SCHEER and J. VAN LAAR, *Solid State Comm.*, 1965, **3**, 189.
139a. R. L. BELL and W. E. SPICER, *Proc. I.E.E.E.*, 1970, **58**, 1788, have reviewed the field of 3–5 compound photocathodes.

140. A. Y. Cho and J. R. Arthur, *Phys. Rev. Letters*, 1969, **22**, 1180.
141. M. J. Sparnaay, *Proc. Kon. Nederland. Akad. Wetenschap.* B, 1968, **71**, 387.
142. B. V. Derjaguin and V. P. Smilga, *J. Appl. Phys.*, 1967, **38**, 4609.
143. W. H. Brattain and C. G. B. Garrett, *Bell Syst. Tech. J.*, **34**, 1955, 129.
144. E. J. W. Verwey, *Proc. Kon. Nederland. Akad. Wetenschap.* B, 1950, **53**, 376.
145. J. F. Dewald, *Bell Syst. Tech. J.*, 1960, **39**, 615.
146. K. Bohnenkamp and H. J. Engell, *Z. Elektrochem.*, 1957, **61**, 1184; see also same authors in: *The Surface Chemistry of Metals and Semiconductors* (ed. H. C. Gatos), Wiley & Sons, 1960, p. 225.
147. M. Hoffmann-Perez and H. Gerischer, *Z. Elektrochem.*, 1961, **65**, 771.
148. W. H. Brattain and P. J. Boddy, *J. Electrochem. Soc.*, 1962, **109**, 574.
149. P. J. Boddy and W. H. Brattain, *J. Electrochem. Soc.*, 1962, **109**, 812.
150. R. Memming, *Surface Sci.*, 1964, **2**, 436.
151. R. Memming and G. Neumann, *Surface Sci,* 1968, **10**, 1.
152. K. M. Hund and P. T. Wrotenbery, *Ann. N.Y. Acad. Sci.,* 1963, **101**, 876.
153. R. Memming and G. Schwandt, *Surface Sci.*, 1966, **4**, 109; 1966, **5**, 97.
154. H. U. Harten and R. Memming, *Phys. Letters*, 1962, **3**, 95 (surface conduction).
155. W. W. Harvey, *Ann. N.Y. Acad. Sci.,* 1963, **101**, 904.
156. (a) B. Lovrecek and J. O'M. Bockris, *J. Phys. Chem.*, 1959, **63**, 1368.
     (b) I. V. Borovkov, *Russ. J. Phys. Chem.*, 1960, **34**, 1263; N. D. Tomashev, E. N. Paleolog and A. Z. Fedotova, *ibid.*, pp. 396, 488.
157. G. Brouwer, *J. Electrochem. Soc.*, 1967, **114**, 743.
158. M. J. Sparnaay, *Rec. Trav. Chim. Pays-Bas,* 1960, **79**, 950; *J. Phys. Chem. Solids,* 1960, **14**, discussion facing p. 136 (Proceedings Second Conference Semiconductor Surfaces).
159. *Oxidation–Reduction Potentials* (Selected Constants), Tables of Constants and Numerical Data, Vol. 8 (ed. G. Charlot *et al.*), Pergamon Press, 1958. In American literature the sign is negative.
160. M. J. Sparnaay, *Surface Sci.*, 1964, **1**, 102.
161. E. Schwarz and E. Huf, *Z. Anorg. u. Allg. Chem.*, 1931, **203**, 188; W. A. Roth and O. Schwartz, *Ber. d. deutschen Chem. Gesellschaft*, 1926, **59**, 338.
162. *The Electrochemistry of Semiconductors* (ed. P. J. Holmes), Academic Press, 1962, contains sufficient references to Gatos' work.
163. S. G. E. Ellis, *J. Appl. Phys.*, 1957, **28**, 1262.
164. M. J. Sparnaay, *Surface Sci.*, 1964, **1**, 213.
165. H. U. Harten, *J. Phys. Chem. Solids*, 1960, **14**, 220.
166. H. Gerischer, *Adv. Electrochemistry and Electro Engineering*, 1960, **1**, 139 (ed. P. Delahay).
167. D. R. Turner, *J. Electrochem. Soc.*, 1956, **103**, 252; and in *The Electrochemistry of Semiconductors* (ed. P. J. Holmes), Academic Press, 1962, chap. 4.
168. R. Memming, *J. Electrochem. Soc.*, 1969, **116**, 85.
169. F. Beck and H. Gerischer, *Z. Elektrochem.*, 1959, **63**, 943.
170. Yu. V. Pleskov and B. N. Kabanov, *Doklady Akad. Nauk U.S.S.R.,* 1958, **123**, 884.
171. H. Gerischer, A. Mauerer and W. Mindt, *Surface Sci.*, 1966, **4**, 431.

172. H. Gerischer and W. Mindt, *Surface Sci.*, 1966, **4**, 440.

173. H. Tributsch and H. Gerischer, *Ber. d. Bunsengesellschaft f. Phys. Chem.*, 1969, **73**, 850 (*Z. Elektrochem.*).

174. H. Gerischer, *Surface Sci.*, 1969, **18**, 97. See also contributions to the 18th CITCE Meeting, Elmenau, April 1967, published in *Electrochim. Acta*, 1968, **13**.

175. B. J. Mulder and J. de Jonge, *Proc. Kon. Nederland Akad. Wetenschap. B*, 1963, **66**, 303; B. J. Mulder, *Rec. Trav. Chim. Pays-Bas*, 1965, **84**, 713; *Philips Res. Repts Suppl.*, 1968, no. 4.

CHAPTER 7

COMMON ASPECTS

THIS final chapter stresses those aspects of double layers which are common to all the systems dealt with in this book. In Section 7.1 this is done by discussing corrections to the simple models and concepts used so far. In Section 7.2 the Langmuir adsorption theory is compared with Fermi statistics and with chemical equilibria of the type $HA \leftrightarrows H^+ + A^-$. Finally, advantages and disadvantages of the use of models are briefly discussed in Section 7.3.

7.1. Corrections

It is useful to distinguish between three ways of approach which have been used in double-layer theory:

(a) The statistical approach.[1-11] The principles of statistical mechanics are applied to double-layer problems, these problems being formulated in as detailed a manner as possible.

(b) The thermodynamic approach.[12-17] A precise analysis, mostly carried out on the basis of local thermodynamic equilibrium conditions, provides correction terms to the usual equations.

(c) The elementary approach.[18-20] The used models are refined. The refinements, leading to correction terms to the original equations, are investigated quantitatively.

It is impossible to do justice here to all the aspects of these investigations, in particular to those belonging to the statistical approach, where rapid progress is being made and can be expected to continue in the future. We mainly restrict ourselves to the elementary approach, except for stating some results of the statistical theory and attempting to make them understandable. As a general rule there should be no conflicting conclusions from the three ways of approach. However, in one instance (the effect of finite ionic size) the results of the elementary approach and those of the thermodynamic approach seem to be at variance, and this point is therefore considered in more detail.

7.1.1. *Diffuse region*

For dilute 1–1 solutions ($\sim 10^{-3}$ M) and for most semiconductors the Poisson–Boltzmann equation is expected to be a good approximation, at least if the absolute value of the wall potential is lower than about 100 mV. This expectation is based on the fact that in the bulk solution the Debye–Hückel theory applies fairly well and that for small wall potentials the concentration near the wall is still fairly small. At higher concentrations and/or wall potentials a number of secondary, hitherto neglected effects may become important. These are enumerated below. In the elementary approach of these corrections each effect is dealt with separately and the corrections upon the space charge parameters are assumed additive. It is realized that this approach can be correct only for moderate concentrations ($\sim 10^{-2}$ M 1–1 electrolytes; Buff and Stillinger[7] arrived at about the same value for the limiting concentration). Whereas for concentrations lower than 10^{-2} M the theory of the diffuse double layer is essentially similar to the one outlined in Section 2.3, the situation for high concentrations may approach that of a crystal. Thus, Stillinger and Kirkwood[3] found, on the basis of the theory of molecular distribution functions, that a z–z electrolyte with ions of finite size (ionic radius a) near a charged wall produced a Gouy layer in which the charge density alternated in sign in a direction normal to the plane of the interface. In their calculations this occurred when $\kappa a > 1\cdot03$. This is reminiscent of a crystal in which exist ordered planes of positive and negative ions.

It is relevant to remember that in a 1–1 10^{-2} M electrolyte there are about 3000 water molecules per ion. This ratio seems too large to affect seriously the structure of the solvent. This problem will be discussed in an elementary way in Subsection 7.1.2.

We now list those topics which must be considered in more detail in any development of the simple treatments outlined earlier in this book.

(a) The statistical basis of the Poisson–Boltzmann equation.

(b) Image effects due to a nearby medium with different dielectric constant.

These two points are of importance to both electrolytes and semiconductors. They are non-specific, i.e. they are based upon general principles not depending on the particular kind of charge carrier. It has proved possible, as shown by Buff and Stillinger,[7] that the effects (a) and (b) can be dealt with in one theory.

The two following points are specific. They are of importance only to electrolytes.

(c) The finite hydration volume of the ions.

(d) The dependence of the dielectric constant of the solution upon the ionic concentration and upon the electric field strength in the Gouy layer.

For semiconductors the following points should be added:

(e) Incomplete ionization of donor and acceptor atoms.

(f) Validity of the Fermi distribution rather than that of the Boltzmann distribution.

(g) The physical meaning of the classical concept of bandbending in a crystal with a finite number of energy states for electrons and holes.

We begin by writing the Poisson equation as

$$\text{div } \varepsilon \text{ grad } \psi_P = -4\pi\rho \qquad (7.1.1)$$

and the Boltzmann equation in the following somewhat modified form:

$$n_i = nf_{bi} \, e^{-z_i e \psi_B/kT}. \qquad (7.1.2)$$

In eqs. (7.1.1) and (7.1.2) ψ_P and ψ_B are two potentials, which are, as a general rule, not equal. In eq. (7.1.2) the factor f_{bi} accounts for the finite hydration volume of the ions, In the preceding chapters it has been supposed that $\psi_P = \psi_B = \psi$, grad $\varepsilon = 0$ and $f_{bi} = 1$ (zero hydration volumes). Moreover, the operator div grad was usually identified with d^2/dx^2 (flat plate approx.).

We now examine (a)–(d) in the light of these remarks:

(a) STATISTICAL BASIS

Onsager pointed out that ψ_P is the average potential at a position in the space charge where the average charge density is ρ, whereas ψ_B is the potential of the average force acting on an ion i at that position. The difference may be expressed as $\psi_B = \psi_P + \varphi_i$ where according to Onsager and to Kirkwood φ_i is a fluctuation term or alternatively a term accounting for the self-atmosphere of the ions in the space charge. Each ion, although under the influence of a potential which is due to a charged wall in the neighbourhood, will still have the tendency to surround itself with ions of the opposite sign. Therefore the contribution φ_i has the same sign for positive and negative ions. It should be noted that from the point of view of the rigorous cluster theory Onsager and Kirkwood followed only an "intuitive" approach.[†] However, this approach provides us with satisfactory numerical estimates, which will now be given.

---

† Private communication of H. Friedman.

It was proved by Kirkwood[2] that if the Debye–Hückel approximations applied, φ_i was small and could be neglected. Thus, identification of ψ_p and ψ_B is consistent with the Debye–Hückel treatment. Estimates by Casimir[3a] showed that φ_i may also be relatively small in the neighbourhood of a charged wall, where ψ_P and ψ_B are large. The condition here for φ_i being small is that the self-atmosphere of an ion near a charged wall can be described by using the Debye–Hückel approximation. The self-atmosphere makes the ions near a charged wall less dependent on the potential due to the wall. The local ionic concentration is somewhat larger than the one estimated on the basis of an uncorrected theory and the average potential is somewhat smaller in absolute value. Calculations were given by Loeb for two parallel plates and by Williams for a single plate. Williams, largely confirming Loeb, found that the correction varied between 3% and 9% for practical cases. These values did not depend much on the distance from the wall or on the concentrations. This behaviour is feasible, because it is characteristic for fluctuations that their relative importance decreases if the concentration is increased. Owing to this property the correction near the wall or in a high concentration is still small in spite of the enhanced electrostatic effects.

(b) IMAGE EFFECTS

These depend on the ratio of the dielectric constants of the adjoining media. Since aqueous solutions have a high dielectric constant, a conducting wall will provide almost no image effect. It is only if the adjoining phase has a low dielectric constant that an image effect (a repulsion) can be expected. Loeb and Williams included image effects in their calculations and found that the repulsive image force led to a lowering of the absolute value of the correction.

For the effects (c) and (d) we mainly follow the reasonings given by Sparnaay (elementary approach) and compare them with those followed by Sanfeld et al.[12] (thermodynamic approach).

(c) HYDRATION VOLUMES

In the elementary approach the non-zero hydration volumes of the ions reduce the free volume which the ions in the solution have at their disposal.[18-23] For cations we have a reduction factor g_+ and for anions we have a reduction factor g_-.

$$g_+ = 1 - B_{++}n_+ - B_{+-}n_- \qquad (7.1.3)$$

N*

where B_{++} and B_{+-} are the excluded volumes of two cations and of a cation and an anion respectively.

$$g_- = 1 - B_{+-}n_+ - B_{--}n_-. \tag{7.1.4}$$

The reduction factors g_+ and g_- are, in fact, Van der Waals-type corrections. The parameters B_{++}, B_{+-} and B_{--} have the same physical meaning as the "Van der Waals-b".

Values of the excluded volumes may be estimated as follows. The number of water molecules in a hydration shell is estimated of the order of 6. The volume of such a hydrated ion which will be considered as a hard sphere ("hard-sphere model") is then about 2×10^{-22} cm³. The excluded volume of a pair of ions i is 16×10^{-22} cm³, or 8×10^{-22} cm³ per ion i. This is the value of B_{ii}. For a 10^{-2} M 1–1 solution (where $n_i = 6 \times 10^{18}$ cm⁻³) the value of $B_{ii}n_i$ may be estimated to be about $0 \cdot 5 \times 10^{-2}$.

It is well known that the hard-sphere model appears already in the original Debye–Hückel (D–H) theory, although it was not handled there in a consistent way. Thus, Wicke and Eigen[21] introduced a Van der Waals-b correction and replaced κ^2 by $\kappa^2[1 - n/2(B_{++} + B_{--} - 2B_{+-})]$. They assumed that B_{++} and $B_{--} \gg B_{+-}$. Only if this assumption is correct will there be an effect in the D–H theory. Hückel and Krafft[6] derived, on the basis of Kirkwood's analysis,[2] additional corrections and by doing so remedied further the inconsistencies of the D–H theory. Their corrections can be interpreted in terms of the formal charge of a hydrated ion. Thus, in the D–H theory each ion is accompanied by a "smeared-out" charged cloud. Hückel and Krafft in fact showed that also within the excluded volumes this smeared out charge must be considered so that the absolute value of the charge e of a monovalent ion must be replaced by[24]

$$e \rightarrow e + \int \rho_{r \leq a} \, d\tau \tag{7.1.5}$$

where $\rho_{r \leq a}$ is the formal charge density within the excluded volumes, a being a brief notation for $r_+ + r_+$, $r_+ + r_-$ or $r_- + r_-$, and $d\tau$ is a volume element of the excluded volume. For monovalent ions, irrespective of the sign one has,†

$$\rho_{r \leq a} = \frac{en}{kT}\left(\frac{e}{\varepsilon r} - \frac{e\kappa}{\varepsilon(1 + \kappa a)}\right). \tag{7.1.6}$$

For cations $d\tau = dB_{++} + dB_{+-}$ and for anions $d\tau = dB_{+-} + dB_{--}$. The

---

† This relation can be derived only if Debye–Hückel approximations are applied.[24] For our purpose here this means that in the final ion size correction terms appear of magnitude $(e\varphi_{\text{cav}}/kT)Bn$ (where $e\varphi_{\text{cav}}/kT < 1$) together with terms Bn (see below). φ_{cav} is the cavity potential, i.e. the potential inside the excluded volume.

bracketed part of eq. (7.1.6) is typically a difference of two potentials, a Coulomb potential and the potential of a conducting sphere in the D–H theory. Returning now to Gouy layers it will be seen that this difference no longer plays a role, at least to a first approximation. The Coulomb potential in eq. (7.1.6) arises from the consideration of a central ion in the D–H theory. This is no longer relevant in the Gouy theory. For this reason Hückel–Krafft corrections can only play a role by modifying the activity coefficient of the ions in the Gouy layer. On the other hand, the Wicke–Eigen correction may be small in the bulk of a solution, namely if B_{++}, B_{--} and B_{+-} are all of the same magnitude, whereas the correction may be important in Gouy layers, where ions of one sign are almost absent.

Finally an important step, made on the basis of the hard sphere model, is mentioned. It is probable that the dielectric constant within the hard sphere is lower than around it. Levine and Wrigley[20] pointed out that in that case there is a tendency towards a repulsion irrespective of the signs of the charges.

We now return to expressions for f_{b+} and f_{b-} (see eq. (7.1.2)). These can be written as

$$f_{b+} = g_+/g_+(\text{bulk}) = 1 - B_{++}(n_+ - n) - B_{+-}(n_- - n), \qquad (7.1.7)$$

$$f_{b-} = g_-/g_-(\text{bulk}) = 1 - B_{+-}(n_+ - n) - B_{--}(n_- - n). \qquad (7.1.8)$$

In later developments (of the correction terms) we use the uncorrected Boltzmann equations for n_+ and n_-. This is consistent with the fact that the series expansions of $g_+^{-1}(\text{bulk})$ and $g_-^{-1}(\text{bulk})$ were cut off after the first term. Such a procedure is characteristic for the elementary approach.

The thermodynamic approach provides expressions of f_{b+} and f_{b-} which are analogous to eqs. (7.1.7) and (7.1.8) although the physical interpretation is different as is seen from the following discussion.

The differential dG of a Gibbs function of a bulk phase is

$$dG = -S\, dT + V\, dp + \sum \mu_i\, dN_i. \qquad (7.1.9)$$

The use of dG as a total differential leads to

$$\left(\frac{\partial \mu_i}{\partial p}\right)_{T,N} = \left(\frac{\partial V}{\partial N_i}\right)_{T,p,N_{j \neq i}} \equiv \bar{v}_i \qquad (7.1.10)$$

where $\bar{v}_i$ denotes the specific molecular volume of component i in the solution at given T, p and composition. The integration may be carried out with the assumption that $\bar{v}_i$ is not a function of the pressure p. The chemical potential then contains a contribution $\bar{v}_i p$, where it is understood that the integration has taken place at constant temperature and composition. It will also be assumed that $\bar{v}_i$ is independent of the composition. We specify the components i as monovalent ions in solution. In a volume element in the Gouy layer the electrochemical potentials can be written as

$$\tilde{\mu}_+ = \mu_+^0 + \bar{v}_+ p + kT \ln n_+ + e\psi, \tag{7.1.11}$$

$$\tilde{\mu}_- = \mu_-^0 + \bar{v}_- p + kT \ln n_- - e\psi. \tag{7.1.12}$$

If the volume element is chosen in the bulk of the solution, where $\psi = 0$, then

$$\tilde{\mu}_+ = \mu_+^0 + \bar{v}_+ p_0 + kT \ln n, \tag{7.1.13}$$

$$\tilde{\mu}_- = \mu_-^0 + \bar{v}_- p_0 + kT \ln n. \tag{7.1.14}$$

Combination of eqs. (7.1.11) and (7.1.13) and of (7.1.12) and (7.1.14) leads to

$$\frac{n_+}{n} = e^{-\bar{v}_+ (p-p_0)/kT} e^{-y} \quad \text{and} \quad \frac{n_-}{n} = e^{-\bar{v}_- (p-p_0)/kT} e^{+y} \quad \left(y = \frac{e}{kT} \psi \right) \tag{7.1.15}$$

Sanfeld *et al.* showed that apart from correction terms arising from the polarization of the medium by the electrostatic field the pressure difference $p-p_0$ is the same as that derived by Verwey and Overbeek[25] (see also Subsection 4.3.2):

$$p - p_0 = kT(n_+ - n + n_- - n). \tag{7.1.16}$$

So far polarization contributions have been ignored in the treatment, both in the bulk of the solution and in the Gouy layer. As is pointed out below (at (d)) polarization effects would have led to slightly different values of n_+ and n_-. Sanfeld *et al.* have discussed this point extensively together with the problem of pressure definitions in solutions.

If eq. (7.1.16) is inserted in eq. (7.1.15) it is seen that n_+/n and n_-/n contain contributions $\exp[-\bar{v}_+(n_+ + n_- - 2n)]$ and $\exp[-\bar{v}_-(n_+ + n_- - 2n)]$. After expansion, thereby retaining only the first term, the ion size correction can be written in a form which is analogous to eqs. (7.1.7) and (7.1.8). We now use the symbols f_{v+} and f_{v-} instead of f_{b+} and f_{b-}. Then we have

$$f_{v+} = 1 - \bar{v}_+(n_+ - n) - \bar{v}_+(n_- - n), \tag{7.1.17}$$

$$f_{v-} = 1 - \bar{v}_-(n_+ - n) - \bar{v}_-(n_- - n). \tag{7.1.18}$$

The only difference between eqs. (7.1.7), (7.1.8) and eqs. (7.1.17) and (7.1.18) is that the excluded volume parameters B_{++}, B_{--} and B_{+-} are replaced by specific molar volume parameters $\bar{v}_+$ and $\bar{v}_-$. However, the physical meaning underlying this difference is important. Thus, the sum $\bar{v}_+ + \bar{v}_-$ is the measured change of the total volume if 1 mole (or, depending on the definition of $\bar{v}$, 1 molecule) of the salt is dissolved. Values of $\bar{v}_+ + \bar{v}_-$ are less than 20 cm³/mole. This means a volume change per ion of about 10–23 Å³. In contrast, the concept of excluded volumes arises from a model, a Van der Waals model. The excluded volume per ion is four times the volume of an ion. The volume of an ion in solution has only little bearing upon $\bar{v}$ because the concept of an hydration shell does not arise in the measurement of the specific molar volume, whereas it may provide an important contribution to the excluded volumes. Comparing the estimates given here for $\bar{v}$ and those given above for B, it is seen that

$$\bar{v} \ll B. \tag{7.1.19}$$

A ratio $B \sim 50\bar{v}$ seems not unreasonable.

Actually, the "true" expression ("true" because so many simplifications are still involved in the corrected theories) of the ion size effect is found by the multiplication of f_{b+} and f_{v+} and of f_{b-} and f_{v-}, but owing to eq. (7.1.19) this will not be undertaken. We see that the elementary and the thermodynamic approaches are not in conflict. They are, so to speak, complementary.

Let us now assume that the wall potential is negative. Then positive ions accumulate in the Gouy layer and it is advantageous to write the charge density as:

$$\frac{\rho}{-2n\,e} = [1 + (B_{++} + B_{+-})n]\sinh y - B_{++}n \sinh 2y$$

$$- (B_{--} - B_{++})n\,e^{\frac{3}{2}y}\sinh \frac{y}{2}. \tag{7.1.20}$$

Inspection shows that the term $-B_{++}n \sinh 2y$ represents the main correction contribution. Consequently we only consider this term. It is seen that, in contrast to the corrections (a) and (b), it is linearly proportional to n and that it is strongly dependent on the value of ψ. Its influence upon the differential capacity of the Gouy layer is

$$C_{db} = C_d[1 - B_{++}n(1 + 3 \sinh^2 \tfrac{1}{2}z)]; \quad \left(z = \frac{e}{kT}\psi_0\right) \tag{7.1.21}$$

where the same integration procedures are followed as in Chapter 3 and where C_{db} now indicates the "corrected" Gouy capacity. The total capacity C_0 can be replaced by

$$C = \frac{C_0 C_d}{C_0 + C_d}\left(1 - \frac{C_0}{C_0 + C_d}\,\omega\right) \tag{7.1.22}$$

where ω indicates the correction term of eq. (7.1.21). It is difficult to check directly on this correction term because at high concentrations the Gouy capacity is large and only C_0 is measured, whereas at low concentrations ω is small. However, it is worth while to reconsider the problem of two colloidal particles in close proximity. Here the total interaction was found as the difference between a repulsive and an attractive energy. The hydrated volume correction will affect the repulsive energy. Although the correction of the repulsive energy may be small, the difference between repulsive and attractive energy may

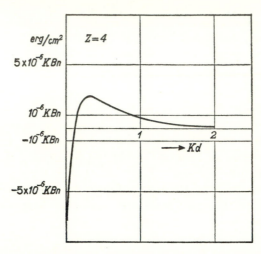

Fig. 7.1. Contribution to the potential energy vs. κd due to non-zero volumes of the ions (accepting the simple hard-sphere model). B is of the order of 10^{-21} cm³; $n = 6 \times 10^{18}$ cm⁻³ for a 0·01 normal 1–1 solution. Wall potential ($z = 4$) is 100 mV.

be relatively important. It has proved to be possible to incorporate the correction term $-B_{++}n \sinh 2y$ in the calculation of the repulsive energy, the methods to be followed being the same as those outlined in Chapter 3 (flat plate approximation). Its contribution is shown in Fig. 7.1 for a potential at the Gouy–Stern interface of 100 mV. The energy–distance curve has a peak which for $\kappa = 3 \times 10^6$ cm⁻¹ and $B_{++}n = 10^{-2}$ (a 10^{-2} M solution) is at 6×10^{-9} J cm² ($= 6 \times 10^{-2}$ erg/cm²). This amounts to about $15kT$ for a particle with an edge length of 300 Å. This certainly is of importance to the overall picture which emerges from the stability theory. The hydrated volume effect increases the height of the barrier between the particles. Thus the ionic volume correction may explain why different ions of the same valency lead to different flocculation concentrations. These differences are as expected: ions where higher hydration volumes can be expected have higher flocculation concentrations. Also the "secondary minimum" will be affected. It is somewhat deepened (about $5kT$), which is of importance to rheology problems.[26] The (non-specific) self atmosphere correction is the same for all the ions. It should be remembered that ions of different size affect the potential drop across the Stern layer in different ways and that this fact led Lijklema (Chapter 4) to an alternative interpretation of the different flocculation values.

The shape of the energy–distance curve of Fig. 7.1 indicates a decrease of the repulsion at short distances and an increase at larger distances. This remarkable consequence of the hydrated volume effect may be visualized as follows. If the distance is small the number of negative ions between the plates will be smaller than that found when the ions are considered as point charges. The positive charge on the particles will then be smaller too than is provided for in the uncorrected theory. Accordingly the repulsion will be smaller. When the distance is increased, the volumes of the ions will tend to increase the thickness of the Gouy layer. This results in a stronger tendency to keep the particles apart than provided in the uncorrected theory. This picture is similar to that given by Falkenhagen[27] for the explanation of the ionic activities of concentrated solutions (see also Subsection 7.1.3).

(d) THE DIELECTRIC CONSTANT

Sanfeld et al.[13] have given a more refined treatment than we are giving here. However, our purpose is only to indicate orders of magnitude. Experiments have indicated that the dielectric constant of an aqueous solution is a function of the ionic concentration and of an applied electric field. It may be written as follows:

$$\varepsilon = \varepsilon_0 + \alpha E^2 + \delta_+ \frac{n_+ - n}{N} \times 10^3 + \delta_- \frac{n_- - n}{N_v} \times 10^3 \qquad (7.1.23)$$

where α and δ_+ and δ_- are negative constants and ε_0 is the relative dielectric constant of the bulk of the solution containing n cations and n anions per cm^3. Finally, E is the field strength. According to old measurements (of pure water) by Malsch[28,29] $\alpha = 10^{-6}$–10^{-7} when E is expressed in e.s.u. (or $0\cdot1$–$0\cdot01$ when E is expressed in V/cm). Hasted et al.[30] were the first[31] to measure for a number of salt solutions the sum $(\delta_+ + \delta_-)$. They assumed that δ_+ for the Na$^+$-ion was about -8 and they calculated accordingly the other δ-values.† (They accounted for the Debye–Falkenhagen effect.) They noted that usually $\delta_+ > \delta_-$. This relation is understandable because positive ions interfere more seriously with the structure of water than do negative ions.

---

† The factor 10^3 in eq. (7.1.23) serves to express δ_- and δ_+ in units applied in the literature.

Estimates of the magnitude of the effects of the field strength and of the dissolved ions upon the Poisson–Boltzmann equation are obtained as follows. For E^2 write

$$E^2 = \left(\frac{\kappa kT}{e}\right)^2 \left(\frac{dy}{d\xi}\right)^2 \quad \text{(flat plate approximation)}.$$

For a single double layer one has

$$\alpha E^2 = 4\alpha \left(\frac{\kappa kT}{e}\right)^2 \sinh^2 \tfrac{1}{2}y = 4\alpha \frac{8\pi nkT}{\varepsilon_0} \sinh^2 \tfrac{1}{2}y \qquad (7.1.24)$$

where again the uncorrected functions are used. The parameter to be used will be $(\alpha/\varepsilon_0)8\pi nkT/\varepsilon_0$ which is about -10^{-4} for a 10^{-3} normal solution. For this same concentration one obtains for $(n\,\delta_+/\varepsilon\,dN) \times 10^3$ (in which $\delta_+ = -10$ is taken) also a value of about -10^{-4}. Therefore on the whole we expect that these effects are somewhat less important than the effect of the non-zero ionic volumes. Returning to eq. (7.1.23) and combining it with the modified Poisson equation, eq. (7.1.1), one has for the flat plate approximation :

$$\frac{d^2\psi}{dx^2} = -\frac{4\pi\rho}{\varepsilon_0} - \frac{d}{\varepsilon_0\,dx}\left[(\varepsilon - \varepsilon_0)\frac{d\psi}{dx}\right] \qquad (7.1.25)$$

where the necessary expansion involves only first-order terms and where a term with $(\delta_- - \delta_+)$ is, owing to its smallness, discarded. The charge per unit area, compensating the charge in the Gouy layer, is

$$\sigma = -\int_\delta^d \rho\,dx \qquad (7.1.26)$$

where in the case of one single double layer $d = \infty$.

Integration of eq. (7.1.26), after inserting eq. (7.1.25), gives

$$\sigma = -\frac{\varepsilon_0}{4\pi}\left(\frac{d\psi}{dx}\right)_{x=\delta} - \frac{1}{4\pi}\left[(\varepsilon - \varepsilon_0)\frac{d\psi}{dx}\right]_{x=\delta} \qquad (7.1.27)$$

In order to obtain explicit expressions for the Gouy capacity of a single double layer and for the repulsive energy between flat plates, it should be noted that not only the Poisson equation is modified, but also the Boltzmann equation. The reason for this modification is that in the diffuse layer the ions, which reduce the dielectric constant, are subject to the influence of the field. They are tending away from positions where the field strength is high to positions where the field strength is low. According to Prigogine *et al.*[16] the electrochemical potential is:

$$\tilde{\mu}_i = \mu_i^0 + kT \ln n_i + z_i e \psi - \frac{E^2}{8\pi} \left(\frac{\partial \varepsilon}{\partial n_i} \right)_{T,V,n,D} \tag{7.1.28}$$

where

$$\left(\frac{\partial \varepsilon}{\partial n_i} \right)_{T,V,n,D} = 10^3 \delta_i$$

in which $\delta_i = \delta_+$ for cations and $\delta_i = \delta_-$ for anions. The derivation of the last term on the right-hand side is given on p. 37.

We see that eq. (7.1.23) leads to corrections of both the Poisson equation and the Boltzmann equation as they are commonly used. In both equations correction terms can be introduced containing the parameter δ_i. It is not difficult to construct a Poisson–Boltzmann equation, which contains δ_i-correction terms. These terms are largely determined by the correction appearing in the Poisson equation.[18]

When the polarization term is small compared to kT, then, near a wall with positive wall potential, the charge density $\rho = e(n_+ - n_-)$ can to a good approximation be given by:

$$\rho = -2ne \left\{ \sinh y + \frac{E^2}{8\pi nkT} \times 10^3 \delta_- \sinh y \right\}. \tag{7.1.29}$$

The differential Gouy capacity finally turns out to be:

$$C_d = C_d \left[1 + 3\pi \left(2 \frac{n\delta_-}{N_0 \varepsilon_0} \times 10^3 + \frac{8\pi nkT}{\varepsilon_0^2} \alpha \right) \sinh^2 \tfrac{1}{2} z \right]. \tag{7.1.30}$$

As expected, the correction is small, the sum of the correction parameters being of the order of -10^{-3} for a 10^{-2} M 1–1 solution. The correction to the repulsive energy between two particles (flat plates) can again be calculated along the same lines as indicated in Chapter 3. The result is less interesting than in case (c): the effect is smaller and it is a decrease which goes monotonically to zero at increasing interparticle distance.

(e) INCOMPLETE IONIZATION OF DONORS AND ACCEPTORS

According to the Fermi distribution the number n_A of electrons per unit volume which is occupied by N_A acceptor atoms per unit volume is

$$n_A = \frac{N_A}{1 + 2 \exp \dfrac{-(E_F - E_A)}{kT}}. \tag{7.1.31}$$

o

In Chapter 6 it was assumed that $(E_F - E_A) \gg kT$, so that $n_A \ll N_A$. Likewise we have for the number n_D of empty donors

$$n_D = \frac{N_D}{1 + \frac{1}{2} \exp \dfrac{-(E_D - E_F)}{kT}} \qquad (7.1.32)$$

where we assumed that $(E_D - E_F) \gg kT$. Thus, in a modified version of the space charge theory the energies E_A and E_D should appear. Calculations have been carried out by Seiwatz and Green.[32] They incorporated case (f) in their calculations.

(f) FERMI DISTRIBUTION OF HOLES AND ELECTRONS

The classical form of the sum-over-states, as used in Section 6.1.3, is only valid when the state density per energy unit is high. This approximation is certainly dangerous if the exponential in the sum-over-states is relatively large, which is the case if the energies involved lie close to the Fermi level. Each special case should be calculated numerically. Seiwatz and Green, following a somewhat different but physically identical approach, found that on the whole for $(E_C - E_D)$, $(E_D - E_P)$, $(E_F - E_A)$ and $(E_P - E_V)$ all $> 3kT$, the usual approximations are valid. They considered the case of germanium doped with one single donor impurity to various concentrations with $E_C - E_D = 0 \cdot 01$ eV $(= 4kT$ at room temperature) and found that the Kingston and Neustadter approximation was valid for most practical cases. For highly degenerate space charges other approximations, given by Stratton,[33] were found to apply.

(g) PHYSICAL MEANING OF BAND BENDING

As pointed out in Subsection 6.2.4, the classical theory of band bending breaks down when the field strength becomes of the order of 10^6 V/cm. Indications pointing to quantization effects at room temperature have been found.[34,35]

7.1.2. *Planar region. Discreteness-of-charge effect*

The Stern layer has been looked upon as a correction of the original double layer theory initiated by Gouy. Thus, the finite dimensions of the ions served to introduce a non-zero minimum distance between the ions and the wall. The decreasing effect upon the dielectric constant of water provided by strong electrostatic fields served to introduce the relatively low dielectric constant in the Stern layer. Refinements are pertinent to the kind of interface under consideration and were dealt with in Chapters 3, 4, and 5.

However, there is one concept, which is of more general validity: the discreteness-of-charge concept. As mentioned in Chapter 3, Frumkin and Ershler and later Grahame, Parsons, Ross Macdonald and Levich used the concept to discuss the Esin and Markov effect and in fact, to explain it. In Chapter 4 the work of Levine *et al.* was mentioned. They interpreted data by Lijklema and Parfitt concerning the stability limits of AgI sols by invoking the discreteness-of-charge concept. Discreteness-of-charge effects have been discussed in connection with other interfaces as well and it was probably De Boer[36] who first invoked the concept for solid surfaces and used it for the explanation of adsorption data of Cs$^+$ ions on Wo-surfaces.[38] The well-known increase of the adsorption energy of Cs$^+$ ions (rather of the Cs atoms which after adsorption are assumed to be ionized) upon increasing degree of coverage can be understood on the basis of the discreteness-of-charge effect. To see this, reconsider the micro-potential versus distance curves of Fig. 3.12. The reason for the curvature already outside the double layer is that the shielding is imperfect. The significance of imperfect shielding is that in the neighbourhood of the plane of positive charges the influence is felt of the negative charges (and, of course, vice versa). This influence is enhanced when more positive charges are brought to the plane of already present positive charges. This apparent paradox is explained by the requirement of electroneutrality: an increase of the positive charge density should be accompanied by an increase of the negative charge density. The reasoning given here is only valid for low charge densities, less than about 10^{13} electrons per cm^2. If the densities are higher than 10^{13} electrons per cm^2, then the shielding becomes almost perfect. (It may be noted that it is fairly immaterial whether in the situation used here the negative charges are assumed localized in a two-dimensional array or distributed over a space charge region.)

Further possible applications of the discreteness-of-charge effect may be found in the discussion of the physical nature of the energies E_D and E_A of surface donor- and acceptor-states. These energies should contain contributions of the type $\frac{1}{2}e(\psi - \psi_A)$ which is the energy difference of electrons in a plane with smeared-out charges and a plane with localized charges. However, concerning state energies there is, apart from the case of Cs$^+$ ions on a Wo-surface, where one evidently could consider the adsorbed ions as empty donors, no convincing experimental evidence for this influence upon E_D and E_A in other cases such as that of the semiconductor/aq. soln. interfaces or semiconductor surfaces, other interpretations always being possible.

Finally, it is worth noting that according to the discreteness-of-charge concept the heat of adsorption depends on the degree of coverage. This means, according to Chapter 2, that virial terms should appear in the equation of state of the adsorbed species. For ionized, adsorbed surfactant molecules, where of course the concept should equally hold true, attempts have been undertaken to derive such an equation of state,[39] and to check on it experimentally.[40]

Vrij[41] has given a quantitative treatment. He pointed out that the discreteness-of-charge effect of adsorbed ions A^- at a liquid surface led to a redistribution of these ions compared with the random distribution of the "smeared-out" model. This redistribution led to a lowering of the interfacial pressure. This lowering was expressed in the form of a virial equation of state in which the second virial coefficient was a function of the bulk ion concentration.

As already pointed out in Chapter 2 there is a non-zero field (field strength F) near the planes of localized charges. Neutral atoms will be polarized by the field and be subject to an attractive polarization energy $V = -\frac{1}{2}\alpha F^2$ where now α is the polarizability of the neutral atom. The direction of the polarization depends upon the sign of the field strength, but F^2 is always positive. There should be a change of the contact potential, ΔV, the value and sign of depending on the average value of F and the degree of coverage. One may expect ΔV to be a complicated function of the degree of coverage. However, it may be estimated that ΔV will never extend beyond 10 mV or thereabouts. This estimate should be contrasted with the very large values, which have been found experimentally ($\Delta V = 1$ V), for the case of heavy gas atoms on tungsten surfaces.[42] These should be explained along other lines than those indicated here.

7.1.3. *Possible effects of the molecular structure of water*

The hard sphere model considered in Subsection 7.1.1 requires an abrupt transition between a rigid hydration shell and a structureless medium. This requirement may not be in agreement with the facts, because "molten zones" (Chapter 1) are probably present around the ions. Moreover, there is an exchange between water molecules adjacent to an ion and water molecules in the bulk.[43] An alternative to the hard sphere model has been offered by the "soft sphere" model, characterized by a soft sphere potential $z_i e/\varepsilon r \, (1 - e^{-\alpha r})$. Here α is a parameter whose value can be adapted to experimental data. The physical meaning of α is not quite clear. We regard α^{-1} as a mean value of the radius of the soft sphere. In the interior of the sphere where $\alpha r \ll 1$ the potential becomes $(z_i e/\varepsilon)\alpha$. The exponential in the expression of the soft-sphere potential crudely describes the deficiencies of the hard-sphere model and of the ionic solution theory. Owing to the minus sign the Coulomb interaction between pairs of ions will be weakened. Therefore the activity coefficients of ionic solutions will have higher values than predicted on the basis of the simple D–H theory, especially at high concentrations. Falkenhagen[27] worked out a theory on the basis of the soft-sphere potential and found with $\alpha^{-1} \sim 4$ Å good agreement between his theory and the experimentally determined values of the activity coefficients of such electrolytes as NaBr in water up to 10^{-1} M. We note that α is regarded as a characteristic of the salt, i.e. its specificity with regards to the ions is ignored. The value of about 4 Å for a "soft sphere" suggests that an ion affects the water structure beyond the range of the adjacent water molecules. The "molten zone" concept suggests the same thing. Therefore we investigate in this subsection the consequences for thermodynamic potentials of dropping the assumption of a structureless medium.

Such an estimate may be important: if two ions are placed at a distance R of, say, 10 Å, the molecular structure of a region in which a hundred or more water molecules are present may be affected. The free energy difference ΔF between such a region and a region of the same size in pure water contributes to the free energy of the interaction between the ions. The force between the ions is affected by an amount $-\partial \Delta F / \partial R$.

To arrive at an estimate a simple model of liquid water must be made which is suitable for calculations. The model, which we outline here, lies in between that of Lennard-Jones and Pople[44] and that of Némethy and Scheraga.[45] In both these models liquid water is considered as a distorted solid and we will do the same in our model. In the (tetrahedral) ice structure each water molecule can form a hydrogen bond (O...H—O) with four neighbouring water molecules. (This means that there are two H-bonds per molecule.) The L-J. and P. theory makes the assumption that in ice these bonds are rigid whereas they are bendable but not breakable in liquid water. As pointed out, the bendability of the four bonds taken together is then characterized by one single constant g and this constant appears to be the only parameter in the L-J. and P. theory. In contrast, the H-bonds are considered non-bendable in the N. and S. theory. Here the bonds can be broken. Configurations are considered of clusters of water molecules with one, two, three or four broken bonds for each water molecule. If four bonds are broken, the water molecule is considered as "free". In order to arrive at thermodynamic potentials, a partition function or sum-over-states is written down in which naturally appears a configurational part. Such a configurational part is missing in the L-J. and P. theory. In our model it is assumed, just as in the L-J. and P. theory, that the H-bonds are bendable. However, if a certain bending angle θ is reached we assume that the H-bond is broken. Therefore we recognize "bound" and "free" molecules. The configuration concept of the N. and S. theory is now adopted on the basis of "bound" and "free" molecules and this concept is used in a simple order–disorder theory. An order–disorder parameter s is introduced, s being zero if "disorder" is complete. To define more quantitatively: if there are n water molecules ("bound" and "free" together) than $\frac{1}{2}n(1-s)$ are bound and $\frac{1}{2}n(1+s)$ are free. If the H-bond energy is high and/or the temperature is low then the mean value of s is positive and may eventually approach the value unity (complete order). If the H-bond energy is very weak and/or the temperature is high then the mean value of s is close to zero. This characterizes disorder. Thermal movements lead in this case to $\frac{1}{2}n$ "bound" and $\frac{1}{2}n$ "free" molecules. There is an over-simplification in this definition of s: it is assumed that the number of positions that a "bound" molecule can occupy is the same as the number of positions of a "free" molecule. However, it is likely that a free molecule can find more positions than a bound molecule. It has translational freedom and, moreover, the value of the angle θ' defined above is of importance. If θ' is small, then the number of positions to be occupied by a bound molecule is relatively small. These positions are only found at bending angles $\theta < \theta'$. In the theory it was proved useful to define a parameter p:

$$p = \frac{\text{number of possible positions of a molecule, free or bound}}{\text{number of possible positions of a bound molecule}}.$$

(7.1.33)

Then the number of bound molecules at a certain moment is $n/p(1+s)$ and the number of free molecules is $n/p(p-1-s)$. Evidently we had so far considered the case $p = 2$. For all p-values complete disorder is characterized by zero s, and complete order by $s = p-1$. Negative s-values are disregarded. They would mean that H-bonds are avoided.

We want to obtain curves of the free energy F versus equilibrium values of r. To this purpose the sum-over-states Z is written down.

$$Z = e^{-F/kT} = \sum g(r)\, e^{-E(s)/kT}.$$

(7.1.34)

As is usual in order–disorder theories,[46] only the largest term of the summation will be considered. Furthermore, in our crude model we follow the "zeroth-order" approximation (Guggenheim).[46] Finally we consider only that part of the sum-over-states, that is directly relevant to the various configurations. Contributions due to (intramolecular) vibrations, etc., are assumed independent of s and are discarded. For $g(s)$ we write

$$g(s) = \frac{\left[\frac{n}{p}\right]!\left[\frac{n}{p}(p-1)\right]!}{\left[\frac{n}{p}(1+s)\right]!\left[\frac{n}{p}(p-1-s)\right]!}(p-1)^{(n/p)(p-1-s)}(p-1)^{-(n/p)(p-1)}$$
(7.1.35)

which is normalized such that $g(0) = 1$. Thus $g(s)$ is the (normalized) number of configurations at given energy $E(s)$. It is a generalization of the well-known configuration factor for $p = 2$:

$$g(s)_{p=2} = \frac{[\frac{1}{2}n]![\frac{1}{2}n]!}{[\frac{1}{2}n(1+s)]![\frac{1}{2}n(1-s)]!}.$$
(7.1.36)

Concerning the energy $E(s)$, this will be of the order of $-(n/p)(1+s)g$ where g ($\approx 10kT$) is the parameter in the theory of Lennard-Jones and Pople. To avoid confusion and also because the model of Lennard-Jones and Pople is different from the one considered here, we use the symbol E_0 instead of g. Another contribution to $E(s)$ may be considered: if a molecule is "free", it may move to a cavity in the water "lattice". In ice the lattice is very open and we assume the same to be true, though to a lesser extent in liquid water. In a cavity, the free molecule is subject to Van der Waals interactions with the surrounding molecules. If the surrounding molecules are all "bound", then in the zeroth-order approximation the contribution to $E(s)$ is equal to

$$2znp^{-2}(1+s)(p-1-s)E_2.$$

Here E_2 is the interaction energy between a "bound" and a "free" molecule and z is the number of nearest neighbours of a molecule. In ice at 0°C $z = 4$ but in liquid water z is assumed to be somewhat greater ($z \approx 4-5$). The point is that it will here be assumed that $E(s)$ contains a contribution which is proportional to s and a contribution which is proportional to s^2.

The Helmholtz free energy is

$$F = -kT \ln Z = -kT \ln g(s) + E(s).$$
(7.1.37)

where

$$E(s) = -\frac{n}{p}(1+s)E_0 + \frac{2zn}{p^2}(1+s)(p-1-s)E_2.$$

For further calculations the introduction of

$$E_1 = E_0 - 2z\left(1-\frac{2}{p}\right)E_2$$

and of

$$E_3' = \frac{2z}{p} E_2$$

has appeared to be useful. We apply the equilibrium condition

$$\frac{\partial F}{\partial s} = 0. \tag{7.1.38}$$

Curves of F versus r can then be constructed[47] for various values of E_1/kT, E_2/kT and p, each value of F being characterized by one value of s. Figure 7.2 gives some examples for $p = 6$. It is seen that if E_1 is

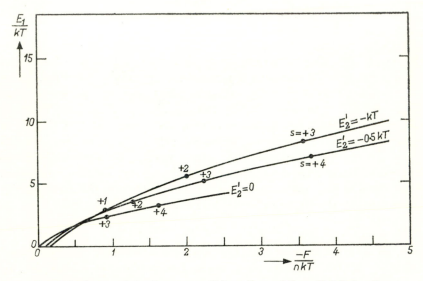

Fig. 7.2. Curves of the free energy F in units $-nkT$ vs. E_1/kT for three values of E_2 ($E_2' = (2z/p) E_2$). For p the value 6 is taken. Values of the order–disorder parameter s are given. The temperature T is held constant.

made smaller at constant E_2 and temperature, i.e. if the H-bonds are weakened, F becomes less negative and s tends to smaller values (more disorder). A reason for lower values of E_1 can be provided by ions in the solution. If two ions are present, a distance R apart, a contribution to the interionic force is $-(\partial \Delta F/\partial R)$ (see above). If the influence of the ionic charges is such as to lower E_1, then this is a repulsive force contribution. This contribution can be considerable. If the value of E_1 is diminished by $0.02 kT$ the free energy per molecule is increased (is made less negative) by about $0.05 kT$. If 20 water molecules are involved the total free energy effect is about $1 kT$. The force effect is (tentatively) $1 kT$ per 10 Å. Concerning the energy E_2, if this is appreciable and negative, the free molecules tend to the cavities in the lattice. The case considered by Némethy and Scheraga is relevant here.[45] These authors considered the hydrophobic part of a molecule in contact with water. They assumed that (in our language) free molecules tend to cavities in pure water rather than to cavities formed partly by water and partly by hydrophobic groups. This assumption seems not unreasonable. Therefore the fraction of free molecules in the neighbourhood of a hydrophobic group will be smaller than in the bulk of the solvent, in agreement with the "frozen-zone" concept. This is also shown in Fig. 7.2: if E_2 is made less negative, then at constant E_1 and T the value of s is removed farther from zero.

Although it seems premature to give quantitative values concerning structure effects in the solvent we are led to believe that their investigation is of great importance in the discussion of corrections to the original theory.

7.1.4. *Review*

NON-SPECIFIC EFFECTS

These effects are independent of the nature of the system under consideration. The most important non-specific effects are the discreteness-of-charge effects of adsorbed charges, the self-atmosphere effect in a space charge (Gouy layer) and image effects. Corrections to the elementary theory, arising from these effects, may dominate all the others. As we have seen, the discreteness-of-charge effect is often rather more than merely a correction: it is required as the unique explanation of such phenomena as the Esin and Markov effect (Chapter 3), of certain coagulation phenomena (Chapter 4), of adsorption and work function phenomena (Subsection 7.1.2). The self-atmosphere effect leads to a smaller absolute value of the average potential and to a lower value of the interparticle repulsion. The effect may amount to 7%.

SPECIFIC EFFECTS

These effects are dependent upon the physical nature of the system. Thus, the medium plays in aqueous systems a role which is entirely different from that in semiconductors. In the latter case we have the problem of the physical meaning of band bending. Furthermore, the mean free path of the conductivity electrons and of holes is of the order of 10^{-5} cm. This should have no influence upon equilibrium properties but, especially when the free mean path is large compared with the Debye length, problems may arise concerning kinetic properties such as mobility. The molecular structure in this case is hardly affected by the presence of conductivity electrons or holes. In contrast, the molecular structure of the solvent in aqueous solutions may be drastically changed as the result of the dissolution of salts. A first approach which accounts for such a change was the "hard-sphere" model and the related concept of a rigidly bound water layer at an interface. The ion size correction is probably the most important "specific" correction. For a 10^{-2} M 1–1 solution the value of the characteristic parameter Bn is here of the order of 10^{-2}–10^{-3}. The ion size correction is opposed in sign to the self-atmosphere correction, but we see that it is usually smaller in absolute value. It is only upon comparing two systems which only differ in the choice of the ions that the ion-size effect appears. Then lyotropic influences may be investigated. These conclusions are essentially the same as those obtained by Levine and Bell.[9] Although their final conclusions (the Gouy layer will be somewhat contracted if compared with the elementary theory, the total effect amounting to about 5%) are the same as those obtained here, their reasoning is somewhat different. The present author considers the "cavity" potential only as a correction upon the self-atmosphere correction; Levine and Bell give the cavity potential a more central place.

On second thoughts this approach appears to be an over-simplification. The structure of the solvent will play a more involved role than suggested by the hard-sphere model. We have indicated this in the previous section and such concepts as "icebergs", "molten zones", soft spheres" are already suggestive. Long-range structured water layers have now been directly detected by using nuclear magnetic resonance techniques.[48] Such layers were already suggested by Derjaguin,[49] but the usefulness of his "anomalous water" concept is doubtful.

Owing to structure effects of the solvent it can be expected that in addition to the Coulomb forces there is a repulsive force contribution to the total interaction force between the ions. Its consequence for modifications in the Gouy theory remain to be investigated.

7.2.   Surface states. Langmuir adsorption

The Fermi and the Langmuir distribution laws have a common statistical basis. In the Langmuir case one considers N sites on a uniform surface, available for the adsorption of n $(n \leqslant N)$ indistinguishable particles. In the Fermi case one considers N eigenstates in the phase space, available for n indistinguishable particles with antisymmetric wave functions. Then, in both cases the sum-over-states Z is reduced to one term which is written as

$$Z = \frac{N!}{n!(N-n)!} Z_i{}^n \qquad (7.2.1)$$

where Z_i is the sum-over-states of one particle at the surface or in an eigenstate. The chemical potential becomes, assuming Stirlings' approximation for the factorials,

$$\mu = -kT \frac{\partial}{\partial n} \ln Z = \mu_0 + kT \ln \frac{\theta}{1-\theta} \qquad (7.2.2)$$

where $\mu_0 = -kT \ln Z_i - nkT(\partial \ln Z_i/\partial n)$ and where $\theta = n/N$. It is seen that $\mu = \mu_0$ if $\theta = \frac{1}{2}$.

In the Langmuir case μ_0 is the free energy of adsorption of a particle. In the Fermi case, where $\mu = E_F$ is the electrochemical potential of the electrons, the quantity μ_0 is the energy of a state. (In the Fermi case, instead of Z_i, one often writes $a \times Z_i$, where a is the number of ways that an electron can choose to occupy a state.) Thus in the usual expressions of the population of donor and acceptor states the energies E_D and E_A have the character of *free* energies per particle.

THE REACTION $AB \rightleftarrows A+B$

As shown in Chapter 2, the chemical potential of one of the dissociation products may be written in such a way that it shows much resemblance to eq. (7.2.2). If the degree of dissociation is α, then the concentration of AB is $c_{AB}(1-\alpha)$ and that of A is $c_{AB}\alpha$, if no A is added from elsewhere. Then, rewriting eq. (2.2.59) (we disregard electrostatic terms):

$$\mu_B = \mu_{AB0} - \mu_{A0} - kT \ln \frac{\alpha}{1-\alpha}. \qquad (7.2.3)$$

If α is identified with $(1-\theta)$ and μ_0 with $\mu_{AB0} - \mu_{A0}$, eqs. (7.2.2) and (7.2.3) are identical. This justifies the following equilibrium reactions for donors and acceptors:

$$(\text{Donor}) \rightleftarrows (\text{Donor})^+ + e^-, \qquad (7.2.4)$$

$$(\text{Acceptor})^- \rightleftarrows (\text{Acceptor}) + e^-. \qquad (7.2.5)$$

Considering eq. (7.2.4), this allows writing

$$\mu = E_F = (\mu_{(\text{Donor})})_0 - (\mu_{(\text{Donor})^+})_0 - kT \ln \frac{\alpha}{1-\alpha} \qquad (7.2.6)$$

where $\alpha = 1 - \theta$ and where

$$E_D = (\mu_{(\text{Donor})})_0 - (\mu_{(\text{Donor})^+})_0. \qquad (7.2.7)$$

An analogous equation can be written down for acceptors where

$$E_A = (\mu_{(\text{Acceptor})})_0 - (\mu_{(\text{Acceptor})^-})_0. \qquad (7.2.8)$$

THE REACTIONS $H_2O \leftrightarrows H^+ + OH^-$ AND $(sc) \leftrightarrows (sc)^+ + e^-$

In neutral, pure, water the ionized molfraction at room temperature is well known to lead to $p_H = 7$ and therefore $\alpha = 0.018 \times 10^{-7}$. Thus

$$\mu_{H_2O} = (\mu_{H_2O})_0 = (\mu_{H^+})_0 + (\mu_{OH^-})_0 + 2kT \ln 18 \times 10^{-10} \qquad (7.2.9)$$

and one has

$$\Delta\mu_0 = (\mu_{H^+})_0 + (\mu_{OH^-})_0 - (\mu_{H_2O})_0 = 40 \, kT,$$
$$= 1 \, eV. \qquad (7.2.10)$$

The heat of neutralization, comparable to Δh_0, has appeared to be $23kT$. It follows that there is a large entropy effect:

$$\Delta s_0 = (s_{H^+})_0 + (s_{OH^=})_0 - (s_{H_2O})_0 = T^{-1}(\Delta h_0 - \Delta\mu_0) = -17k. \qquad (7.2.11)$$

This entropy effect may be explained by a more ordered structure of the water molecules around H^+ and OH^- ions compared with the structure around neutral water molecules. Thus much entropy is lost upon ionization.

The equilibrium $(sc) \leftrightarrows (sc)^+ + e^-$ (where (sc) stands for a semi-conductor atom, for instance Ge or Si) can be dealt with in the same way and leads to :

$$\Delta\mu_0 = (\mu_{sc^+})_0 + (\mu_{e^-})_0 - \mu_{sc} = -kT \ln npV^2$$
$$= -2kT \ln n_i V \qquad (7.2.12)$$

However, in the case of a semiconductor the band picture is usually invoked to find an expression for $-kT \ln np$ (see Chapter 6) and one has (assuming $a = 1$)

$$\Delta\mu_0 = E_c - E_v - 3kT \ln (2\pi\sqrt{(m_e^* m_h^*)}kT \, h^{-2}) = -kT \ln np. \qquad (7.2.13)$$

The entropy contribution to $\Delta\mu_0$ is

$$\Delta s_0 = 3kT \ln (2\pi\sqrt{(m_e^* m_h^*)}kT \, h^{-2}) + 3k + T^{-2} \frac{\partial(E_c - E_v)}{\partial T^{-1}} \qquad (7.2.14)$$

where the first term on the right is the normalization constant which is similar to that used for gases and where the second term, $3k$, accounts for the gain of 6 degrees of freedom, 3 for the released electrons and 3 for the holes. The third term is zero, if the value of the bandwidth is independent of temperature.

For the change of the enthalpy upon ionization one has

$$\Delta h_0 = E_c - E_v + 3kT + T^{-1} \frac{\partial(E_c - E_v)}{\partial T^{-1}}. \qquad (7.2.15)$$

The two ionization reactions, both producing important mobile charge carriers, can be linked by placing a semiconductor, which can act as a hydrogen electrode, in an aqueous solution. This may be done in such a way that the following equation is obeyed:

$$H^+ + e^- \leftrightarrows H. \qquad (7.2.16)$$

The equilibrium condition is:

$$\mu_{H^+} + E_F = \mu_H \qquad (7.2.17)$$

where $\mu_H \ (= \tfrac{1}{2}\mu_{H_2})$ is to be held constant.

A "bandgap" in solutions, which refers to electrons, may be introduced if redox systems are considered

$$X^{n+} + e^- \leftrightarrows X^{(n-1)+} \qquad (7.2.18)$$

where X^{n+} and $X^{(n-1)+}$ are cations capable of releasing and trapping an electron. Anions instead of cations may be used too. Gerischer et al.[51] have studied these systems in connection with semiconductors.

Adsorption energies, state energies, and (the logarithms of) ionization constants are closely related properties. Their value can, in all the widely varying systems, be affected by the same physical influences. In the first place, of course, there is the contribution $z_i\, e\psi_0$, if the charged particles under consideration (ions, electrons, holes) are under the influence of the potential ψ_0. Furthermore, the three energy quantities are all subject to the same modifications arising from the discreteness-of-charge effect. We also want to mention image effects. As pointed out in Chapter 6, image effects can explain the negative adsorption of ions at the surface of an aqueous solution, and it may be held responsible for the differences of the ionization constant of a surfactant molecule dissolved in water and of the same molecule adsorbed at the water surface. The image effect may now also have a bearing upon state energies and may be (partly) responsible for differences of properties of surface states, which have been found before and after gas adsorption, adsorbed gas layers having a higher dielectric constant than has vacuum.

7.3. Future work. Models in science and philosophy

In the future, progress will be made by means of steadily improved experimental techniques, and in conjunction with other theoretical disciplines which thus far have been hardly touched upon.

Thus, concerning experimental techniques, a vacuum of 10^{-12} torr can be obtained and a working pressure of 10^{-9} torr is quite common.

Force measurements can be carried out down to changes of 10^{-5} dynes; displacements of 1–10 Å can be measured; currents of 10^{-17} A and potential changes of 0·1 mV (difference of work functions) are well within the range of electronic techniques. The combination of these various techniques in one experimental equipment of course often provides difficulties, but much progress has been made in recent years.

Electron-, ion-, and molecular-beam techniques as well as esr and nmr techniques, field ion microscopy and optical techniques such as ellipsometry are now in an advanced state. To date, these techniques provide information concerning the molecular structure of surface layers and the exact location of atoms at surfaces. Furthermore, the chemical composition and the state of cleanliness of surfaces can be given with great precision. Methods to obtain clean surfaces are now available. With this information, long-standing problems can be approached on a new level. Theoreticians may be encouraged by this course of things.

Concerning the theoretical disciplines, which have hardly been touched upon to date, the following may be mentioned:

(a) Molecular physics, to be used for the elucidation of the nature of surface states, of adsorption energies and ionization constants of chemical equilibria in bulk phases and interfacial phases, and also for such problems as the behaviour of hydration shells around ions in solution, of "hydrophobic bonding", of the outermost layer of molecules at a liquid surface and of the solvent itself.

(b) Molecular biology, to be used for the study of the highly ordered molecular structures in nerve and muscle cell membranes.

(c) Modern statistical developments (cluster theory), to be used for obtaining a better expression for the potential distribution in space charge regions in connection with better models of the physical systems involved.

Application of these three disciplines (they should of course often be used in combination) will invalidate a number of the models which have been used thus far, and which we have tried to explain in the preceding chapters.

A model can be defined as an intentionally simplified version of what is considered at a certain stage as a scientific truth. The simplification is intentional because this facilitates the attainment of further scientific achievements. Then new models become necessary, but often features of a former model are contained in a new model.

It is remarkable to note that often in elementary models either an "ideal" orderliness, or an "ideal" disorder is assumed. Thus, the well-ordered Helmholtz layer proved to be insufficient to explain the experimental data and Gouy invoked a diffuse region in which ions carried out thermal, i.e. disordered, movements. Another example of this sort is provided by crystals. Here first an ideal structure was assumed and later on imperfections had to be introduced.

On the other hand, we have the well-known example in the classical theory of the ideal gas. Elements of orderliness had to be introduced in this concept: Van der Waals interactions and, apart from that, the quantum theory according to which entropy values are lower than predicted by the classical theory. Also the example may be mentioned of the early models of cell membranes, which were based on a purely electrostatic theory in a structureless medium. These models are certainly invalid in the light of the more recently revealed highly ordered structures, but they have served, and still serve, as a tool to accomplish further progress.

In some cases the change from one model to another to handle a certain problem is quite drastic. Thus, Frenkel[52] considered a liquid, which is not too close to the critical point, as a distorted solid rather than as a dense imperfect gas, as Van der Waals did. Onsager has remarked[53] that the problems connected with highly concentrated ionic solutions may be approached from the "crystal side" rather than from the "dilute-solution side". Also the example is relevant here of the various equations of state of an adsorbed phase. Mobile and localized adsorption represent two opposite views, although the same limiting adsorption law is usually obtained for low adsorption. Models in which too little, or too much order was assumed, have been used outside science as well, notably in philosophy. Some of these models will now be mentioned. The reason why we shall stress the concept of order in these models is that the development of this concept is one of the greatest achievements of scientific thinking of the past 300 years and that it has had, ever since the discussions between Newton and Leibniz,[54] a considerable bearing on present-day philosophy.[55]

The philosopher David Hume[56] published a book in 1738, entitled *Treatise of Human Nature*, in which ideas were considered as (in the words of modern scientists) gas atoms. These gas atoms did not form an ideal gas: "Were ideas entirely loose and unconnected, chance alone would join them." In fact, Hume's "ideas" were not unconnected. He assumed (weak) ordering principles present in the mind, the most important of them being the "custom". Two ideas may often be experienced to have a certain relation. If one idea occurs to us, the other will follow in our mind. If we let go a stone (first idea), it drops on the floor (second idea). After some experience we predict what happens if we let go a stone. This was Hume's explanation of causality. The theory has often been criticized because the mind will have more powerful and also more hidden ordering principles at its disposal. One of the opponents, Husserl,[57] replaced the gas model by a liquid model. The ideas (i.e. the atoms or molecules of the liquid) were now in closer contact. By comparing the liquid with a river, Husserl imposed a certain "Intentionalität" (the direction of the river) upon the ideas. A model in which too much order was assumed is that of Hegel.[58] He assumed a perfect obedience not only of the ideas but, more generally, of the history of mankind, to the "Weltgeist". Phenomena not fitting the model had to be considered as "accidental", as disordering. Although Hegel attempted to incorporate these disordering phenomena in his model,[58] it seems appropriate to say that Hume assumed too little order in his model and Hegel too much.[59]

REFERENCES

1. L. ONSAGER, *Chem. Revs.,* 1933, **13**, 73.
2. J. G. KIRKWOOD, *J. Chem. Phys.,* 1934, **2**, 767.
3. F. H. STILLINGER and J. G. KIRKWOOD, *J. Chem. Phys.,* 1960, **33**, 1282.
3a. H. B. G. CASIMIR, Contribution to a Symposium held in Utrecht by the Kon. Ned. Chem. Vereeniging, 1944.
4. A. L. LOEB, *J. Coll. Sci.,* 1951, **6**, 75.
5. W. E. WILLIAMS, *Proc. Roy. Soc.* A, 1953, **66**, 372.
6. E. HÜCKEL and G. KRAFFT, *Z. Physik. Chemie* N.F., 1955, **3**, 135.
7. F. P. BUFF and F. H. STILLINGER, *J. Chem. Phys.,* 1963, **39**, 1911.
8. Papers in: B. V. DERYAGIN (DERJAGUIN) editor, *Research in Surface Forces 2, Three-dimensional Aspects of Surface Forces,* Plenum Press (consultants Bureau), N.Y., 1966 (translated from the Russian).
9. S. LEVINE and G. M. BELL, *Disc. Faraday Soc.,* Sept. 1966.
10. H. L. FRIEDMAN, *Ionic Solution Theory,* Interscience, 1962.
11. S. LEVINE and G. M. BELL, *J. Phys. Chem.,* 1960, **64**, 1186.
12. A. SANFELD, A. STEINCHEN-SANFELD and R. DEFAY, *J. Chim. Physique,* 1962, no. 15, p. 132.
13. A. SANFELD, A. STEINCHEN-SANFELD, H. HURWITZ and R. DEFAY, *J. Chim. Physique,* 1962, no. 16, p. 139.
14. R. DEFAY and A. SANFELD, *J. Chim. Physique,* 1963, **60**, 635.
15. H. D. HURWITZ, A. SANFELD and A. STEINCHEN-SANFELD, *Electrochim. Acta,* 1964, **9**, 929.
16. I. PRIGOGINE, P. MAZUR and R. DEFAY, *J. Chim. Physique,* 1953, **50**, 156.
17. A. SANFELD and R. DEFAY, *Physica,* 1964, **30**, 2232.
18. M. J. SPARNAAY, *Rec. Trav. Chim. Pays-Bas,* 1958, **77**, 871; 1962, **81**, 395.
19. H. BRODOWSKY and H. STREHLOW, *Z. Elektrochem.,* 1959, **63**, 262.
20. S. LEVINE and H. E. WRIGLEY, *Disc. Faraday Soc.,* 1957, **24**, 43.
21. E. WICKE and M. EIGEN, *Naturwissenschaften,* 1951, **38**, 453; 1952, **39**, 108; *Z. Elektrochem.,* 1951, **55**, 352; 1952, **56**, 551; 1952, **56**, 836; 1953, **57**, 319; *Z. Naturforschung,* 1953, **8a**, 161.
22. J. J. BIKERMAN, *Phil. Mag.,* 1942, **33**, 384.
23. F. B. GRIMLEY and N. F. MOTT, *Disc. Faraday Soc.,* 1947, **1**, 3.
24. M. J. SPARNAAY, *Z. Phys. Chem.* N.F., 1957, **10**, 156.

25. E. J. W. VERWEY and J. TH. G. OVERBEEK, *Theory of the Stability of Lyophobic Colloids*, Elsevier, 1948, p. 92. See also ref. 13.
26. *Rheology of Emulsions*, ed. P. SHERMAN (Proceedings of a Symposium held by the British Society of Rheology, October 1962), Pergamon Press, 1963.
27. H. FALKENHAGEN, *Die Elektrolytarbeiten von Max Planck und ihre weitere Entwicklung*, Max-Planck-Festschrift, 1958, p. 11 (especially p. 22). This paper contains a lucid review of modern Russian work in the field of ionic solution theory. H. FALKENHAGEN and G. KELBG, *Disc. Faraday Soc.*, 1957, **24**, 20.
28. J. MALSCH, *Phys. Z.*, 1928, **29**, 770; 1929, **30**, 837.
29. J. ROSS MACDONALD and C. A. BARLOW, *J. Chem. Phys.*, 1962, **36**, 3062.
30. J. B. HASTED, D. M. RITSON and C. H. COLLIE, *J. Chem. Phys.*, 1948, **16**, 1.
31. F. E. HARRIS and C. T. O'KONSKI, *J. Phys. Chem.*, 1957, **61**, 310.
32. R. SEIWATZ and M. GREEN, *J. Appl. Phys.*, 1958, **29**, 1034.
33. R. STRATTON, *Proc. Phys. Soc.* B, 1955, **68**, 746.
34. J. F. DEWALD, *Ann. N.Y. Acad. Sci.*, 1963, **101**, art. 3, p. 872.
35. N. ST. J. MURPHY, *Surface Science*, 1964, **2**, 86.
36. J. H. DE BOER and C. F. VEENEMANS, *Physica*, 1934, **1**, 960.
37. A. N. FRUMKIN invoked the concept for liquid interfaces (see Chap. 3, ref. 6).
38. N. S. RASOR and C. WARNER, *J. Appl. Phys.*, 1964, **35**, 2589.
39. R. P. BELL, S. LEVINE and B. A. PETHICA, *Trans. Faraday Soc.*, 1962, **58**, 904.
40. J. MINGINS and B. A. PETHICA, *Trans. Faraday Soc.*, 1963, **59**, 1892.
41. A. VRIJ, Grenslaagverschijnselen, *Coll. Royal Flemish Acad. Sci. Brussels*, 1965, p. 13 (in English).
42. See for example: G. EHRLICH and F. H. HUDDA, *J. Chem. Phys.*, 1959, **30**, 493; evidence was first given by J. C. P. MIGNOLET, *ibid.*, 1953, **21**, 1298.
43. O. YA. SAMAILOV, *Disc. Faraday Soc.*, 1957, **24**, 141 and book: *Structure of Aqueous Electrolytes and the Hydration of Ions* (Engl. transl., Consultants Bureau, New York, 1965).
44. J. LENNARD-JONES and J. A. POPLE, *Proc. Roy. Soc.* A, 1951, **205**, 155; J. A. POPLE, *Proc. Roy. Soc.* A, 1951, **205**, 163.
45. G. NÉMETHY and H. A. SCHERAGA, *J. Chem. Phys.*, 1962, **36**, 3382; 1962, 3401.
46. E. A. GUGGENHEIM, *Mixtures*, Clarendon Press, Oxford, 1952.
47. M. J. SPARNAAY, *J. Coll. Interface Sci.*, 1966, **20**, 23.
48. G. A. JOHNSON, S. M. A. LECCHINI, E. G. SMITH, J. CLIFFORD and B. A. PETHICA, *Disc. Faraday Soc.*, Sept. 1966.
49. B. V. DERJAGUIN, *Soc. Expt. Biol. Symp.*, 1964, **19**, 55.
50. T. L. HILL, *J. Chem. Phys.*, 1949, **17**, 762.
51. H. GERISCHER in *Electrochemistry*, **1** (ed. P. DELAHAY), Interscience, 1961, p. 139.
52. J. FRENKEL, *Kinetic Theory of Liquids*, chap. III, Russian ed. 1943, English ed. Dover, 1946.
53. Improvised, unpublished speech at the Discussion of the Faraday Soc., "Interaction in Ionic Solutions", Oxford, 1957. Onsager referred to *J. Phys. Chem.*, 1939, **43**, 189.
54. A. ROBINET, *La Correspondence Leibniz–Clarke*, Paris, 1957 (Clarke represented the opinions of Newton).
55. L. BRUNSCHVICG, *L'Expérience humaine et la Causalité physique*, Paris, 1948.

56. D. HUME, *Treatise of Human Nature*, published in Everyman's Library, Dent, London.
57. E. HUSSERL, in *Husserliana* VI, M. Nijhoff, Den Haag, 1954.
58. G. W. F. HEGEL, in Geschichte der Philosophie.
59. M. J. SPARNAAY, *Alg. Ned. Tijdschrift voor Wijsbegeerte en Psychologie*, 1959, **51**, 98 (in Dutch).

AUTHOR INDEX

Names of authors are followed by numbers of pages where these names appear. References listed after each chapter are given in the Index by Roman numerals indicating chapter number followed by reference number in italics.

SUBJECT INDEX